Special External Effects on Fluvial System Evolution

Special External Effects on Fluvial System Evolution

Special Issue Editors

Jef Vandenberghe
David R. Bridgland
Xianyan Wang

MDPI • Basel • Beijing • Wuhan • Barcelona • Belgrade

MDPI

Special Issue Editors
Jef Vandenberghe
VU University Amsterdam
The Netherlands

David R. Bridgland
Durham University
UK

Xianyan Wang
Nanjing University
China

Editorial Office
MDPI
St. Alban-Anlage 66
4052 Basel, Switzerland

This is a reprint of articles from the Special Issue published online in the open access journal *Quaternary* (ISSN 2571-550X) from 2018 to 2019 (available at: https://www.mdpi.com/journal/quaternary/special_issues/fluvial_system_evolution)

For citation purposes, cite each article independently as indicated on the article page online and as indicated below:

LastName, A.A.; LastName, B.B.; LastName, C.C. Article Title. *Journal Name* **Year**, *Article Number*, Page Range.

ISBN 978-3-03921-544-7 (Pbk)
ISBN 978-3-03921-545-4 (PDF)

Cover image courtesy of Martin Gibling.

Contents

About the Special Issue Editors

Jef Vandenberghe graduated from the Catholic University of Leuven (Belgium) and continued his career at the Vrije Universiteit Amsterdam. He is Visiting Professor at the Universities of Nanjing (China) and Novi Sad (Serbia) and member of the Leopoldina Academy. Vandenberghe's general scientific expertise is in the field of geomorphology and sedimentology of river and wind deposits and inferred palaeoclimatological reconstructions. A main part of his research activities focuses on (former) periglacial (permafrost) and monsoonal environments. His research has extended over many regions in western, central, and eastern Europe, as well as in Asia and South America. Characteristic of his research is the multidisciplinary approach involving the specializations of palaeoecology, sedimentology, geomorphology, and climatic modeling. Besides having served as Guest Editor on 19 occasions for various international journals, he has been or is currently a member of the Editorial Board of 7 international journals and has published 245 peer-reviewed papers and 4 monographs. He has been a member of review committees for scientific quality both abroad and in the Netherlands. He has served as Chair and Full Member of several commissions of international scientific organizations.

David R. Bridgland completed his BSc and PhD studies at the City of London Polytechnic, and then spent the first decade of his career with the conservation agency of the British Government, the Nature Conservancy Council, selecting and providing justification for Earth science sites representing the quaternary of the River Thames, which remains an important research theme for him. He moved to Durham University in the early 1990s where he is now Professor in the Geography Department. Extending his interests to the worldwide evidence for changing rivers during the Quaternary, he co-founded the Fluvial Archives Group in 1996. He has also been active in the London-based Quaternary Research Association and Geologists' Association. In addition to his broad interests in Quaternary fluvial archives, David specializes in evidence for the human occupation of the Old World as represented within fluvial sequence, mainly from stone artefacts, as well as Earth science conservation, especially of Quaternary sites. He was co-leader of successive International Geoscience Programme (IGCP) projects: IGCP 449 "Global Correlation of Late Cenozoic Fluvial Deposits" (2000–2004) and IGCP 518 "Fluvial Sequences as Evidence for Landscape and Climatic Evolution in the Late Cenozoic" (2005–2007). He has published numerous books and >100 journal articles.

Xianyan Wang graduated from the institute of Earth Environment (Chinese Academy of Sciences) and the Vrije University Amsterdam, and he continued his career at Nanjing University where he is now Professor in the School of Geography and Ocean Science. Xianyan Wang's general scientific expertise is in the field of fluvial geomorphology and sedimentology, tectonic geomorphology, and Quaternary climate and environment. A main part of his research activities focuses on the impacts of drainage and fluvial evolution on human settlement as a response to climate change and tectonic activity on the Tibetan Plateau and in Southeast Asian Monsoon area. Characteristic of his research is the multidisciplinary approach involving the specializations of sedimentology, geomorphology, and dating. His ORCID ID is 0000-0002-8281-5734.

quaternary

MDPI

Editorial

Specific Exogenetic (External) and Endogenetic (Internal) Effects on Fluvial System Evolution

Jef Vandenberghe [1,*], David Bridgland [2] and Xianyan Wang [3]

[1] Institute of Earth Sciences, VU University Amsterdam, 1081HV Amsterdam, The Netherlands
[2] Department of Geography, Durham University, South Road, Durham DH1 3LE, UK; d.r.bridgland@durham.ac.uk
[3] School of Geography and Ocean Science, Nanjing University, Nanjing 210023, China; xianyanwang@nju.edu.cn
* Correspondence: jef.vandenberghe@vu.nl

Received: 18 November 2018; Accepted: 21 November 2018; Published: 26 November 2018

Abstract: A collection of papers appears under the title "Special External Effects on Fluvial System Evolution" in the journal, *Quaternary*. This is a new Special Issue under the aegis of the Fluvial Archives Group (FLAG), illustrating the recent progress made in paleo-fluvial research. These papers highlight the high complexity of the external forcing of fluvial dynamics, and especially, the combined results of several interfering variables. In addition, it appears that the study of fluvial archives cannot be limited to the general and direct effects of external variables, but it also has to include the indirect influences that are regionally variable.

Keywords: FLAG; paleo-fluvial; fluvial forcing; fluvial archives

Rivers are excellent witnesses of the dynamics affecting the Earth's surface environments. Their activity is highly sensitive to the external impacts of climate change, base-level change, crustal movements, and human influence, as evidenced over variable temporal and spatial scales. Because these strong relationships, fluvial geomorphology and sedimentology have, in turn, often been applied in order to infer information related to the reconstructions of climate, tectonic (and atectonic) crustal movements, base-level changes, and human activity. The specification of these relationships in terms of the processes involved has always been a crucial point in the study of fluvial archives, and thus a focal point of interest for Fluvial Archives Group (FLAG) research (e.g., [1]). In addition, it was shown several decades ago that the fluvial system is not static, but is affected by its own endogenetic processes, and undergoes its own internal evolution in dynamic equilibrium (e.g., [2,3]). This means that the complexity of the external forcing of fluvial dynamics has to be supplemented with the effects of internal mechanisms. Furthermore, it has been found that, both in internal and external impacts, the delay and feedback effects and threshold crossing often lead to particular results and to more complexity in fluvial evolution (e.g., [4,5]).

Until now, there has been focus on evaluating the general impacts of the individual external factors, and more specifically, on disentangling their individual contributions. For instance, there is a recurrent question about how to identify the separate effects of climate and crustal movement on fluvial systems, or how to evaluate the impact of human action in comparison with climate changes. Such basic topics still need further investigation, as is shown by the contributions to this Special Issue. However, it remains a challenge to investigate the combined results of several interfering variables. Furthermore, several cases in the present issue show that we cannot continue to limit the study of fluvial archives to the general and direct effects of external variables. For instance, the climate has an immense impact on the fluvial system, not only in terms of temperature or precipitation, but also indirectly, by way of vegetation and frozen ground as intermediaries (e.g., [6,7]). Another example is the impact of crustal movement, which is often specified indirectly through its effects on the steepness of the river gradient, the topographic shape of the river catchment, or the physical properties of the substratum (e.g., [8]). An important result from the earlier years of FLAG research has been the realization that the effects of the various influences, and the style and pattern of the resultant fluvial archives, can vary according to the crustal type, particularly in terms of the susceptibility of different crustal provinces to vertical movement [9,10]. In addition, of practical importance, is the impact of human activity, such as land-use change and demographic pressure, in comparison with climatic change (e.g., [11–13]).

Finally, the controlling influences, because of their own complexity, may lead to the opposite, or at least different effects on the fluvial system. For example, cold- and warm-climate conditions have to be differentiated in order to cope with the effects of seasonality, the role of snow cover, and the duration of frost in the soil, while the relative magnitude and frequency of precipitation and temperature greatly affects vegetation. Similarly, crustal movements may have different effects on the fluvial processes as a function of the variety and rate of tectonic movement, or the regionally variable erodibility of the substratum. The regional differences of these indirect impacts may be very important for fluvial evolution.

In summary, the fluvial archive is a rich but also complex source of information. Rivers are reliable recorders of complex dynamics via their sedimentary products and morphological expression, and it is a challenge to contribute step-by-step to the understanding of the operation of the system and its influencing factors. But, in addition to the importance of regional differentiation and the effect of the indirect impacts outlined above, it is desirable to extract the validity of the rules that globally govern the evolution of fluvial systems and are reflected in the fluvial archives.

The present collection of papers under the title "Special External Effects on Fluvial System Evolution" illustrates the recent progress in the direction outlined above. It originally arose from a conference organized by Dr Zhenbo Hu (Lanzhou University, Lanzhou, China) and Prof Xianyan Wang (Nanjing University, Nanjing, China) in the summer of 2017, under the aegis of the Fluvial Archives Group (FLAG), the Quaternary Research Association (QRA), and the Geologists' Association (GA) in the upper and middle reaches of the Yellow River. In their paper, Daley and Cohen stress the significant impacts of regional climatic change above the relatively minor, and rather local-scale effects of intrinsic factors. More specifically, the channel incision at the beginning of the Holocene coincided with a precipitation maximum widely recognized in the otherwise tectonically stable area of subtropical Australia. In contrast, Gao et al. show the internal steering mechanisms apparent in the detailed evolution of an alluvial fan (the Huangshui river, NE Tibetan Plateau), clearly related to the position at the entrance of a subsiding basin, in combination with climatic effects. Other papers illustrate a less equivocally forced fluvial evolution, in which climate impact is interfering with other forcing factors. For instance, Stokes et al. discuss the evolution of the surface of the intramontane Sorbas basin in southern Spain. That surface, originally forming the top of the basin infill, was successively eroded towards a pediment by autogenic processes in the Early Pleistocene, followed by a (external) base-level lowering in the Middle Pleistocene. The next case deals with the tectonic impact of relatively local significance overprinted onto climate changes of a much wider extent in shaping a characteristic example of fluvial morphology. It describes the evolution of the catchment of the Tis(z)a river in the Pannonian Basin, where the differential tectonic subsidence has had a direct impact on the river gradient, and thus on the energy conditions (Vandenberghe et al.). Demir et al. provide a review of the inferred influence of the crustal type within the Eurasian continent on patterns of fluvial-archive preservation over Quaternary timescales, an influence that is shown to have affected landscape evolution in different crustal provinces. At the opposite end of the geographical scale, Cunha et al. attempt to interpret and date a fluvial archaeological context within the lowest (T6) terrace of the Lower Tejo (Tagus), just downstream from the Spain–Portugal border. This concerns a basal gravelly bed beneath silty sands and sandy silts, with the archaeology occurring, in conjunction with the remains of large mammals, at the interface between the gravel and the overlying finer-grained sequence. Indeed, the mammals here include some of the youngest examples of megafauna, such as the straight-tusked elephant. Finally, Gibling et al. focus on the impact of different kinds of human activities (for instance, as expressed by land or water management and agriculture) on the modification of channels and floodplains augmented since the beginning of the Holocene.

Funding: This research received no external funding.

Acknowledgments: Financial support for the conference was obtained from the Nanjing and Lanzhou Universities, and from the QRA.

Conflicts of Interest: The authors declare no conflict of interest.

References

1. Cordier, S.; Bridgland, D. De la géomorphologie fluviale aux archives fluviales (Introduction au numéro spécial). *Géomorphol. Relief Process. Environ.* **2012**, *4*, 391–404. [CrossRef]
2. Schumm, S.A.; Parker, R.S. Implications of complex response of drainage systems for Quaternary alluvial stratigraphy. *Nat. Phys. Sci.* **1973**, *243*, 99–100. [CrossRef]
3. Vandenberghe, J.; Cordier, S.; Bridgland, D.R. Extrinsic and intrinsic forcing on fluvial development: Understanding natural and anthropogenic influences. *Proc. Geol. Assoc.* **2010**, *121*, 107–112. [CrossRef]
4. Schumm, S.A. Geomorphic thresholds: The concept and its applications. *Trans. Inst. Br. Geogr.* **1979**, *4*, 485–515. [CrossRef]
5. Vandenberghe, J. A typology of Pleistocene cold-based rivers. *Quat. Int.* **2001**, *79*, 111–121. [CrossRef]
6. Vandenberghe, J.; Woo, M.K. Modern and ancient periglacial river types. *Prog. Phys. Geogr.* **2002**, *26*, 479–506. [CrossRef]
7. Verstraeten, G.; Broothaerts, N.; Van Loo, M.; Notebaert, B.; D'Haen, K.; Dusar, B.; de Brue, H. Varaibility in fluvial geomorphic response to anthropogenic disturbance. *Geomorphology* **2017**, *294*, 20–39. [CrossRef]
8. Rixhon, G.; Demoulin, A. The picturesque Ardennian valleys: Plio-Quaternary incision of the drainage system in the uplifting Ardenne. In *Landscapes and Landforms of Belgium and Luxembourg*; Demoulin, A., Ed.; Springer: Berlin, Germany, 2018; pp. 159–176.
9. Westaway, R.; Bridgland, D.R.; Mishra, S. Rheological differences between Archaean and younger crust can determine rates of Quaternary vertical motions revealed by fluvial geomorphology. *Terra Nova* **2003**, *15*, 287–298. [CrossRef]
10. Bridgland, D.R.; Westaway, R. Preservation patterns of Late Cenozoic fluvial deposits and their implications: Results from IGCP 449. *Quat. Int.* **2008**, *189*, 5–38. [CrossRef]
11. Thorndycraft, V.R.; Benito, G. Late Holocene fluvial chronolgy of Spain: The role of climatic variability and human impact. *Catena* **2006**, *66*, 34–41. [CrossRef]
12. Hoffmann, T.; Thorndycraft, V.R.; Brown, A.G.; Coulthard, T.J.; Damnati, B.; Kale, V.S.; Middelkoop, H.; Notebaert, B.; Walling, D. Human impact on fluvial regimes and sediments flux during the Holocene: Review and future research agenda. *Glob. Planet. Chang.* **2010**, *72*, 87–98. [CrossRef]
13. Wolf, D.; Seim, A.; Faust, D. Fluvial system response to external forcing and human impact—Late Pleistocene and Holocene fluvial dynamics of the lower Guadalete River in western Andalucía (Spain). *Boreas* **2014**, *43*, 422–449. [CrossRef]

quaternary

MDPI

Review

The Influence of Crustal Properties on Patterns of Quaternary Fluvial Stratigraphy in Eurasia

Tuncer Demir [1], Rob Westaway [2] and David Bridgland [3,*]

1   Department of Geography, Akdeniz University, 07070 Konyaaltı/Antalya, Turkey;
    tuncerdemir@akdeniz.edu.tr
2   School of Engineering, University of Glasgow, Glasgow G12 8QQ, UK; Robert.Westaway@glasgow.ac.uk
3   Department of Geography, Durham University, Durham DH1 3LE, UK
*   Correspondence: d.r.bridgland@durham.ac.uk

Academic Editors: Xianyan Wang, Jef Vandenberghe and Valentí Rull
Received: 3 August 2018; Accepted: 19 November 2018; Published: 5 December 2018

Abstract: Compilation of empirical data on river-terrace sequences from across Eurasia during successive International Geoscience Programme (IGCP) projects revealed marked contrasts between the records from different crustal provinces, notably between the East European Platform (EEP) and the Caledonian/Variscan/Alpine provinces of western/central Europe. Well-developed terrace staircases, often indicative of hundreds of metres of Late Cenozoic uplift/fluvial incision, are preserved in many parts of the European continent, especially westward of the EEP. In contrast, rivers within the EEP have extensive sedimentary archives that are not preserved as terrace staircases; instead, they form sets of laterally accreted sediment packages, never more than a few tens of metres above or below modern river level. There are parallels in Asia, albeit that the crust of the Asian continent has a greater proportion of tectonically active zones, at one extreme, and stable platforms/cratons at the other. The observed patterns point strongly to the mobility of lower-crustal material within younger provinces, where the continental crust is significantly hotter, as a key part of the mechanism driving the progressive uplift that has led to valley incision and the formation of river terraces: a process of erosional isostasy with lower-crustal flow as a positive-feedback driver. The contrast between these different styles of fluvial-archive preservation is of considerable significance for Quaternary stratigraphy, as such archives provide important templates for the understanding of the terrestrial record.

Keywords: river terraces; fluvial archives; uplift; crustal properties; craton; sedimentary basins

1. Introduction

This review highlights the enhancement of terrestrial Quaternary stratigraphy that has been made possible by the widespread evaluation of fluvial sequences, achieved as a result of successive International Geoscience Programme (IGCP) projects (449, 518: see acknowledgements) and the continuation of this work under the auspices of the Fluvial Archives Group (FLAG). This has allowed the comparison of similarities and differences in preservation style and has provided new insight into the relation between the patterns discerned and different types of crust. The importance of the physical properties of different crustal provinces, which were inherited from the ancient geological past, in governing Late Cenozoic geomorphic evolution, (e.g., [1–6]), is further assessed here on the scale of the Eurasian continent.

This work has founded a view of long-timescale fluvial sequences, particularly (but not solely) river terraces, as regional templates for Quaternary terrestrial stratigraphy, on the basis that they represent semi-continuous sedimentary archives representative of the period of existence of the river system to which they pertain [6–9]. Fluvial sedimentary sequences (both Quaternary and pre-Quaternary) can be valuable repositories of fossils of many types (e.g., [10]) and can also yield Palaeolithic artefact assemblages [7,11–15]. They also represent valuable data-sources for investigating palaeoclimate and landscape evolution, the joint themes of the second of the above-mentioned International Geoscience Programme projects [6]. Indeed, it is widely believed that the climatic fluctuation that has characterized the Quaternary has driven the changes in river activity, essentially between sedimentation and (downward) erosion, that have given rise to terrace sequences (e.g., [3,4,7,16]). In comparison with fluvial sequences, the sedimentary archives from other environments, such as caves and lakes, often represent shorter intervals of time; it is useful to cross-correlate these with fluvial sequences, using biostratigraphy and other proxies. Long sequences of marine terrace deposits are found in some coastal regions (e.g., [17]) and they can usefully be compared with nearby fluvial sequences, often representing interglacials, whereas the latter represent cold-stage aggradations [8,18]. In areas where loess sequences are well developed, often as 'overburden' above fluvial archives, these can be sources of complementary evidence, particularly from intercalated warm-climate palaeosols; as well as the famous occurrences in China [19–21], such sequences are well known from central Europe [22–24] and northern France (e.g., [25]).

In recent decades, the formation of Quaternary fluvial terrace staircases has been associated with 'epeirogenic' vertical crustal motions within continental interiors [3,4,6,7,16,26–29]. Such uplifting areas occur worldwide, but not everywhere. They are commonplace in the interiors of tectonic plates, suggesting that the uplift cannot be attributed to plate-tectonic activity, despite contrary suggestions (e.g., [30]). Indeed, there are several examples that show that the effects of tectonic activity are very different from those of epeirogenic movement (e.g., [31,32]) Terrace staircases are widespread in Europe, although two of the continent's largest rivers have reaches that show clear evidence that they have not been uplifting. Thus the Lower Rhine in the Netherlands is well known to coincide with a subsiding depocentre (Figure 1), with the terraces from upstream in Germany passing downstream into a stack of accumulated sediments with conventional 'layer-cake' stratigraphy (Figure 2) [33]. The Danube flows across a similar large subsiding basin, the Pannonian Basin, which formed as a result of subduction of the Carpathian Ocean in association with the formation of the Carpathian fold-mountains [30]; its infilled surface forms the Great Hungarian Plain (Figure 1) [34]. There are also smaller fault-bounded basins within the upstream parts of Rhine Graben that have been subsiding relative to adjacent terraced valley reaches, such as the Neuwied Basin [35,36]; small structural basins of this sort have generally been omitted from Figure 1.

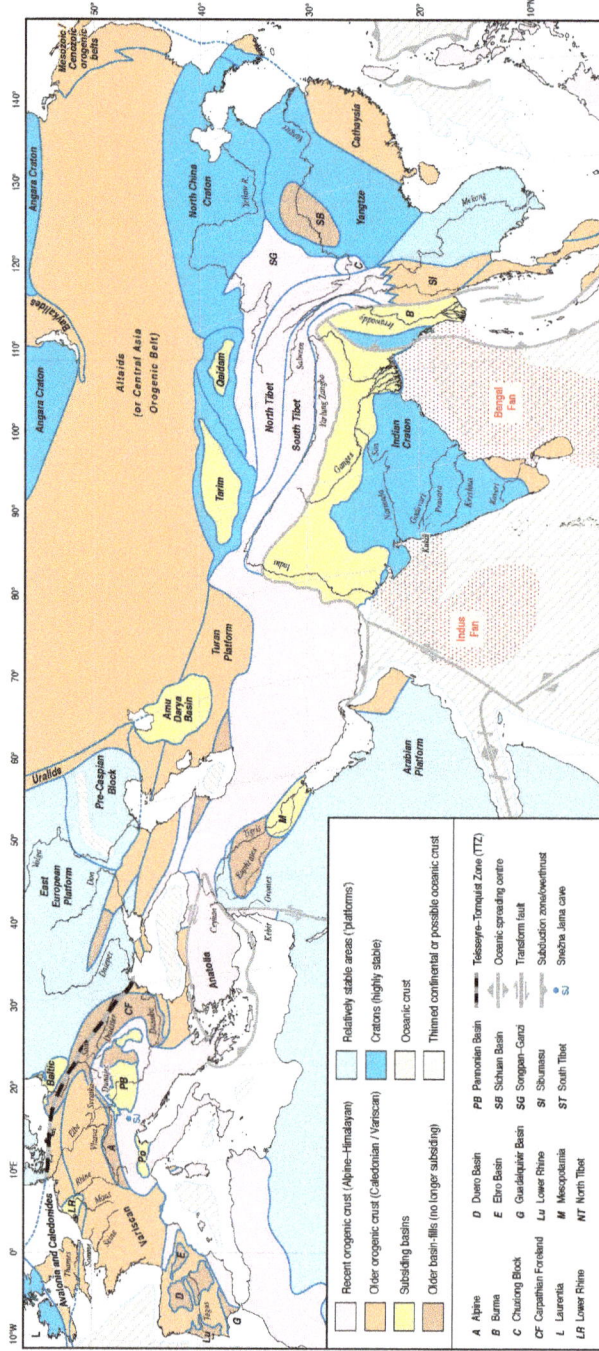

Figure 1. Crustal provinces of the Eurasian continent. Modified from numerous sources, including [37–48]; for a larger version of this figure, see Supplementary Material.

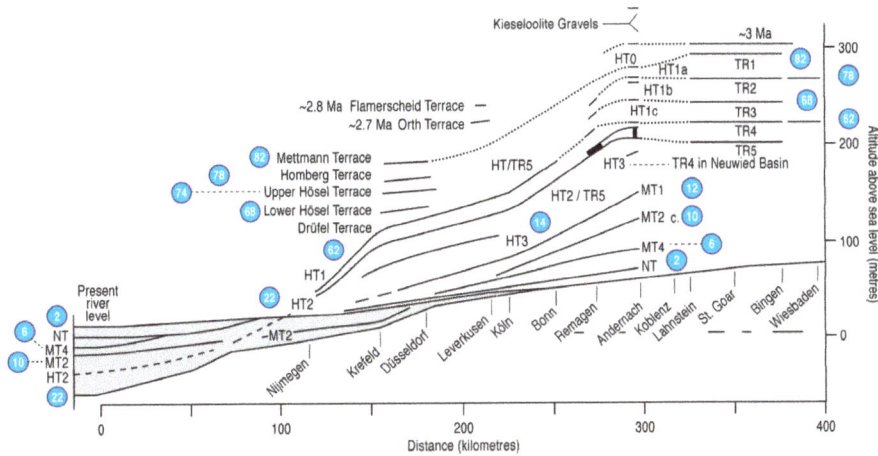

Figure 2. Longitudinal profiles of the Middle and Lower Rhine terraces and stacked sediments, from the uplifting Rhenish Massif into the subsiding coastal region of the Netherlands adapted from [49,50]; redrawn from [3]. The numbers in blue roundels are suggested marine oxygen isotope stage (MIS) correlations.

The contrast between uplifting areas, with terraces, and subsiding areas, with subterranean stacked sequences, is not the only pattern apparent. Indeed, in non-subsiding areas, there are three characteristic patterns discernible, which can be related to different types and ages of continental crust (Figure 1) [5,51]. First, in Archaean cratons, there is evidence of considerable long-term stability, in that the pattern and disposition of fluvial sediment-body preservation implies minimal vertical crustal movement during the Quaternary. Second, in most other regions, including western and central Europe, there is evidence of progressive vertical motion, upwards in the case of areas that are not loaded by accumulating sediment. Many parts of Eurasia display combined sedimentary and geomorphological records that match this scenario, these being the areas in which the world's best-documented long-timescale Late Cenozoic river terrace sequences are located, in some cases extending back for several millions of years (e.g., [52–54]). The third pattern of preservation can perhaps be regarded as intermediate between the other two, with a mixture of sediment stacking and terrace-staircase formation, with the latter never amounting to long continuous sequences, giving rise overall to little net change in valley-floor level during the latest Cenozoic–Quaternary. This pattern essentially shows alternation between uplift and subsidence on timescales of hundreds of thousands of years or more and it is typical of Early Proterozoic crustal provinces [3,4,6,55].

It has been shown that, in uplifting areas, the rates of vertical motion have varied over time, in phases that suggest correlation with patterns of climate change, leading to the suggestion that climatically induced changes in the rates of surface processes have influenced the crustal motion [3–6,56–60]. Any suggested mechanism for this effect needs also to factor in the differences in crustal movements observed from fluvial archives in contrasting crustal provinces. Erosional isostasy is a well known potential driver of progressive uplift [61]. Varying effectiveness of erosional isostasy in different crustal provinces can perhaps be accounted for by the differential operation of an important positive feedback mechanism: lower crustal flow, whereby mobile lower crust is squeezed from loaded, subsiding areas into unloaded uplifting areas [17,56,57,62,63]. The concept of lower crustal flow stems from the 'jelly sandwich' model of the continental lithosphere, which holds that the lower crust is weaker than either the upper crust or the mantle lithosphere [64–66]. This mechanism can explain the observed enhancement of uplift following significant deterioration of climate (cooling), such as occurred globally late in the Mid-Pliocene and again with the transition to 100 ka climate cycles at the Mid-Pleistocene Revolution (MPR). This works in terms of the increased rates of erosion and concomitant accelerated sedimentation in adjacent depocentres that would have occurred, thus accelerating the displacement of lower-crustal material from beneath sediment-accumulation areas to beneath the sediment sources [5,6].

An essential difference between modern ideas regarding erosional isostasy and the original form of this concept [61] concerns the relative magnitudes of forcing (i.e., erosion rates) and the associated response (i.e., uplift rates). Calculations [61], which are based on the assumption of conventional Airy isostasy, indicate that, although erosion will drive uplift, this is inevitably accompanied by a reduction in the spatially averaged altitude of the eroding landscape. On the contrary, when the effect of erosion-induced lower-crustal flow is taken into account, erosion can indeed drive uplift at a rate that exceeds the spatially averaged erosion rate of an eroding landscape [56]. As a result, erosion can drive not only uplift, but also surface uplift, meaning an increase in the spatially averaged altitude of an eroding landscape, or an increase in relief (if parts of the landscape remain at sea-level). Furthermore, if the lower-crustal rheology is linear, with viscosity independent of strain rate, then the flow rate in the lower crust will be proportional to the rate of forcing by erosion and/or sedimentation, even within the limit of the very low rates that have been observed in many intraplate regions (e.g., [56]). This mechanism can therefore cause surface uplift in such regions and the behaviour of isostatic models of this type [56] thus replicates the observed systematic worldwide increase in altitude and relief of landscape during the Late Cenozoic. These are represented by many examples, including those that are covered in the present review, in a manner that follows naturally, strongly suggesting that this general approach, incorporating lower-crustal flow, is on the right track (e.g., [59]). However, others have been slow to recognize this reasoning [58], accepting that erosion rates have increased as a result of climate change during the Late Cenozoic, and that relief has correspondingly increased, but expressing no view as to whether these changes have been associated with surface uplift and offering no opinion regarding the causal mechanism coupling landscape response to climate change. Signals have been sought in the remote, secondary situations of offshore fans (e.g., [67]), overlooking the availability of direct evidence from numerous fluvial archives across the world, which can provide much better age-control for phases of uplift that can be inferred to result from increased erosion (as discussed above). Others (e.g., [68]) have even disputed the view that erosion rates have increased during the late Cenozoic, but nonetheless accept that uplift has occurred, implying that they consider uplift and erosion to be independent processes with no causal connection.

The understanding of these mechanisms and the patterns of landscape evolution that they produce are of considerable importance to Quaternary stratigraphy. The case studies of fluvial-system stratigraphies that will be presented below, and which have provided the evidence from which the main insights have come, are well-dated exemplars amongst numerous other well-developed sedimentary archives that, for various reasons, are less well constrained temporally, and, as a result, are less well understood. Some may lack the bedrock geology that would have provided lime-rich groundwater, promoting the long-term (Quaternary-timescale) preservation of the vertebrate, molluscan and other calcareous fossils that are important for biostratigraphical age constraint (e.g., [10,69–72]). Another important bedrock-related factor is the presence or otherwise of lithologies that are suitable as raw material for Lower and Middle Palaeolithic tool-making, without which well-developed artefact assemblages and related stratigraphies are unlikely to exist. Most river systems will have sediments rich in the quartz sand that provides the basis for optically stimulated luminescence (OSL) dating, which is the most extensively used geochronological technique for fluvial archives, although infrared-stimulated luminescence and cosmogenic nuclide methodologies are fast gaining ground (e.g., [73]); some rivers have been affected by Quaternary volcanism, which has produced interbedded tephra and lava that can be dated with the reliable argon–argon geochronological method [74–78]. In systems with well-developed fluvial sedimentary archives but a dearth of dating evidence, it is possible to apply uplift–incision computer-modelling techniques [49,56] to generate an age model for the terrace (morpho) stratigraphy, which can be calibrated by any available indication of age for particular elements within the system, and, failing that, can be matched to patterns of terrace development in systems that are better dated and which occur in comparable crustal provinces and climatic zones (cf., [3–5]). It is important to note, in addition, that the examples expounded upon in this paper are dominantly from intraplate regions, where the only viable mechanism for driving and sustaining vertical crustal motions is the lower-crustal displacement described above, giving rise to 'regional' or 'epeirogenic' uplift rather than the increases in relief that can occur in plate-boundary zones as a result of plate-tectonic processes or the effects of active faulting. The latter can of course also give rise to localized vertical crustal motions, often overprinted onto a regional (epeirogenic) pattern of movement (e.g., [31,77,79]). A discussion of mechanisms of tectonic/atectonic uplift is not fundamental to this review paper, which is not the place to expand upon the various models. Readers are instead referred to the relevant papers in the Reference List.

2. Example Eurasian Records from Areas of Dynamic Crust

Western Europe boasts many of the world's best documented and most significant Quaternary river terrace sequences, amongst which are the classic records from which the modern understanding of Pleistocene landscape evolution and considerable advances in terrestrial stratigraphy have arisen. In the north-west of the continent, these include the sequences from the Thames, UK, with its biostratigraphical and Palaeolithic richness (Figure 3) [7,80–82], the Somme and Seine–Yonne systems in France, with similar riches and the addition of loess–soil overburden sequences [25,83], and the Maas, in the southernmost Netherlands, which can claim the largest number (31) of terraces of any world river, with a sequence within which the effects of both the mid-Pliocene cooling and the MPR are well illustrated and can be dated with some precision (Figure 4) [6,49,50,84,85]. The Rhine and its tributaries also have important river-terrace archives (e.g., [3,49,86–89]). Further south in western Europe, river-terrace sequences are well developed on the Iberian Peninsula [90–98], where again they are often the context for important Palaeolithic artefact records [98–102]. Parts of this peninsula are sufficiently close to the African–European plate boundary for an influence from tectonic activity to have been invoked in explanation of the disposition of fluvial archives, e.g., in the Almeria, Sorbas, Tabernas, and Vera basins of SE Spain [103]. Nonetheless, the calculated amounts of late Quaternary uplift in those systems compare closely with those from a similarly dynamic crust in NW Europe [3,4].

From the number and disposition of their terraces it can be inferred that many rivers in western Europe have, since the MPR, formed terraces at approximately one per 100 ka climatic cycle; again the Thames is a good example (Figure 3), as are the Somme and Seine [25]. In the case of the Maas, it has been suggested that more than a single terrace was formed during some Middle Pleistocene cycles (Figure 4) [50,104], whereas the (now-drowned) River Solent in southern England appears to have produced roughly two terraces in each of the last four 100 ka cycles [63]. Similarly, the Vltava, in the Czech Republic, has formed more than one terrace per 100 ka cycle during the Middle and Late Pleistocene (Figure 5) [105]. Further east, the Svratka, a Danube tributary that flows through Brno, has fewer Middle–Late Pleistocene terraces than one per 100 ka cycle. This system is dated with reference to the loess–soils overburden sequence that covers the fluvial deposits at the famous 'Red Hill' in Brno (Figure 6) [22,24]. It has been suggested [3] that the Svratka terraces correspond with the most pronounced climatic oscillations, those identified as 'supercycles' [106]; essentially those containing the marine oxygen isotope stages (MIS) that correspond with the 'Donau', 'Günz', 'Mindel', 'Riss', and 'Würm' alpine glaciations ([107] cf. [108]).

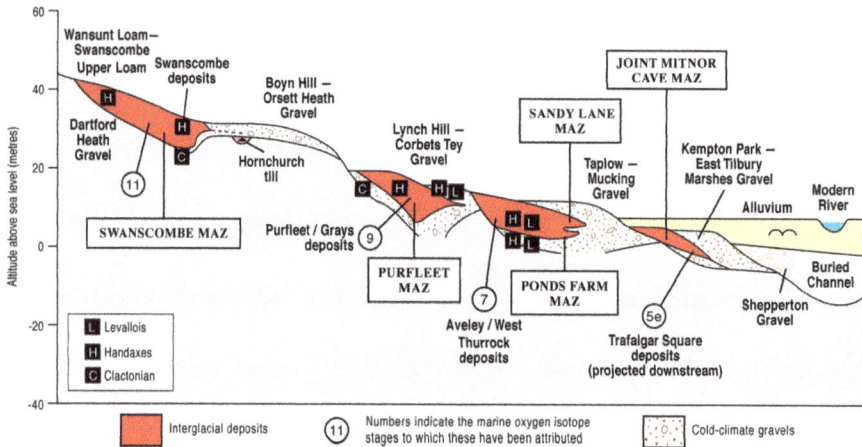

Figure 3. Idealized transverse section through the terrace sequence of the Lower Thames (specifically, the valley between central London and the estuary), showing the occurrence of Palaeolithic industries and of Mammal Assemblage Zones (MAZ). Note that this and similar diagrams are not straightforward transverse sections across the valley but are typically synthetic sections that draw information from different points along a fluvial reach, with heights compensated for upstream or downstream projection, as necessary. Modified from [82], with additions from [71,81], from which details of mammalian assemblages can be accessed.

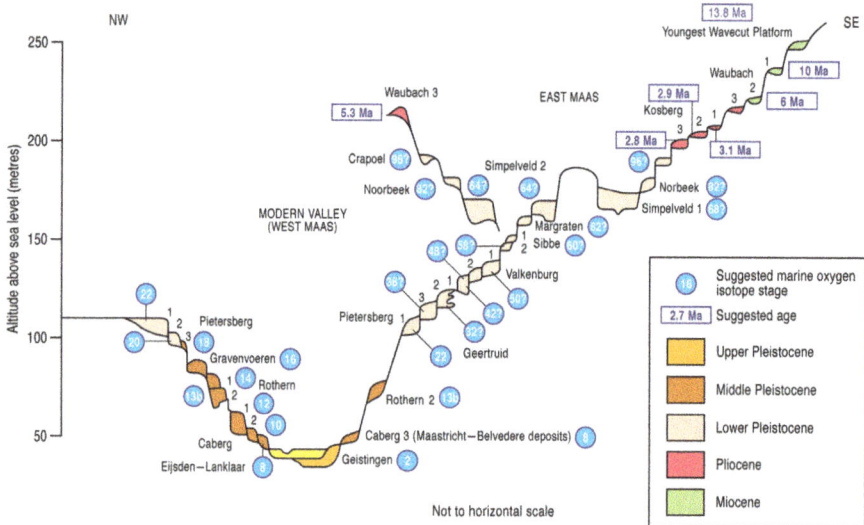

Figure 4. Idealized transverse sequence through the terraces of the River Maas, in the area of Maastricht, Netherlands. Modified from [50,84]; reproduced from [5].

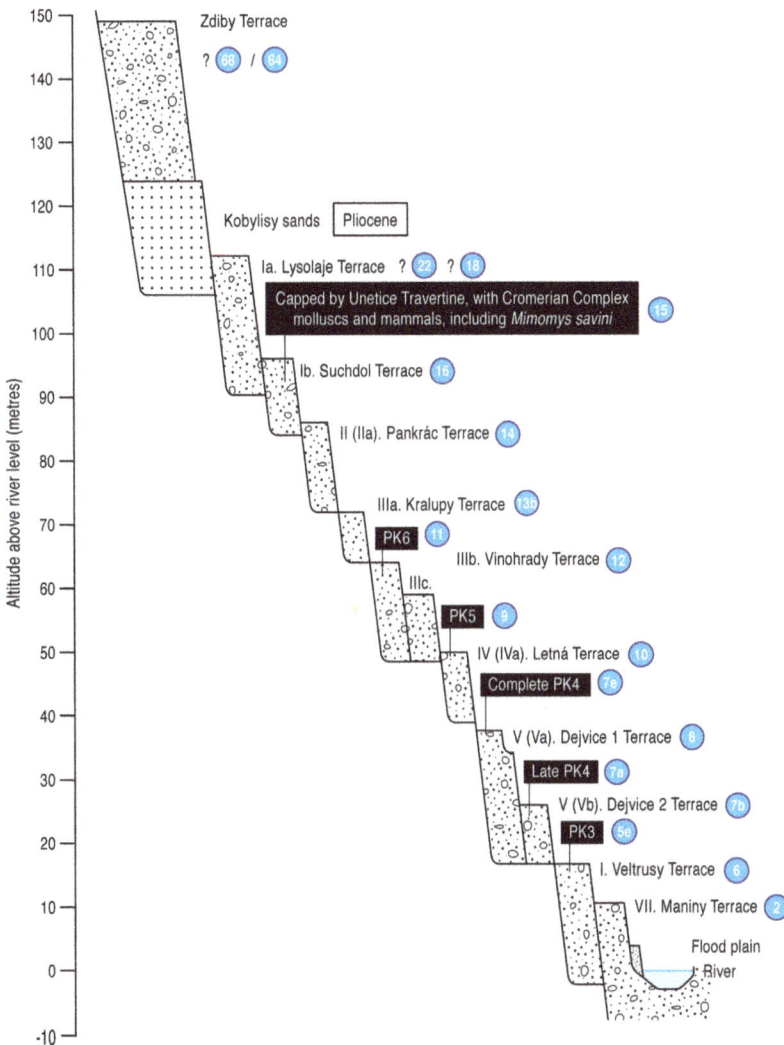

Figure 5. Idealized transverse section through the terrace sequence of the River Vltava at Prague, Czech Republic modified from [105], from which details can be accessed. The numbers in blue roundels are suggested MIS correlations.

The Danube, Europe's second largest river, flows through diverse crustal blocks that have experienced contrasting Quaternary evolution, some forming subsiding basins in which sediment has been accumulating (e.g., the Great Hungarian Plain: see above), whereas others have been uplifting and have staircases of terraces (see [34] for a review). In the uplifting reach upstream of the Great Hungarian Plain, through the Transdanubian Mountain Range, there seems to have been a terrace formed in most but not all late Middle–Late Pleistocene climate cycles, with both MIS 8 and 6 represented [3,109], as well as being marked by karstic levels within the limestone mountains [110]. Other rivers in the central–eastern part of Europe with well-developed terraces systems, in which most post-MPR climate cycles are separately represented, include several draining from the Carpathian Mountains, such as the Dniester, San, and Dunajec [32,52,111–115].

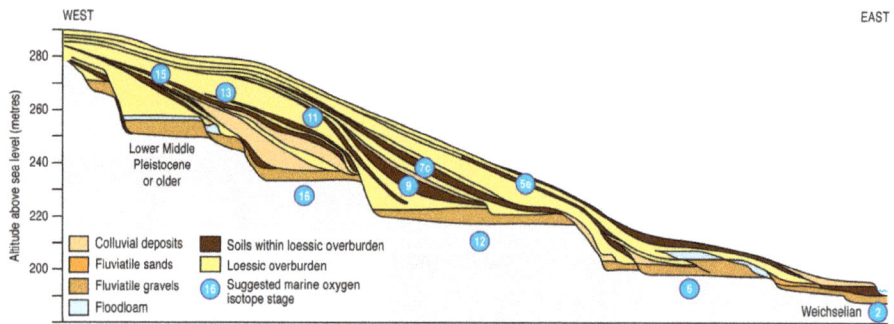

Figure 6. Idealized transverse sequence through the terraces of the River Svratka and their loessic overburden, the Red Hill, Brno, Czech Republic From [22–24]; suggested MIS ages of fluviatile gravels and soils from [7].

Young, dynamic crust also extends throughout the tectonically active Mediterranean region, where terrace staircases, both marine and fluviatile, are found in Italy [17,116], Bulgaria–Greece [5,117,118], and Turkey–Syria [18,119,120]. Turkey straddles the boundary between Europe and Asia, and terrace sequences indicative of particularly rapidly uplifting crust have been detected here in the Mediterranean coastal region from the İskenderun Gulf, east of Adana, through Hatay and into the NW corner of Syria, in the region of Latakia [18]. The dynamic crust of the wider Mediterranean region results from its deformation in response to the convergence of the African and Eurasian plates and subduction of the Tethys Ocean (e.g., [121,122]), although the reason for the rapidly deforming crust of the Adana–Latakia area, with an uplift rate of ~0.1–0.4 mm a^{-1} during the latest Middle and Late Pleistocene [18], is less clear. The evidence for this exceptional uplift comes from the terrace sequences of the Ceyhan and the lowermost Orontes in Turkey and the Nahr el Kebir in Syria, the first of which is well constrained by the Ar–Ar dating of basaltic lava emplaced above fluvial deposits down to the fourth terrace (of seven), ~90 m above floodplain level, to 278 ± 7 ka, i.e., within MIS 9 (Figure 7A) [77]. Indeed, this same volcanism has armoured the landscape around the Ceyhan course through the Amanos Mountains, preventing the complete destruction by erosion of earlier Middle Pleistocene fluvial deposits, as has occurred in the Orontes and Kebir, where the combination of rapid uplift and concomitant denudation has resulted in only the youngest few climate cycles being represented in the preserved terrace staircases (Figure 7B) [18]. It has been suggested (e.g., [123]) that terrace formation will have been limited in areas of very rapid uplift, such as New Zealand [124–127].

Another case study that has received attention as part of FLAG activity, in this case under the auspices of IGCP 518 (see above) [6], concerns an area of China with relatively rapid fluvial incision, in response to rapid uplift: the Middle Yangtze region in Yunnan. This was visited during the IGCP 518 conference/field meeting to Nanjing and the Yangtze in October 2006. Evidence from river terraces near the entrance of Tiger-Leaping Gorge, where the Yangzte traverses the Yulong mountain range, taken in combination with geomorphological and thermochronological techniques, has been used to estimate the rates of uplift in this region [128]. The area in question can be defined as the western part of the Yangtze crustal province, also known as the Chuxiong Block, which is located to the SE of the Tibetan Plateau and east of the eastern Himalayan syntaxis (Figure 1). The Tibetan plateau is a region of ~3 million km^3, which is characterized by high topography (~5–6 km above sea level (a.s.l.)) and with extreme crustal thickness (up to ~70 km), with the latter thought to result from inflow of mobile lower crust from the south in response to the convergence between the Indian and Eurasian plates (e.g., [129,130]). This extreme build-up of crustal thickness has been facilitated by Tibet being surrounded by crust with dominant properties of relative coldness and high strength (such as the Indian craton to the south, the Yangtze craton to the east, and the Tarim crustal province to the north and west (Figure 1)); these surroundings have helped to constrain the thickening crustal block and promote the elevation of the plateau. The uplift of the Tibetan Plateau has been related to the intensification of the Indian monsoon from ~8 Ma [131] and it has been calculated [130] that if all the crustal volume flux arising from the convergence between India and Eurasia had been converted into crustal thickening beneath Tibet (with none escaping), the resultant uplift rate of the land surface would have been ~0.3 mm a^{-1}.

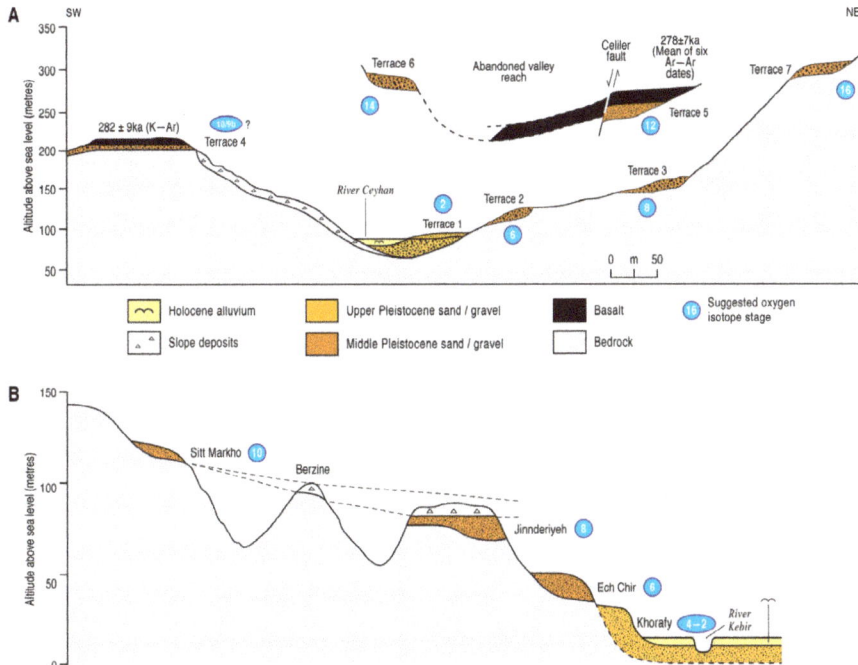

Figure 7. Cross sections (with vertical scales equalized) through example terrace sequences from the area of rapid uplift in the NE corner of the Mediterranean [18]. (**A**) The River Ceyhan valley in the vicinity of the Aslantaş Dam, near Düziçi (Turkey), showing the relation of river terrace deposits to dated basalt lavas. (**B**) The Kebir valley ~10 km upstream from Latakia, Syria, showing the relative disposition of river terraces, marine terraces (raised beaches), and slope deposits, the last-mentioned giving rise to a bogus terrace named after a hilltop at Berzine (for explanation see [18]).

3. Contrasting Records from Less Dynamic Crust: The East European Plain, Arabian Platform and Cratonic Regions Such as India

Important differences in the patterns of river-terrace preservation are observed in areas eastwards from those described above. In northern and central Europe, these relate to crustal differences that are associated with the crossing of the Teisseyre–Tornquist zone (TTZ), also called the Trans-European Suture Zone (marking the suture of the former Tornquist ocean (e.g., [132]), which runs NW–SE through Poland and western Ukraine and separates the relatively young and dynamic crust of the area described above (Variscan and younger) from the much older (Proterozoic) crust of the East European Platform to the NE (Figure 1). This important geological boundary, which marks a considerable difference in heat-flow and temperature at the mantle–crust transition (e.g., [55,133]), thus separates regions with significantly different Quaternary landscape evolution histories, as demonstrated by their river-terrace records. To the west of the suture, such records invariably point to monotonic Late Cenozoic vertical crustal motion, generally progressive uplift, as is consistent with the younger, dynamic crustal type, which has a mobile lower layer >10 km thick, beneath ~20 km of upper-crust (cf. [3–5,50,55]). To the east of the suture, there is evidence of alternations of uplift and subsidence over timescales of many hundreds of thousands of years, with little net crustal motion in either direction, as exemplified by the sedimentary archives of rivers such as the Vistula in central Poland [32], in the north, and the Dnieper and Don, rivers flowing southwards to the Black Sea [52]. Further east the East European Platform extends to the Urals, largely drained southwards by tributaries of the Volga, the latter entering the Caspian Sea across the ultra-stable pre-Caspian Block, which is thought to include a component of oceanic crust (Figure 1) [1,134].

The contrast between the two sides of the TTZ is well illustrated by the stark differences between the fluvial archives of the large south-flowing rivers of the Black Sea region, which, fortuitously, are also important repositories of stratigraphical data that is well constrained in terms of age, thanks to a combination of biostratigraphy (vertebrates and Mollusca) and geochronology (multiple techniques, including magnetostratigraphy); these are, from west to east, the Dniester, Dnieper, and Don (Figure 8) [3–5,32,52,55,113]. There are also important loess–palaeosol sequences occurring as overburden above the fluvial archives, providing a further check on the age constraint for the latter, and two important glacial episodes are also recognized, named after two of the rivers (Don and Dnieper), their sediments interbedded with the fluvial archives (Figure 8B,C).

The River Dniester has its course immediately west of the East European Platform and has a staircase of terraces that records continuous uplift since basin inversion during the Mid-Pliocene, before which there had been accumulation of fluvial sediments within the northern part of a larger Black Sea (or 'Paratethys') Basin: the Upper Miocene Balta Series (Figure 8A) [3,5,52,135]. Although this would seem to represent sedimentary isostatic subsidence, the Balta Group was accumulating as the 'coastline' of the landlocked 'Paratethys Sea' retreated southwards towards the present Black Sea. Thus, the ancestral Dniester was probably forced to aggrade to maintain its gradient, forming a stacked sequence that does not necessarily imply concomitant subsidence [5].

Neither of the other two rivers (Dnieper or Don), both situated well east of the TTZ, show this staircase pattern of river terrace preservation. Instead their sedimentary records imply that there have been alternations in the sense of vertical crustal motions, with little or no net uplift during the Quaternary (Figure 8B,C) [3,52]. Each of these rivers has a different pattern of fluvial-archive preservation. The Dnieper, ~300 km east of the Dniester, has a depositional archive in which sediment bodies of various ages, many of which can be correlated with the regimented terraces of the Dniester, occupy positions in the landscape that lie between ~40 m below and ~50 m above the modern valley floor (Figure 8B). These show no clear relation between age and elevation, but instead would appear to record alternating episodes of uplift and subsidence, with preservation also predicated by lateral migration of the river and its tributaries, the horizontal extent of the terrain represented by the summary diagram (Figure 8B) exceeding 200 km [5,55,113]. This type of preservation pattern, suggestive of little net vertical crustal motion during the Quaternary, is typical of the ancient 'cratonic' crust of continental cores, dating from the Archaean; similar evidence has been observed from cratons in South Africa and India, where early Pleistocene or even pre-Quaternary fluvial sediments are close to, or even below, modern river levels [4,51]. In the case of the Dnieper, the craton in question is the Ukrainian Shield [136]. It has been noted [5], when comparing the records from Dniester and Dnieper systems, that the marked contrast in valley evolution they record can only have resulted from the different crustal stability on either side of the TTZ, there being little difference in the climate and hydrological properties of their catchments. The crustal properties of the Ukrainian Shield and their effect on landscape evolution has been discussed at length [55]; in fact, the course of the lowermost Dniester, with its terrace staircase, lies ~50–200 km NE of the TTZ (Figure 1), seemingly within the stable zone of the East European Platform. However, the crust in that area experienced later regional metamorphism and igneous intrusion, and so it was not fully cratonized during the Archaean. Furthermore, the ancient crust in this area is deeply buried by later sediments, which has the effect of raising its temperature and therefore its potential for lower-crustal mobility [55].

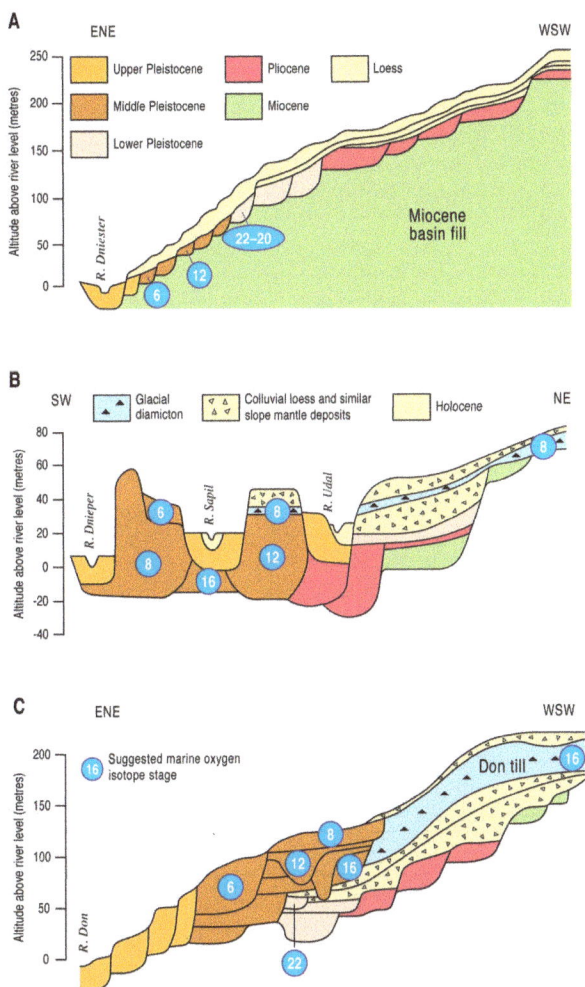

Figure 8. The Rivers of the Northern Black Sea region, showing suggested MIS correlations. Modified from [5,52]. (**A**) Idealized transverse section through the middle to lower Dniester in the Ukraine–Moldova border region. (**B**) Transverse section through the deposits of the Dnieper in the area of Kiev, central Ukraine. The SW–NE distance depicted is 240 km. (**C**) Idealized transverse section through the terraces of the River Don in the vicinity of Voronezh, Russia. Rivers located in Figure 1.

The final river in the west to east transect across this region is the Don, which flows through SW Russia, traversing the 'Voronezh Shield' or the 'Lipetsk–Losev crustal domain', representing the Early Proterozoic (~2300–1900 Ma), with no evidence of older Archaean material. The Don sequence dates back to the late Middle Miocene (Sarmatian Stage of the Paratethys realm). The summary diagram (Figure 8C) shows evidence for lengthy periods of uplift, represented by terraces, interspersed with episodes of subsidence during which the earlier terraces were buried. Thus, an initial terrace staircase, spanning the late Middle Miocene to the latest Pliocene, was formed during a period of steady uplift, but was subsequently buried beneath younger deposits, which accumulated during a phase of subsidence that began at ~2 Ma. Renewed uplift, as indicated by incision, was followed by further aggradation and the accumulation of a stacked fluvial succession interbedded with loess layers, palaeosols, and the deposits of the MIS 16 Don glaciation (Figure 8C). This accumulation culminated in MIS 8 (250 ka), when renewed incision left the capping sediments as the fourth terrace and progressed to the modern valley depth, forming the lowest terraces of the of the Don. This sequence has been explained in terms of interactions between conventional Airy isostatic compensation involving flow in the mantle and the aforementioned isostatic compensation by lower-crustal flow; these have different characteristic response times [55], with the repeated alternations of uplift and subsidence, as evidenced by the Don fluvial sequence, considered to be characteristic of crustal types in which the lower mobile layer is of highly restricted depth. This is typical of Lower Proterozoic crust, as here in the Voronezh region, but also occurs where the crust has experienced significant mafic underplating, as in rather younger crust of the Arabian Platform (see below). Indeed, seismic profiling has indicated that crust of the Voronezh Shield has a basal layer with a P-wave velocity of 6.95–7.8 km s^{-1} [137], which is consistent with mafic underplating. There are various problems in determining the crustal characteristics of this region [55], which probably has a thin mobile layer, in approximate agreement with the previous calculations (cf. [3]).

The comparison of these three fluvial sequences thus shows stark differences. The Dniester has a standard river-terrace staircase, with most 100 ka climate cycles since the MPR being represented, as well as high-level terraces representing the Pliocene and Early Pleistocene (Figure 8A). The sedimentary records of the Dnieper and Don have considerably more restricted altitude ranges, with the Dnieper being the more restricted of the two. Its succession (Figure 8B) is also less complete, notably having no deposits between the earliest Pliocene (the Parafiivka Formation) and the early Middle Pleistocene (the Traktemyriv Formation). In the Chornobyl district, ~200 km upstream of the sequence illustrated in Figure 8B, a ~35 m stacked fluvial succesion (the Chornobyl Formation) overlies the Parafiivka Formation and it is thought to represent the late Early Pliocene and Middle–Late Pliocene, although the Early Pleistocene is still lacking [52]. Downstream from the illustrated sequence, the Dnieper flows for ~400 km ESE along the Dnieper Basin, and then turns to the SW to reach the Black Sea by way of its ~300 km long 'Lower Dnieper' course, where the sedimentary sequence is even more closely spaced altitudinally [52], suggesting even greater crustal stability. This has been attributed to the Archaean crust beneath the lowermost Dnieper course [55], which has a lower heat-flow (cf. [138]) and an evidently constricted the mobile lower layer, perhaps completely missing in some places (cf. [51]).

To the south and south-east, there are valuable comparisons to be made between the highly dynamic crust that has been deformed by Alpine orogenic activity, which extends from western Syria across most of Turkey and southern-central Asia towards the Himalayas, separating the East European Platform from a comparable stable region further south, the Arabian Platform, which underlies central-eastern Syria and the southern fringe of central Turkey (Figure 1). This latter province is essentially similar to the East European Platform in terms of its properties, with mafic underplating and/or a narrow depth of mobile lower crustal material, often of limited effect. Thus the river terrace records from the Rivers Euphrates and Tigris, where they flow through this province, show evidence of alternating uplift and subsidence, or uplift and stability (Figure 9) (e.g., [5,18,54,78,139–141]), in a manner that is reminiscent of the Don (see above). The Orontes, already seen in its lowermost reach, flows northwards from Lebanon through Syria and Turkey, crosses into the stable interior of the Arabian Platform in its middle reach (Figures 1 and 9A) before entering that region of rapidly uplifting crust described above, around and downstream of Antakya in Turkey [18,119]. In north-western Syria and immediately downstream of the border with Turkey, the river passes through two small pull-apart basins that have experienced long-term subsidence, possibly throughout the Quaternary [119], another example of small structural basins of insufficient size to be depicted in Figure 1. The Arabian Platform covers the entire Arabian Peninsula to the south (Figure 1), so comparable sedimentary archives might be anticipated there, although the history of long-term aridity and resultant absence of modern-day perennial rivers explains the paucity of data, despite the importance of the region as an archive of Quaternary environmental change [72,142,143].

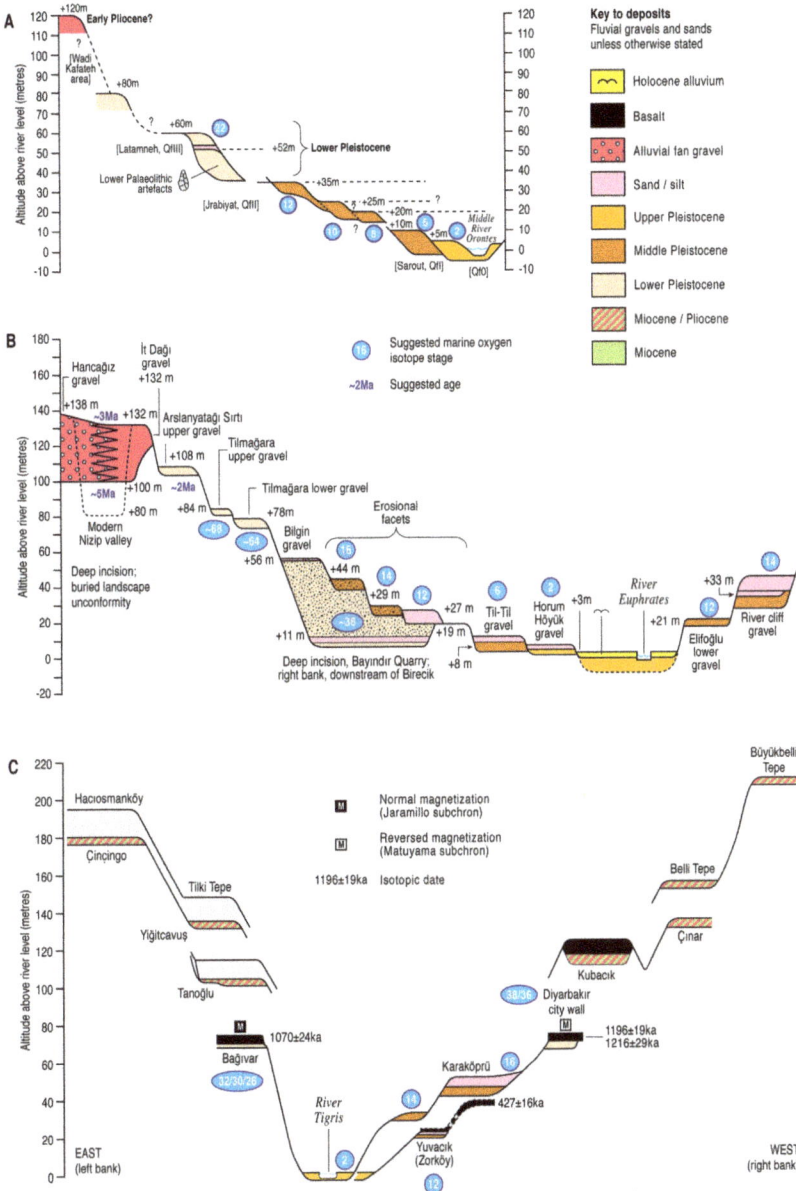

Figure 9. Fluvial archives from rivers flowing over the Arabian Platform, showing evidence for alternations between accumulation and subsidence or stability (see text) modified from [18]. (**A**) Idealized transverse section through the terrace sequence of the Middle Orontes, in the Hama–Latamneh area of Syria, showing suggested MIS correlations [119]. (**B**) Idealized transverse profile through the sequence of the Euphrates in the Birecik area, southern Turkey; Holocene overbank deposits that cover the terraces assigned to MIS 6 and 2 (cf. [144]) are omitted. (**C**) Idealized transverse profile across the River Tigris at Diyarbakır, SE Turkey, showing the disposition of terrace gravels and dated basalts, the latter revealing a lengthy period of crustal stability during the early Pleistocene.

Much of the crust further to the east belongs to the orogenic belts of the Himalayan massif, the Tibetan Plateau, and, further north, the Altaid collage (Figure 1). To the south of these orogenic regions, however, is a further ultrastable peninsula, that of the Indian subcontinent. The Indian peninsula is an Archaean craton that was one of the first to be recognized as ultrastable in terms of its Late Cenozoic fluvial archive record, with pre-Quaternary alluvial sediments at near river level in the valleys of rivers such as the Kukdi, Narmada, Pravara, and Son [3,4,51,145–148]. The main line of evidence for dating these ancient fluvial deposits, apart from weathering of their components and of bedrock below them, was Palaeolithic artefacts, although dated tephra deposits from Toba in Sumatra, which are also found near modern river level, have provided support [148–150], albeit controversial (cf. [151,152]).

4. Anomalous Records from Orogenic Belts in Central Asia

The orogenic belts of the Himalayan massif and the Tibetan Plateau are associated with the India–Eurasia collision, whereas the Altaid collage, also known as the Central Asian Orogenic Belt, is a huge area that experienced large-scale plate movement (including ocean-crust subduction) during the Palaeozoic, creating an agglomeration of different terranes, including island arcs (Figure 1) (e.g., [48,134]). There is a somewhat patchy documentation of fluvial archives from such regions, although the high mountain ranges are associated with deep foreland basins, such as that to the south of the Himalayas (Figure 1), which have experienced prolonged sediment accumulation, much of it fluvial or fluvio-deltaic (e.g., [45]). The recent plate-tectonically generated Himalayan orogenic belt comprises accreted terranes that are associated with the convergence of the Indian and Eurasian plates. These include the Tibetan Plateau (visited during the FLAG meeting that generated this publication), where a complex history of crustal accretion has given rise to west–east crustal extension on north–south striking normal faults, creating the present 'basin and range' landscape (see other contributions to this special issue). The accreted terranes, derived from the Gondwana supercontinent, are aligned west–east to the north of the Himalayan mountain range, turning towards north–south and extending, in attenuated form, into southern China (Figure 1). Here, they give rise to the 'three rivers' region, immediately east of the collision zone, in which the Salween, Mekong and Yangtze flow in deep parallel gorges within a few tens of kilometres of each other, and incised several kilometres beneath an uplifted low-relief palaeosurface [153–156]. This attenuation, which is presumed to have pre-dated gorge initiation, has been attributed to orogen-perpendicular crustal shortening and/or orogen-parallel stretching [156,157], the deeply incised valleys having been brought into greater proximity as a result of the attenuation of the crustal terranes to which they belong. Such drainage systems, being highly influenced by this tectonism, are anomalous in that they are not part of systematic patterns, as described in this paper, although the uplift that has led to the formation of these deep valleys is perhaps as much the result of crustal properties and lower-crustal processes as of the orogenic thickening of the crust; analysis of the Yangtze [128] indicates that the incision of these valleys is a consequence of regional crustal thickening in the absence of shortening, which is potentially attributable to movement of mobile lower crustal material beneath the region. Despite flowing in deep gorges, it has been suggested that the three rivers and their near neighbours and tributaries have also been involved in a complex history of drainage diversion and capture, part of the progressive enlargement westwards of the Yangtze, largely at the expense of the Red River [156].

Further west two other examples of the world's great rivers are worthy of mention, although their records are equally anomalous. First, the Yarlung–Zangbo, which drains from Tibet and, becoming the Bhramaputra, through India to the Bay of Bengal in Bangladesh, having turned sharply southward from its original west–east course and flowed through a deep and steeply graded gorge reach; this course has long been attributed to drainage diversion by river capture, with the original downstream continuation of the modern Zangbo suggested to have been the lower Irrawaddy, or, further back in time, the Red River (Figure 1) [153,155,156]. In its middle reaches in Tibet, the Yarlung–Zangbo valley preserves a remarkable sequence of terraces and evidence for transitory glacially impounded lakes, with some of the higher terraces taking the form of thick (>200 m) sedimentary sequences akin to 'perched basin-fills' [158]. These early deposits, clearly representing the river in ancestral form, date back to pre-Quaternary times, leading to the claim [158] that this reach of the Yarlung-Zangbo has existed since the Miocene, presumably when it drained to the Irrawaddy or the Red River. The second noteworthy great river of the Himalayan orogenic region is the Indus, which has also been affected by river capture during the Late Cenozoic, gaining left-bank tributary rivers draining from the Punjab [159,160]. As it crosses the Himalayan mountain range, the Indus forms what is arguably the world's deepest gorge, >5000 m deep, in the vicinity of the Nanga Parbat massif [155]. This provides a clear indication that this river has been 'locked' in its present course to the Arabian Sea throughout the Himalayan orogenesis, as is also evidenced from the provenance of sediments that were retrieved by offshore drilling in the Indus Fan (Figure 1) [159,160].

5. The Effects of Glaciation

There is little potential for comparison with areas further north from those described, as these have been repeatedly glaciated and long-timescale fluvial records are, as a result, generally absent, although again, this generalization is less well-established in Asia than in Europe. Given that sufficient precipitation is a requirement for the formation of glaciers, some northern and/or upland areas that have been sufficiently cold have seemingly escaped, or largely escaped, the effects of glaciation on account of their aridity, a prime example being large parts of the Tibetan Plateau (e.g., [161,162]). The influence upon river systems of interactions with glaciation, in both upstream and downstream areas, was reviewed recently, with the important role of deglaciation, with its (temporary) boosting of discharge and sediment supply, particularly noted [163]. Such fluvial–glacial interactions can also have a profound effect on the evolution of drainage patterns, since rivers can be diverted and even obliterated by glaciation; the recently elucidated evolution of the River Trent, in Midland England, provides a good example [164–166].

Notwithstanding the emphasis thus far given to variations in crustal properties as an influence on uplift rates, it is important to note for comparison the considerable difference in uplift rates that can result from the contrast between upland glacial erosion and extra-glacial surface processes within a single crustal province. By way of example, such a contrast has been revealed by recent research the SE margin of the Alps in Slovenia ([167,168]; cf. [59]). Here, in a high-altitude, formerly glaciated region, cave sediments that have been dated by magnetostratigraphy and cosmogenic dating have provided an insight into Late Cenozoic landscape evolution. The studied cave, Snežna Jama (Figure 1), is situated ~1530 m a.s.l. and ~950 m above the adjacent River Savinja, which is a tributary of the Sava, itself a right-bank Danube tributary. From analysis of its sediments, and by analogy with other caves, it can be inferred that Snežna Jama cave would have formed and been filled while close to local fluvial base level. Dating results have led to the conclusion that the cave was isolated from the local fluvial system [167], presumably by erosional valley-floor lowering, by ~1.9 Ma. It has been estimated [59] that the local post-MPR uplift rate might have been as high as ~1 mm a^{-1}. In contrast, in the lowlands of the Ljubljana Basin <40 km from Snežna Jama, outside the maximum limit of Pleistocene glaciation [169], the oldest river terrace deposit (of another Sava tributary) is at 564 m a.s.l., 119 m above the modern river, and is likewise dated cosmogenically and from magnetostratigraphy to ~1.9 Ma [168]. The time-averaged local uplift rate thus indicated has therefore been ~0.06 mm a^{-1}, with a maximum upper bound to the local uplift rate post-MPR of ~0.13 mm a^{-1}. Both localities are in the same crustal province of Alpine age (Figure 1), the difference between them being that one has experienced glacial erosion and the other has not.

There is thus evidence from the above case study in support of the view that glaciation has had an important effect on erosion rates in glaciated regions [72]. Nonetheless, it also provides evidence for an increase in uplift and erosion rates during the Late Cenozoic in unglaciated regions, as demonstrated by numerous examples in this paper, albeit that these rates have been typically lower than in glaciated regions of equivalent crustal properties [59]. Furthermore, the contrasting localities in Slovenia are separated by the active Sava Fault [168], with downthrow towards the Ljubljana Basin, although it is currently unclear to what extent the dramatic variation in Quaternary uplift has been accommodated by slip on this fault or by tilting or warping of the adjoining crustal blocks. Moreover, it is also currently unclear whether this fault is accommodating plate motions (a small part of the relative motion between the African and Eurasian plates) or is slipping to relieve build up of stress as a result of the warping of the adjoining crustal blocks, in response to the different rates of surface processes that are being imposed on them [170].

6. Discussion: The Influence of Crustal Province on Erosion Rates and Patterns of Stratigraphical Disposition and Preservation of Fluvial Sediments

The different styles of fluvial archive preservation in the different parts of the Eurasian continent are an important consideration in the understanding of Quaternary stratigraphy in these regions, given that fluvial sequences provide valuable templates for the Late Cenozoic terrestrial record [5,8,27]. There are significant differences between these patterns of preservation that point to important contrasts in landscape evolution, in particular, relating to the extent of valley incision [6,51]. These have important repercussions for Quaternary stratigraphical data from other environments. Thus, the occurrence of karstic caves, with their important records of fauna, hominin occupation, and datable speleothems, is directly related to valley incision, paralleling underground the records of progressive down-cutting provided above ground by river terraces [59,60,171–175]. Furthermore, the preservation patterns of fluvial archives have a direct bearing on the likelihood of preservation of evidence from ancient lacustrine and other terrestrial environments, since the former document the extent of, and likely loss to, erosional processes, which will be substantially greater in areas of dynamic and rapidly uplifting crust.

From their distribution, especially in Europe, it is clear that the different preservational patterns of fluvial archives are related to crustal type, with progressive and extensive vertical incision being evidenced from fluvial sequences in regions of relatively 'young' dynamic crust, whereas older crust generally has greater stability and has seen much less valley deepening by rivers, if any at all (see above). These different patterns of fluvial archive preservation can be matched to the different crustal provinces. As has been seen, the most stable regions, in which the fluvial archives suggest a complete or near absence of net uplift during the Quaternary, coincide with the most ancient cratonic crustal provinces, such as peninsular India and parts of the East European Platform, in particular the Ukrainian Shield (Figures 1 and 8). Such highly stable regions are the exception in the case of the East European Platform; however, over much of this crustal province, there has been limited net uplift as a result not of ultra-stability but of alternations of vertical crustal movements, resulting in periods of terrace generation with intervening periods of subsidence and burial. Such patterns are seen as far west as eastern Poland, to the north of which is an area of general subsidence, at the margins of the Baltic Basin [32]. The difference between the fluvial records from the East European Platform and those from the youngest and most dynamic crust are quite profound, although many of the comparisons above are with crust of somewhat intermediate age, such as the Variscan and Avalonia provinces (Figure 1). This is because a large part of the youngest crust, such as the Alpine and Carpathian provinces, remains tectonically active and so has fluvial archives that are less clearly related to regional vertical crustal movements.

The patterns are less clear in Asia (excepting India), despite the well-known occurrence of terraces in the valleys of many of the larger rivers, including the Huang He (Yellow River), the inspiration for the meeting from which this special issue arises [53,176]. In the very high relief areas of much of central Asia, the fluvial response to uplift will have been gorge incision, with only sporadic terrace formation, such as described above for the Middle Yangtze [128] and the Yarlung-Zangbo [158]. As already noted for Europe, the active tectonic processes in these areas make the interpretation of the fluvial archives and their relation to other forcing effects problematic. Asia also has a greater proportion of highly stable cratonic crust (Figure 1); the combination of this and the dominantly arid climate regimes of inland area has led to the persistence of endorheic drainage systems, which are probably characterized by subsidence (cf. [177]), such as the cratonic Tarim Basin (e.g., [58]). A feature of the Yellow River system is that its upper reaches pass through the basins and ranges of the high-altitude Tibetan Plateau, many of them endorheic earlier in the Quaternary and several remaining so, in total or in part [176,178–180].

7. Conclusions

It has been shown that crustal properties have a very important influence on the style of fluvial archive preservation, which is linked to their role in determining patterns of landscape evolution. Thus, the staircases of river terraces familiar in many parts of Europe, and from which much stratigraphical information (including biostratigraphy and Palaeolithic artefact sequences) is derived, are largely confined to the younger and more dynamic crustal provinces, whereas more ancient, stable crust has not undergone the progressive Quaternary uplift that is required to generate such flights of fluvial terraces. Some ancient crust, such as in the East European and Arabian Platforms, has been subject to periodic uplift, such that combinations of river terraces and limited stacked sequences are observed. The patterns of differential preservation, which have important implications for understanding and resolution of Quaternary stratigraphy, appear likely to result from the movement of lower crustal material from beneath areas under isostatic load to beneath adjacent areas undergoing Quaternary erosion, thus acting as a positive-feedback driver that sustains the uplift generated by denudational isostasy in these latter areas. This uplift has accelerated through coupling between crustal and surface processes, in response to global cooling in the late pre-Quaternary and again as a result of the greater climatic severity as part of the 100 ka cycles that followed the Mid Pleistocene Revolution. The patterns are less clear-cut in Asia, where much of the crust has been (and/or continues to be) subject to plate tectonic processes.

Supplementary Materials: The following are available online at http://www.mdpi.com/2571-550X/1/3/28/s1, Figure S1: Crustal provinces of the Eurasian continent; large-format version.

Author Contributions: Conceptualization, field data collection (Turkey and Syria) T.D., D.B. and R.W.; Methodology and Writing-and Reviewing Original Draft, D.B. and R.W.; Software, R.W.; Funding Acquisition (dating in Turkey and Syria) and presentation of oral version at INQUA 2015, T.D.

Funding: The dating of basalts in Syria was supported by HUBAK (Harran University Scientific Research Council), project 442 (T.D.). Fieldwork in Syria was supported by the Council for British Research in the Levant (serial small travel grants to D.B.)

Acknowledgments: In addition to the above-mentioned funding, Chris Orton of the Department of Geography, Durham University, is thanked for production of the figures. We would also like to thank two anonymous referees for constructive comments that have led to improvements in the paper. The synthesis included here was facilitated by the successive UNESCO-funded International Geoscience Programme (IGCP) projects, Nos 449 and 518, co-led (under the auspices of FLAG) by DB, respectively entitled 'Global Correlation of Late Cenozoic Fluvial Deposits' (2000–2004) and 'Fluvial sequences as evidence for landscape and climatic evolution in the Late Cenozoic' (2005–2007).

Conflicts of Interest: The authors declare no conflict of interest.

References

1. Nikishin, A.M.; Ziegler, P.A.; Stephenson, R.; Cloetingh, S.; Furne, A.V.; Fokin, P.A.; Ershov, A.V.; Bolotov, S.N.; Korotaev, A.S.; Alekseev, A.S.; et al. Late Precambrian to Triassic history of the East-European Craton: Dynamics of Basin Evolution. *Tectonophysics* **1996**, *268*, 23–63. [CrossRef]
2. Cloetingh, S.A.P.L.; Ziegler, P.A.; Beekman, F.; Andriessen, P.A.M.; Matenco, L.; Bada, G.; Garcia-Castellanos, D.; Hardebol, N.; Dezes, P.; Sokoutis, D. Lithospheric memory, state of stress and rheology: Neotectonic controls on Europe's intraplate continental topography. *Quat. Sci. Rev.* **2005**, *24*, 241–304. [CrossRef]
3. Bridgland, D.R.; Westaway, R. Preservation patterns of Late Cenozoic fluvial deposits and their implications: Results from IGCP 449. *Quat. Int.* **2008**, *189*, 5–38. [CrossRef]

4. Bridgland, D.R.; Westaway, R. Climatically controlled river terrace staircases: A worldwide Quaternary phenomenon. *Geomorphology* **2008**, *98*, 285–315. [CrossRef]

5. Bridgland, D.R.; Westaway, R. Quaternary fluvial archives and landscape evolution: A global synthesis. *Proc. Geol. Assoc.* **2014**, *125*, 600–629. [CrossRef]

6. Westaway, R.; Bridgland, D.R.; Sinha, R.; Demir, T. Fluvial sequences as evidence for landscape and climatic evolution in the Late Cenozoic: A synthesis of data from IGCP 518. *Glob. Planet. Chang.* **2009**, *68*, 237–253. [CrossRef]

7. Bridgland, D.R. River terrace systems in north-west Europe: An archive of environmental change, uplift and early human occupation. *Quat. Sci. Rev.* **2000**, *19*, 1293–1303. [CrossRef]

8. Bridgland, D.R.; Maddy, D.; Bates, M. River terrace sequences: Templates for Quaternary geochronology and marine–terrestrial correlation. *J. Quat. Sci.* **2004**, *19*, 203–218. [CrossRef]

9. Bridgland, D.; Keen, D.; Westaway, R. Global correlation of Late Cenozoic fluvial deposits: A synthesis of data from IGCP 449. *Quat. Sci. Rev.* **2007**, *26*, 2694–2700. [CrossRef]

10. Schreve, D.C.; Keen, D.H.; Limondin-Lozouet, N.; Auguste, P.; Santistebane, J.I.; Ubilla, M.; Matoshko, A.; Bridgland, D.R.; Westaway, R. Progress in faunal correlation of Late Cenozoic fluvial sequences 2000–2004: The report of the IGCP 449 biostratigraphy subgroup. *Quat. Sci. Rev.* **2007**, *26*, 2970–2995. [CrossRef]

11. Bridgland, D.R.; Antoine, P.; Limondin-Lozouet, N.; Santisteban, J.I.; Westaway, R.; White, M.J. The Palaeolithic occupation of Europe as revealed by evidence from the rivers: Data from IGCP 449. *J. Quat. Sci.* **2006**, *21*, 437–455. [CrossRef]

12. Mishra, S.; White, M.J.; Beaumont, P.; Antoine, P.; Bridgland, D.R.; Howard, A.J.; Limondin-Lozouet, N.; Santisteban, J.I.; Schreve, D.C.; Shaw, A.D.; et al. Fluvial deposits as an archive of early human activity. *Quat. Sci. Rev.* **2007**, *26*, 2996–3016. [CrossRef]

13. Bridgland, D.R.; White, M.J. Fluvial archives as a framework for the Lower and Middle Palaeolithic: Patterns of British artefact distribution and potential chronological implications. *Boreas* **2014**, *43*, 543–555. [CrossRef]

14. Bridgland, D.R.; White, M.J. Chronological variations in handaxes: Patterns detected from fluvial archives in NW Europe. *J. Quat. Sci.* **2015**, *30*, 623–638. [CrossRef]

15. Chauhan, P.R.; Bridgland, D.R.; Moncel, M.-H.; Antoine, P.; Bahain, J.-J.; Briant, R.M.; Cunha, P.; Locht, J.-L.; Martins, A.; Schreve, D.; et al. Fluvial deposits as an archive of early human activity: Progress during the 20 years of the Fluvial Archives Group. *Quat. Sci. Rev.* **2017**, *166*, 114–149. [CrossRef]

16. Maddy, D. Uplift-driven valley incision and river terrace formation in southern England. *J. Quat. Sci.* **1997**, *12*, 539–545. [CrossRef]

17. Westaway, R.; Bridgland, D.R. Late Cenozoic uplift of southern Italy deduced from fluvial and marine sediments: Coupling between surface processes and lower-crustal flow. *Quat. Int.* **2007**, *175*, 86–124. [CrossRef]

18. Bridgland, D.R.; Demir, T.; Seyrek, A.; Daoudd, M.; Abou Romieh, M.; Westaway, R. River terrace development in the NE Mediterranean region (Syria and Turkey): Patterns in relation to crustal type. *Quat. Sci. Rev.* **2017**, *166*, 307–323. [CrossRef]

19. Kukla, G.; An, Z. Loess stratigraphy in Central China. *Palaeogeogr. Palaeoclimatol. Palaeoecol.* **1989**, *72*, 203–225. [CrossRef]

20. Fang, X.; Li, J.; Derbyshire, E.; FitzPatrick, E.A.; Kemp, R. Micromorphology of the Beiyuan Loess-Paleosol sequence in Gansu Province, China: Geomorphological and paleoenvironmental significance. *Palaeogeogr. Palaeoclimatol. Palaeoecol.* **1994**, *111*, 289–303.

21. Ding, Z.; Derbyshire, E.; Yang, S.; Yu, Z.; Xiong, S.; Liu, T. Stacked 2.6-Ma grain size record from the Chinese loess based on five sections and correlation with the deep-sea $\delta^{18}O$ record. *Paleoceanography* **2002**, *17*, 1033–1053. [CrossRef]

22. Kukla, G.J. Loess stratigraphy of Central Europe. In *After the Australopithecines: Stratigraphy, Ecology and Culture Change in the Middle Pleistocene*; Butzer, K.W., Isaac, G.L., Eds.; Mouton: The Hague, The Netherlands, 1975; pp. 99–188, ISBN 9027976295.

23. Kukla, G.J. Pleistocene land-sea correlations. I. Europe. *Earth Sci. Rev.* **1977**, *13*, 307–374. [CrossRef]

24. Kukla, G.J. The classical European glacial stages: Correlation with deep-sea sediments. *Trans. Nebraska Acad. Sci.* **1978**, *6*, 57–93.

25. Antoine, P.; Limondin Lozouet, N.; Chaussé, C.; Lautridou, J.-P.; Pastre, J.-F.; Auguste, P.; Bahain, J.-J.; Falguères, C.; Galehb, B. Pleistocene fluvial terraces from northern France (Seine, Yonne, Somme): Synthesis, and new results from interglacial deposits. *Quat. Sci. Rev.* **2007**, *26*, 2701–2723. [CrossRef]

26. Vandenberghe, J. Timescales, climate and river development. *Quat. Sci. Rev.* **1995**, *14*, 631–638. [CrossRef]

27. Vandenberghe, J. The relation between climate and river processes, landforms and deposits during the Quaternary. *Quat. Int.* **2002**, *91*, 17–23. [CrossRef]

28. Vandenberghe, J. Climate forcing of fluvial system development; an evolution of ideas. *Quat. Sci. Rev.* **2003**, *22*, 2053–2060. [CrossRef]

29. Vandenberghe, J. The fluvial cycle at cold-warm-cold transitions in lowland regions: A refinement of theory. *Geomorphology* **2008**, *98*, 275–284. [CrossRef]

30. Cloetingh, S.; Burov, E. Lithospheric folding and sedimentary basin evolution: A review and analysis of formation mechanisms. *Basin Res.* **2011**, *23*, 257–290. [CrossRef]

31. Abou Romieh, M.; Westaway, R.; Daoud, M.; Radwan, Y.; Yassminh, R.; Khalil, A.; Al-Ashkar, A.; Loughlin, S.; Arrell, K.; Bridgland, D.R. Active crustal shortening in NE Syria revealed by deformed terraces of the River Euphrates. *Terra Nova* **2009**, *27*, 427–437. [CrossRef]

32. Krzyszkowski, D.; Bridgland, D.R.; Allen, P.; Westaway, R.; Wachecka-Kotkowska, L.; Czerwonka, J.A. Drainage evolution in the Polish Sudeten Foreland in the context of European fluvial archives. *Quat. Res.* **2018**, in press. [CrossRef]

33. Brunnacker, K.; Löscher, M.; Tillmans, W.; Urban, B. Correlation of the Quaternary terrace sequence in the lower Rhine valley and northern Alpine foothills of central Europe. *Quat. Res.* **1982**, *18*, 152–173. [CrossRef]

34. Gábris, G.; Nádor, A. Long-term fluvial archives in Hungary: Response of the Danube and Tisza rivers to tectonic movements and climatic changes during the Quaternary: A review and new synthesis. *Quat. Sci. Rev.* **2007**, *26*, 2758–2782. [CrossRef]

35. Bosinski, G. The earliest occupation of Europe: Western Central Europe. In *The Earliest Occupation of Europe*; Roebroeks, W., Van Kolfschoten, T., Eds.; University of Leiden: Leiden, The Netherlands, 1995; pp. 103–121, ISBN 90-73368-07-3.

36. Meyer, W.; Stets, J. Pleistocene to recent tectonics in the Rhenish Massif (Germany). *Neth. J. Geosci. (Geologie en Mijnbouw)* **2002**, *81*, 217–221. [CrossRef]

37. Pharaoh, T.C.; England, R.W.; Verniers, J.C.L.; Zelazniewicz, A. Introduction: Geological and geophysical studies in the Trans-European Suture Zone. *Geol. Mag.* **1997**, *134*, 585–590. [CrossRef]

38. Aplonov, S. Oil in 'Holes-in-the-Continent' (Relict Oceanic Basins). In *Paradoxes in Geology*; Briegel, U., Xiao, W.-J., Eds.; Elsevier: Amsterdam, The Netherlands, 2001; pp. 113–130.

39. Garzanti, E.; Gaetani, M. Unroofing history of Late Paleozoic magmatic arcs within the "Turan Plate" (Tuarkyr, Turkmenistan). *Sediment. Geol.* **2002**, *151*, 67–87. [CrossRef]

40. Brunet, M.-F.; Korotaev, M.V.; Ershov, A.V.; Nikishin, A.M. The South Caspian Basin: A review of its evolution from subsidence modelling. *Sediment. Geol.* **2003**, *156*, 119–148. [CrossRef]

41. Şengör, A.M.C.; Natal'in, B.A. Tectonics of the Altaids: An example of a Turkic-type Orogeny. In *Earth Structure*, 2nd ed.; van der Pluijm, B.A., Marshak, S., Eds.; W.W. Norton: New York, NY, USA, 2004; pp. 535–546.

42. Jahn, B.M. The Central Asian Orogenic Belt and growth of the continental crust in the Phanerozoic. In *Aspects of the Tectonic Evolution of China*; Malpas, J., Fletcher, C.J.N., All, J.R., Aitchison, J.C., Eds.; Special Publication 226; Geological Society: London, UK, 2004; pp. 73–100.

43. Khain, V.; Bogdanov, N. (Eds.) *International Tectonic Map of the Caspian Sea Region*; Geological Institute, Russian Academy of Sciences: Moscow, Russia, 2005; 140p.

44. Natal'in, B.A.; Şengör, A.M.C. Late Palaeozoic to Triassic evolution of the Turan and Scythian platforms: The pre-history of the Palaeo-Tethyan closure. *Tectonophysics* **2005**, *404*, 175–202. [CrossRef]

45. Sinha, R.; Kumar, R.; Tandon, S.K.; Gibling, M.R. Late Cenozoic fluvial deposits of India: An overview. *Quat. Sci. Rev.* **2007**, *26*, 2801–2822. [CrossRef]

46. Janik, T.; Grad, M.; Guterch, A.; Vozár, J.; Bielik, M.; Vozárova, A.; Hegedüs, E.; Kovács, C.A.; Kovács, I.; Keller, G.R. Crustal structure of the Western Carpathians and Pannonian Basin: Seismic models from CELEBRATION 2000 data and geological implications. *J. Geodyn.* **2011**, *52*, 97–113. [CrossRef]

47. Cunha, P.P.; de Vicente, G.; Martín-González, F. Cenozoic sedimentation along the piedmonts of thrust related basement ranges and strike-slip deformation belts of the Iberian Variscan Massif. In *The Geology of Iberia: a geodynamic approach*; Quesada, C., Oliveira, J.T., Eds.; Regional Geology Review series; Springer: Berlin, Germany, 2018; (in press). Volume 4, Chapter 5.

48. Stern, R.J.; Li, S.-M.; Keller, G.R. Continental crust of China: A brief guide for the perplexed. *Earth Sci. Rev.* **2018**, *179*, 72–94. [CrossRef]

49. Westaway, R. Flow in the lower continental crust as a mechanism for the Quaternary uplift of the Rhenish Massif, north-west Europe. In *River Basin Sediment Systems: Archives of Environmental Change*; Maddy, D., Macklin, M., Woodward, J., Eds.; Balkema: Abingdon, UK, 2001; pp. 87–167.

50. Westaway, R. Long-term river terrace sequences: Evidence for global increases in surface uplift rates in the Late Pliocene and early Middle Pleistocene caused by flow in the lower continental crust induced by surface processes. *Neth. J. Geosci.* **2002**, *81*, 305–328. [CrossRef]

51. Westaway, R.; Bridgland, D.R.; Mishra, S. Rheological differences between Archaean and younger crust can determine rates of Quaternary vertical motions revealed by fluvial geomorphology. *Terra Nova* **2003**, *15*, 287–298. [CrossRef]

52. Matoshko, A.V.; Gozhik, P.F.; Danukalova, G. Key Late Cenozoic fluvial archives of eastern Europe: The Dniester, Dnieper, Don and Volga. *Proc. Geol. Assoc.* **2004**, *115*, 141–173. [CrossRef]

53. Pan, B.; Su, H.; Hu, Z.; Hu, X.; Gao, H.; Li, J.; Kirby, E. Evaluating the role of climate and tectonics during non-steady incision of the Yellow River: Evidence from a 1.24 Ma terrace record near Lanzhou, China. *Quat. Sci. Rev.* **2009**, *28*, 3281–3290. [CrossRef]

54. Demir, T.; Seyrek, A.; Westaway, R.; Guillou, H.; Scaillet, S.; Beck, A.; Bridgland, D.R. Late Cenozoic regional uplift and localised crustal deformation within the northern Arabian Platform in southeast Turkey: Investigation of the Euphrates terrace staircase using multidisciplinary techniques. *Geomorphology* **2012**, *165–166*, 7–24. [CrossRef]

55. Westaway, R.; Bridgland, D.R. Relation between alternations of uplift and subsidence revealed by Late Cenozoic fluvial sequences and physical properties of the continental crust. *Boreas* **2014**, *43*, 505–527. [CrossRef]

56. Westaway, R. The Quaternary evolution of the Gulf of Corinth, central Greece: Coupling between surface processes and flow in the lower continental crust. *Tectonophysics* **2002**, *348*, 269–318. [CrossRef]

57. Westaway, R. Improved modelling of the Quaternary evolution of the Gulf of Corinth, incorporating erosion and sedimentation coupled by lower-crustal flow. *Tectonophysics* **2007**, *440*, 67–84. [CrossRef]

58. Herman, F.; Champagnac, J.-D. Plio-Pleistocene increase of erosion rates in mountain belts in response to climate change. *Terra Nova* **2016**, *28*, 2–10. [CrossRef]

59. Westaway, R. Feedbacks between climate change and landscape evolution. *Terra Nova Debates* **2016**. Available online: https://terranovadebates.wordpress.com/2015/10/13/debate-articles-have-changes-in-quaternary-climate-affected-erosion/ (accessed on 26 November 2018).

60. Westaway, R. Isostatic compensation of Quaternary vertical crustal motions: Coupling between uplift of Britain and subsidence beneath the North Sea. *J. Quat. Sci.* **2017**, *32*, 169–182. [CrossRef]

61. England, P.C.; Molnar, P. Surface uplift, uplift of rocks, and exhumation of rocks. *Geology* **1990**, *18*, 1173–1177. [CrossRef]

62. Morley, C.K.; Westaway, R. Subsidence in the super-deep Pattani and Malay basins of Southeast Asia: A coupled model incorporating lower-crustal flow in response to post-rift sediment-loading. *Basin Res.* **2006**, *18*, 51–84. [CrossRef]

63. Westaway, R.; Bridgland, D.R.; White, M.J. The Quaternary uplift history of central southern England: Evidence from the terraces of the Solent River system and nearby raised beaches. *Quat. Sci. Rev.* **2006**, *25*, 2212–2250. [CrossRef]

64. Chen, W.P.; Molnar, P. Focal depths of intracontinental and intraplate earthquakes and their implications for the thermal and mechanical properties of the lithosphere. *J. Geophys. Res.* **1983**, *88*, 4183–4214. [CrossRef]

65. Bürgmann, R.; Dresen, G. Rheology of the lower crust and upper mantle: Evidence from rock mechanics, geodesy, and field observations. *Annu. Rev. Earth Planet. Sci.* **2008**, *36*, 531–567. [CrossRef]

66. Westaway, R. A numerical modelling technique that can account for alternations of uplift and subsidence revealed by Late Cenozoic fluvial sequences. *Geomorphology* **2012**, *165–166*, 124–143. [CrossRef]

67. Zhang, P.; Molnar, P.; Downs, W.R. Increased sedimentation rates and grain sizes 2–4 Myr ago due to the influence of climate change on erosion rates. *Nature* **2001**, *410*, 891–897.

68. Willenbring, J.K.; Jerolmack, D.J. The null hypothesis: Globally steady rates of erosion, weathering fluxes and shelf sediment accumulation during Late Cenozoic mountain uplift and glaciation. *Terra Nova* **2016**, *28*, 11–18. [CrossRef]

69. Keen, D.H. Significance of the record provided by Pleistocene fluvial deposits and their included molluscan faunas for palaeoenvironmental reconstruction and stratigraphy: Cases from the English Midlands. *Palaeogeogr. Palaeoclimatol. Palaeoecol.* **1990**, *80*, 25–34. [CrossRef]

70. Keen, D.H. Towards a late Middle Pleistocene non-marine molluscan biostratigraphy for the British Isles. *Quat. Sci. Rev.* **2001**, *20*, 1657–1665. [CrossRef]

71. Schreve, D.C. Differentiation of the British late Middle Pleistocene interglacials: The evidence from mammalian biostratigraphy. *Quat. Sci. Rev.* **2001**, *20*, 1693–1705. [CrossRef]

72. White, T.S.; Bridgland, D.R.; Schreve, D.C.; Limondin-Lozouet, N.; Markova, A.K.; Santisteban, J.I.; Woodburne, M.O. Fossil evidence from Quaternary fluvial archives: Biostratigraphical, biogeographical and palaeoclimatic potential. *Quat. Sci. Rev.* **2017**, *166*, 150–176. [CrossRef]

73. Rixhon, G.; Briant, R.; Cordier, S.; Duval, M.; Jones, A.; Scholz, D. Dating techniques (from methodological improvements to a better knowledge of the evolution of fluvial environments). *Quat. Sci. Rev.* **2017**, *166*, 91–113. [CrossRef]

74. Pastre, J.-F. The Perrier Plateau: A Plio–Pleistocene long fluvial record in the River Allier basin, Massif Central, France. *Quaternaire* **2004**, *15*, 87–101. [CrossRef]

75. Westaway, R.; Guillou, H.; Yurtmen, S.; Beck, A.; Bridgland, D.R.; Demir, T.; Scaillet, S.; Rowbotham, G. Late Cenozoic uplift of western Turkey: Improved dating of the Kula Quaternary volcanic field and numerical modelling of the Gediz river terrace staircase. *Glob. Planet. Chang.* **2006**, *51*, 131–171. [CrossRef]

76. Bridgland, D.R.; Demir, T.; Seyrek, A.; Pringle, M.; Westaway, R.; Beck, A.R.; Rowbotham, G.; Yurtmen, S. Dating Quaternary volcanism and incision by the River Tigris at Diyarbakır, SE Turkey. *J. Quat. Sci.* **2007**, *22*, 387–393. [CrossRef]

77. Seyrek, A.; Demir, T.; Pringle, M.; Yurtmen, S.; Westaway, R.; Bridgland, D.R.; Beck, A.; Rowbotham, G. Late Cenozoic uplift of the Amanos Mountains and incision of the Middle Ceyhan river gorge, southern Turkey; Ar–Ar dating of the Düziçi basalt. *Geomorphology* **2008**, *97*, 321–355. [CrossRef]

78. Westaway, R.; Guillou, H.; Seyrek, A.; Demir, T.; Bridgland, D.R.; Scaillet, S.; Beck, A. Late Cenozoic surface uplift, basaltic volcanism, and incision by the River Tigris around Diyarbakır, SE Turkey. *Int. J. Earth Sci.* **2009**, *98*, 601–625. [CrossRef]

79. Abou Romieh, M.; Westaway, R.; Daoud, M.; Bridgland, D.R. First indications of high slip rates on active reverse faults NW of Damascus, Syria, from observations of deformed Quaternary sediments: Implications for the partitioning of crustal deformation in the Middle Eastern region. *Tectonophysics* **2012**, *538–540*, 86–104. [CrossRef]

80. Bridgland, D.R. *Quaternary of the Thames*; Geological Conservation Review Series 7; Chapman and Hall: London, UK, 1994; 401p.

81. Bridgland, D.R.; Schreve, D.C. Quaternary lithostratigraphy and mammalian biostratigraphy of the Lower Thames terrace system, south-east England. *Quaternaire* **2004**, *15*, 29–40. [CrossRef]

82. Bridgland, D.R. The Middle and Upper Pleistocene sequence in the Lower Thames; a record of Milankovitch climatic fluctuation and early human occupation of southern Britain: Henry Stopes Memorial Lecture. *Proc. Geol. Assoc.* **2006**, *117*, 281–305. [CrossRef]

83. Antoine, P.; Lautridou, J.P.; Laurent, M. Long-term fluvial archives in NW France: Response of the Seine and Somme rivers to tectonic movements, climate variations and sea-level changes. *Geomorphology* **2000**, *33*, 183–207. [CrossRef]

84. Van den Berg, M.W. Neo-tectonics in the Roer Valley Rift System. Style and rate of crustal deformation inferred from syntectonic sedimentation. *Geologie en Mijnbouw* **1994**, *73*, 143–156.

85. Van den Berg, M.W.; van Hoof, T. The Maas terrace sequence at Maastricht, SE Netherlands: Evidence for 200 m of late Neogene and Quaternary surface uplift. In *River Basin Sediment Systems: Archives of Environmental Change*; Maddy, D., Macklin, M., Woodward, J., Eds.; Balkema: Abingdon, UK, 2001; pp. 45–86.

86. Bibus, E.; Wesler, J. The middle Neckar as an example of fluviomorphological processes during the Late Quaternary Period. *Zeitschrift für Geomorphologie* **1995**, *100*, 15–26.

87. Cordier, S.; Frechen, M.; Harmand, D.; Beiner, M. Middle and Upper Pleistocene fluvial evolution of the Meurthe and Moselle valleys in the Paris Basin and the Rhenish Massif. *Quaternaire* **2005**, *16*, 201–215. [CrossRef]
88. Cordier, S.; Harmand, D.; Frechen, M.; Beiner, M. Fluvial system response to Middle and Upper Pleistocene climate change in the Meurthe and Moselle valleys (Eastern Paris Basin and Rhenish Massif). *Quat. Sci. Rev.* **2006**, *25*, 1460–1474. [CrossRef]
89. Cordier, S.; Harmand, D.; Lauer, T.; Voinchet, P.; Bahain, J.-J.; Frechen, M. Geochronological reconstruction of the Pleistocene evolution of the Sarre valley (France and Germany) using OSL and ESR dating techniques. *Geomorphology* **2012**, *165–166*, 91–106. [CrossRef]
90. Santisteban, J.I.; Schulte, L. Fluvial networks of the Iberian Peninsula: A chronological framework. *Quat. Sci. Rev.* **2007**, *26*, 2738–2757. [CrossRef]
91. Cunha, P.P.; Martins, A.A.; Huot, S.; Murray, A.; Raposo, L. Dating the Tejo river lower terraces in the Rodao area (Portugal) to assess the role of tectonics and uplift. *Geomorphology* **2008**, *102*, 43–54. [CrossRef]
92. Martins, A.A.; Cunha, P.P.; Huot, S.; Murray, A.S.; Buylaert, J.-P. Geomorphological correlation of the tectonically displaced Tejo River terraces (Gavião–Chamusca area, central Portugal) supported by luminescence dating. *Quat. Int.* **2009**, *199*, 75–91. [CrossRef]
93. Martins, A.A.; Cunha, P.P.; Rosina, P.; Osterbeck, L.; Cura, S.; Grimaldi, S.; Gomes, J.; Buylaert, J.-P.; Murray, A.; Matos, J. Geoarchaeology of Pleistocene open air sites in the Vila Nova da Barquinha-Santa Cita area (Lower Tejo River basin, central Portugal). *Proc. Geol. Assoc.* **2010**, *121*, 128–140. [CrossRef]
94. Meikle, C.; Stokes, M.; Maddy, D. Field mapping and GIS visualisation of Quaternary river terrace landforms: An example from the Rio Almanzora, SE Spain. *J. Maps* **2010**, *6*, 531–542. [CrossRef]
95. Viveen, W.; Van Balen, R.T.; Schoorl, J.M.; Veldkamp, A.; Temme, A.J.A.M.; Vidal-Romani, J.R. Assessment of recent tectonic activity on the NW Iberian Atlantic margin by means of geomorphic indices and field studies of the lower Miño River terraces. *Tectonophysics* **2012**, *544–545*, 13–30. [CrossRef]
96. Viveen, W.; Braucher, R.; Bourlès, D.; Schoorl, J.M.; Veldkamp, A.; Van Balen, R.T.; Wallinga, J.; Fernandez-Mosquera, D.; Vidal-Romani, J.R.; Sanjurjo-Sanchez, J. A 0.65 Ma chronology and incision rate assessment of the NW Iberian Miño River terraces based on [10]Be and luminescence dating. *Glob. Planet. Chang.* **2012**, *94–95*, 82–100. [CrossRef]
97. Viveen, W.; Schoorl, J.M.; Veldkamp, A.; van Balen, R.T.; Desprat, S.; Vidal-Romani, J.R. Reconstructing the interacting effects of base level, climate, and tectonic uplift in the lower Miño River terrace record: A gradient modelling evaluation. *Geomorphology* **2013**, *186*, 96–118. [CrossRef]
98. Cunha, P.P.; Martins, A.A.; Buylaert, J.-P.; Murray, A.S.; Raposo, L.; Mozzi, P.; Stokes, M. New data on the chronology of the Vale do Forno sedimentary sequence (Lower Tejo River terrace staircase) and its relevance as fluvial archive of the Middle Pleistocene in western Iberia. *Quat. Sci. Rev.* **2017**, *166*, 204–226. [CrossRef]
99. Raposo, L.; Santonja, M. The earliest occupation of Europe: The Iberian Peninsula. In *The Earliest Occupation of Europe*; Roebroeks, W., van Kolfschoten, T., Eds.; University of Leiden Press: Leiden, The Netherlands, 1995; pp. 7–25.
100. Mozzi, P.; Azevedo, T.; Nunes, E.; Raposo, L. Middle terrace deposits of the Tagus River in Alpiarça, Portugal, in relation to early human occupation. *Quat. Res.* **2000**, *54*, 359–371. [CrossRef]
101. Santonja, M.; Pérez-González, A. Mid-Pleistocene Acheulian industrial complex in the Iberian Peninsula. *Quat. Int.* **2010**, *223–224*, 154–161. [CrossRef]
102. Santonja, M.; Pérez-González, A.; Panera, J.; Rubio-Jara, S.; Méndez-Quintas, E. The coexistence of Acheulean and Ancient Middle Palaeolithic technocomplexes in the Middle Pleistocene of the Iberian Peninsula. *Quat. Int.* **2016**, *411*, 367–377. [CrossRef]
103. Stokes, M.; Mather, A.E. Response of Plio–Pleistocene alluvial systems to tectonically induced base-level changes, Vera Basin, SE Spain. *J. Geol. Soc.* **2000**, *157*, 303–316. [CrossRef]
104. Westaway, R. Geomorphological consequences of weak lower continental crust, and its significance for studies of uplift, landscape evolution, and the interpretation of river terrace sequences. *Neth. J. Geosci.* **2002**, *81*, 283–304. [CrossRef]
105. Tyráček, J.; Westaway, R.; Bridgland, D. River terraces of the Vltava and Labe (Elbe) system, Czech Republic, and their implications for the uplift history of the Bohemian Massif. *Proc. Geol. Assoc.* **2004**, *115*, 101–124. [CrossRef]

106. Kukla, G.J. Saalian Supercycle, Mindel/Riss Interglacial and Milankovitch's Dating. *Quat. Sci. Rev.* **2005**, *24*, 1573–1583. [CrossRef]

107. Penck, A.; Brückner, E. *Die Alpen im Eiszeitalter*; Verlag Tauchnitz: Leipzig, Germany, 1901–1909; 3 Volumes; 1396p.

108. Šibrava, V. Correlations of European glaciations and their relation to the deep sea record. *Quat. Sci. Rev.* **1986**, *5*, 433–442. [CrossRef]

109. Gascoyne, M.; Schwarcz, H.P. Carbonate and sulphate precipitates. In *Uranium Series Diseqilibrium: Applications to Environmental Problems*; Ivanovitch, M., Harmon, R.S., Eds.; Clarendon Press: Oxford, UK, 1982; pp. 268–301.

110. Ruszkiczay-Rüdiger, Z.; Fodor, L.; Bada, G.; Leél-Össy, S.; Horváth, E.; Dunai, T.J. Quantification of Quaternary vertical movements in the central Pannonian Basin: A review of chronologic data along the Danube River, Hungary. *Tectonophysics* **2005**, *410*, 157–172. [CrossRef]

111. Zuchiewicz, W. Pozycja stratygraficzna tarasów Dunajca w Karpatach Zachodnich. *Przegląd Geologiczny* **1992**, *40*, 436–444, (In Polish with English summary).

112. Zuchiewicz, W. Quaternary tectonics of the Outer West Carpathians, Poland. *Tectonophysics* **1998**, *297*, 121–132. [CrossRef]

113. Matoshko, A.V.; Gozhik, P.F.; Ivchenko, A.S. The fluvial archive of the Middle and Lower Dnieper (a review). *Neth. J. Geosci.* **2002**, *81*, 339–355. [CrossRef]

114. Starkel, L. Climatically controlled terraces in uplifting mountain areas. *Quat. Sci. Rev.* **2003**, *22*, 2189–2198. [CrossRef]

115. Olszak, J. Evolution of fluvial terraces in response to climate change and tectonic uplift during the Pleistocene: Evidence from Kamienica and Ochotnica River valleys (Polish Outer Carpathians). *Geomorphology* **2011**, *129*, 71–78. [CrossRef]

116. Karner, D.B.; Marra, F. Correlation of fluviodeltaic aggradational sections with glacial climate history: A revision of the Pleistocene stratigraphy of Rome. *Geol. Soc. Am. Bull.* **1998**, *110*, 748–758. [CrossRef]

117. Westaway, R. Late Cenozoic extension in southwest Bulgaria: A synthesis. In *Tectonic Development of the Eastern Mediterranean Region*; Robertson, A.H.F., Mountrakis, D., Eds.; Special Publication 260; Geological Society: London, UK, 2006; pp. 557–590.

118. Zagorchev, I. Late Cenozoic development of the Strouma and Mesta fluviolacustrine systems, SW Bulgaria. *Quat. Sci. Rev.* **2007**, *26*, 2783–2800. [CrossRef]

119. Bridgland, D.R.; Westaway, R.; Abou Romieh, M.; Candy, I.; Daoud, M.; Demir, T.; Galiatsatos, N.; Schreve, D.C.; Seyrek, A.; Shaw, A.; et al. The River Orontes in Syria and Turkey: Downstream variation of fluvial archives in different crustal blocks. *Geomorphology* **2012**, *165–166*, 25–49. [CrossRef]

120. Seyrek, A.; Demir, T.; Westaway, R.; Guillou, H.; Scaillet, S.; White, T.S.; Bridgland, D.R. The kinematics of central-southern Turkey and northwest Syria revisited. *Tectonophysics* **2014**, *618*, 35–66. [CrossRef]

121. Aktaş, G.; Robertson, A.H.F. The Maden Complex, SE Turkey: Evolution of a Neotethyan active margin. In *The Geological Evolution of the Eastern Mediterranean*; Dixon, J.E., Robertson, A.H.F., Eds.; Special Publications; Geological Society of London: London, UK, 1984; Volume 17, pp. 375–402.

122. Allen, M.B.; Armstrong, H.A. Arabia–Eurasia collision and the forcing of mid-Cenozoic global cooling. *Palaeogeogr. Palaeoclimatol. Palaeoecol.* **2008**, *265*, 52–58. [CrossRef]

123. Veldkamp, A.; Van Dijke, J.J. Simulating internal and external controls on fluvial terrace stratigraphy: A qualitative comparison with the Maas record. *Geomorphology* **2000**, *33*, 225–236. [CrossRef]

124. Adams, J. Contemporary uplift and erosion of the Southern Alps, New Zealand: Summary. *Geol. Soc. Am. Bull.* **1980**, *91*, 2–4. [CrossRef]

125. Batt, G.E.; Braun, J. The tectonic evolution of the Southern Alps, New Zealand: Insights from fully thermally coupled dynamical modelling. *Geophys. J. Int.* **1999**, *136*, 403–420. [CrossRef]

126. House, M.A.; Gurnis, M.; Kamp, P.J.J.; Sutherland, R. Uplift in the Fiordland Region, New Zealand: Implications for Incipient Subduction. *Science* **2002**, *297*, 2038–2041. [CrossRef]

127. Claessens, L.; Veldkamp, A.; ten Broeke, E.M.; Vloemans, H. A Quaternary uplift record for the Auckland region, North Island, New Zealand, based on marine and fluvial terraces. *Glob. Planet. Chang.* **2009**, *68*, 383–394. [CrossRef]

128. Westaway, R. Active crustal deformation beyond the SE margin of the Tibetan Plateau: Constraints from the evolution of fluvial systems. *Glob. Planet. Chang.* **2009**, *68*, 395–417. [CrossRef]

129. Zhao, W.; Morgan, W.J.P. Injection of Indian crust into Tibetan lower crust; a two-dimensional finite element model study. *Tectonics* **1987**, *6*, 489–504. [CrossRef]

130. Westaway, R. Crustal volume balance during the India–Eurasia collision and altitude of the Tibetan plateau: A working hypothesis. *J. Geophys. Res.* **1995**, *100*, 15173–15194. [CrossRef]

131. Molnar, P.; England, P.; Martinod, J. Mantle dynamics, uplift of the Tibetan Plateau, and the Indian monsoon. *Rev. Geophys.* **1993**, *31*, 357–396. [CrossRef]

132. Grad, M.; Jensen, S.L.; Keller, G.R.; Guterch, A.; Thybo, H.; Janik, T.; Tiira, T.; Yliniemi, J.; Luosto, U.; Motuza, G.; et al. Crustal structure of the Trans-European suture zone region along POLONAISE'97 seismic profile P4. *J. Geophys. Res.* **2003**, *108*, 2511. [CrossRef]

133. Goes, S.; Spakman, W.; Bijwaard, H. A lower mantle source for Central European volcanism. *Science* **1999**, *286*, 1928–1931. [CrossRef]

134. Şengör, A.M.C.; Natal'in, B.A.; Burtman, V.S. Evolution of the Altaid tectonic collage and Palaeozoic crustal growth in Eurasia. *Nature* **1993**, *364*, 299–307. [CrossRef]

135. Matoshko, A.; Gozhik, P.; Semenenko, V. Late Cenozoic fluvial development within the coastal plains and shelf of the Sea of Azov and Black Sea basin. *Glob. Planet. Chang.* **2009**, *68*, 270–287. [CrossRef]

136. Shchipansky, A.A.; Bogdanova, S.V. The Sarmatian crustal segment: Precambrian correlation between the Voronezh Massif and the Ukrainian Shield across the Dniepr-Donets Aulacogen. *Tectonophysics* **1996**, *268*, 109–125. [CrossRef]

137. Tarkov, A.P.; Basula, I.V. Inhomogenous structure of the Voronezh Shield lithosphere from explosion seismology data. *Phys. Earth Planet. Inter.* **1983**, *31*, 281–292. [CrossRef]

138. Kutas, R.I. A geothermal model of the Earth's crust on the territory of the Ukrainian shield. In *Terrestrial Heat Flow in Europe*; Čermák, V., Rybach, L., Eds.; Springer: Berlin, Germany, 1979; pp. 309–315.

139. Demir, T.; Westaway, R.; Bridgland, D.; Pringle, M.; Yurtmen, S.; Beck, A.; Rowbotham, G. Ar–Ar dating of Late Cenozoic basaltic volcanism in northern Syria: Implications for the history of incision by the River Euphrates and uplift of the northern Arabian Platform. *Tectonics* **2007**, *26*, TC3012. [CrossRef]

140. Demir, T.; Westaway, R.; Seyrek, A.; Bridgland, D. Terrace staircases of the River Euphrates in southeast Turkey, northern Syria and western Iraq: Evidence for regional surface uplift. *Quat. Sci. Rev.* **2007**, *26*, 2844–2863. [CrossRef]

141. Demir, T.; Seyrek, A.; Westaway, R.; Bridgland, D.; Beck, A. Late Cenozoic surface uplift revealed by incision by the River Euphrates at Birecik, southeast Turkey. *Quat. Int.* **2008**, *186*, 132–163. [CrossRef]

142. Breeze, P.S.; Drake, N.A.; Groucutt, H.S.; Parton, A.; Jennings, R.P.; White, T.S.; Clark-Balzan, L.; Shipton, C.; Scerri, E.M.L.; Stimpson, C.M.; et al. Remote sensing and GIS techniques for reconstructing Arabian palaeohydrology and identifying archaeological sites. *Quat. Int.* **2015**, *382*, 98–119. [CrossRef]

143. Breeze, P.S.; Groucutt, H.S.; Drake, N.A.; White, T.S.; Jennings, R.P.; Petraglia, M.P. Palaeohydrological corridors for hominin dispersals in the Middle East ~250–270,000 years ago. *Quat. Sci. Rev.* **2016**, *144*, 155–185. [CrossRef]

144. Kuzucuoğlu, C.; Fontugne, M.; Mouralis, D. Holocene terraces in the Middle Euphrates valley between halfeti and karkemish (Gaziantep, Turkey). *Quaternaire* **2004**, *15*, 195–206. [CrossRef]

145. Mishra, S.; Rajaguru, S. Quaternary deposits at Bhedaghat, near Jabalpur, Madhya Pradesh. *Man Environ.* **1993**, *18*, 7–13.

146. Mishra, S. Prehistoric and Quaternary studies at Nevasa: The last forty years. *Mem. Geol. Soc. India* **1995**, *32*, 324–332.

147. Mishra, S.; Naik, S.; Adhav, U.; Deo, S.; Rajaguru, S. Studies in the geomorphology, Quaternary palaeoenvironments and archaeology of the Vel River, a tributary of the Bhima in western Maharashtra. *Man Environ.* **1999**, *24*, 159–166.

148. Westaway, R.; Mishra, S.; Deo, S.; Bridgland, D.R. Methods for determination of the age of Pleistocene tephra, derived from eruption of Toba, in central India. *J. Earth Syst. Sci.* **2011**, *120*, 503–530. [CrossRef]

149. Mishra, S.; Venkatesan, T.R.; Rajaguru, S.N.; Somayajulu, B.L.K. Earliest Acheulian industry from Peninsular India. *Curr. Anthropol.* **1995**, *36*, 847–851. [CrossRef]

150. Horn, P.; Müller-Sohnius, D.; Storzer, D.; Zöller, L. K–Ar, fission track, and thermoluminescence ages of Quaternary volcanic tuffs and their bearing on Acheulian artefacts from Bori, Kukdi valley, Pune district, India. *Z. Deutschen Geol. Ges.* **1993**, *144*, 327–329.

151. Shane, P.; Westgate, J.; Williams, M.; Korisettar, R. New geochemical evidence for the youngest Toba tuff in India. *Quat. Res.* **1995**, *44*, 200–204. [CrossRef]

152. Westgate, J.; Shane, P.; Pearce, N.; Perkins, W.; Korisettar, R.; Chesner, C.; Williams, M.; Acharyya, S. All the Toba occurrences across peninsular India belong to 75,000 bp eruption. *Quat. Res.* **1998**, *50*, 107–112. [CrossRef]

153. Koons, P.O. Modeling the topographic evolution of collisional belts. *Annu. Rev. Earth Planet. Sci.* **1995**, *23*, 375–408. [CrossRef]

154. Wang, E.; Burchfiel, B.C. Interpretation of Cenozoic tectonics in the right-lateral accommodation zone between the Ailao Shan Shear Zone and the eastern Himalayan syntaxis. *Int. Geol. Rev.* **1997**, *39*, 191–219. [CrossRef]

155. Brookfield, M.E. The evolution of the great river systems of southern Asia during the Cenozoic India–Asia collision: Rivers draining southwards. *Geomorphology* **1998**, *2*, 285–312. [CrossRef]

156. Clark, M.K.; Schoenbohm, L.M.; Royden, L.H.; Whipple, K.X.; Burchfiel, B.C.; Zhang, X.; Tang, W.; Wang, E.; Chen, L. Surface uplift, tectonics, and erosion of eastern Tibet from large-scale drainage patterns. *Tectonics* **2004**, *23*, TC1006. [CrossRef]

157. Hallet, B.; Molnar, P. Distorted drainage basins as markers of crustal strain east of the Himalaya. *J. Geophys. Res.* **2001**, *106*, 13697–13709. [CrossRef]

158. Zhu, S.; Wu, Z.; Zhao, X.; Li, J.; Xiao, K. Ages and genesis of terrace flights in the middle reaches of the Yarlung Zangbo River, Tibetan Plateau, China. *Boreas* **2014**, *43*, 485–504. [CrossRef]

159. Clift, P.D.; Lee, J.I.; Hildebrand, P.; Shimizu, N.; Layne, G.D.; Blusztajn, J.; Blum, J.D.; Garzanti, E.; Khan, A.A. Nd and Pb isotope variability in the Indus River system: Implications for sediment provenance and crustal heterogeneity in the Western Himalaya. *Earth Planet. Sci. Lett.* **2002**, *200*, 91–106. [CrossRef]

160. Clift, P.D.; Blusztajn, J. Reorganization of the western Himalayan river system after five million years ago. *Nature* **2005**, *438*, 1001–1003. [CrossRef] [PubMed]

161. Stroeven, A.P.; Hättestrand, C.; Heyman, J.; Harbor, J.; Li, Y.K.; Zhou, L.P.; Caffee, M.W.; Alexanderson, H.; Kleman, J.; Ma, H.Z.; et al. Landscape analysis of the Huang He headwaters, NE Tibetan Plateau—Patterns of glacial and fluvial erosion. *Geomorphology* **2009**, *103*, 212–226. [CrossRef]

162. Zhou, S.Z.; Li, J.J.; Zhao, J.D.; Wang, J.; Zheng, J.X. Quaternary glaciations: Extent and chronology in China. In *Quaternary Glaciations—Extent and Chronology. A Closer Look*; Ehlers, J., Gibbard, P.L., Hughes, P.D., Eds.; Developments in Quaternary Science No. 15; Elsevier: Amsterdam, The Netherlands, 2011; pp. 981–1002.

163. Cordier, S.; Adamson, K.; Delmas, M.; Calvet, M.; Harmand, D. Of ice and water: Quaternary fluvial response to glacial forcing. *Quat. Sci. Rev.* **2017**, *166*, 57–73. [CrossRef]

164. Bridgland, D.R.; Howard, A.J.; White, M.J.; White, T.S. *The Quaternary of the Trent*; Oxbow Books: Oxford, UK, 2014; 406p.

165. Bridgland, D.R.; Howard, A.J.; White, M.J.; White, T.S.; Westaway, R. New insight into the Quaternary evolution of the River Trent, UK. *Proc. Geol. Assoc.* **2015**, *126*, 466–479. [CrossRef]

166. White, T.S.; Bridgland, D.R.; Westaway, R.; Straw, A. Evidence for late Middle Pleistocene glaciation of the British margin of the southern North Sea. *J. Quat. Sci.* **2017**, *32*, 261–275. [CrossRef]

167. Häeselmann, P.; Mihevc, A.; Pruner, P.; Horáček, I.; Čermák, S.; Hercman, H.; Sahy, D.; Fiebig, M.; Zupan Hajna, N.; Bosák, P. Snežna Jama (Slovenia): Interdisciplinary dating of cave sediments and implication for landscape evolution. *Geomorphology* **2015**, *247*, 10–24. [CrossRef]

168. Mihevc, A.; Bavec, M.; Häuselmann, P.; Fiebig, M. Dating of the Udin Boršt conglomerate terrace and implication for tectonic uplift in the northwestern part of the Ljubljana Basin (Slovenia). *Acta Carsol.* **2015**, *44*, 169–176. [CrossRef]

169. Ferk, M.; Gabrovec, M.; Komac, B.; Zorn, M.; Stepišnik, U. Pleistocene glaciation in Mediterranean Slovenia. In *Quaternary Glaciation in the Mediterranean Mountains*; Hughes, P.D., Woodward, J.C., Eds.; Special Publication 433; Geological Society: London, UK, 2017; pp. 179–191.

170. Westaway, R. Investigation of coupling between surface processes and induced flow in the lower continental crust as a cause of intraplate seismicity. *Earth Surf. Process. Landf.* **2006**, *31*, 1480–1509. [CrossRef]

171. Waltham, A.C.; Simms, M.J.; Farrant, A.J.; Goldie, H.S. *Karst and Caves of Great Britain*; Geological Conservation Review Series 12; Chapman & Hall: London, UK, 1997; 358p.

172. Granger, D.E.; Fabel, D.; Palmer, A.N. Pliocene–Pleistocene incision of the Green River, Kentucky, determined from radioactive decay of cosmogenic ^{26}Al and ^{10}Be in Mammoth Cave sediments. *Geol. Soc. Am. Bull.* **2001**, *113*, 825–836. [CrossRef]

173. Westaway, R. Late Cenozoic uplift of the eastern United States revealed by fluvial sequences of the Susquehanna and Ohio systems: Coupling between surface processes and lower-crustal flow. *Quat. Sci. Rev.* **2007**, *26*, 2823–2843. [CrossRef]

174. Westaway, R. Quaternary uplift of northern England. *Glob. Planet. Chang.* **2009**, *68*, 257–282. [CrossRef]

175. Meyer, M.C.; Cliff, R.A.; Spötl, C. Speleothems and mountain uplift. *Geology* **2011**, *39*, 447–450. [CrossRef]

176. Hu, Z.; Pan, B.; Bridgland, D.R.; Vandenberghe, J.; Guo, L.; Fan, Y.; Westaway, R. The linking of the upper-middle and lower reaches of the Yellow River as a result of fluvial entrenchment. *Quat. Sci. Rev.* **2017**, *166*, 324–338. [CrossRef]

177. Yu, X.; Guo, Z.; Fu, S. Endorheic or exorheic: Differential isostatic effects of Cenozoic sediments on the elevations of the cratonic basins around the Tibetan Plateau. *Terra Nova* **2015**, *27*, 21–27. [CrossRef]

178. Perrineau, A.; Van der Woerd, J.; Gaudemer, Y.; Jing, L.-Z.; Pik, R.; Tapponnier, P.; Thuizat, R.; Zheng, R. Incision rate of the Yellow River in Northeastern Tibet constrained by ^{10}Be and ^{26}Al cosmogenic isotope dating of fluvial terraces: Implications for catchment evolution and plateau building. In *EGU General Assembly Conference Abstracts*; Special Publication 353; Geological Society: London, UK, 2001; pp. 189–219.

179. Craddock, H.W.; Kirby, E.; Harkins, W.N.; Zhang, H.; Shi, X.; Liu, J. Rapid fluvial incision along the Yellow River during headward basin integration. *Nat. Geosci.* **2010**, *3*, 209–213. [CrossRef]

180. Hu, Z.; Wang, X.; Pan, B.; Bridgland, D.R.; Vandenberghe, J. *The Quaternary of the Upper Yellow River and Its Environs: Field Guide*; Quaternary Research Association: London, UK, 2017; 84p.

quaternary

MDPI

Article

The Lowermost Tejo River Terrace at Foz do Enxarrique, Portugal: A Palaeoenvironmental Archive from c. 60–35 ka and Its Implications for the Last Neanderthals in Westernmost Iberia

Pedro P. Cunha [1,*], António A. Martins [2], Jan-Pieter Buylaert [3], Andrew S. Murray [4], Maria P. Gouveia [1], Eric Font [5,6], Telmo Pereira [7], Silvério Figueiredo [8], Cristiana Ferreira [9], David R. Bridgland [10], Pu Yang [1], José C. Stevaux [11] and Rui Mota [1]

[1] MARE—Marine and Environmental Sciences Centre, Department of Earth Sciences, University of Coimbra, Rua Sílvio Lima, Univ. Coimbra—Pólo II, 3030-790 Coimbra, Portugal; mariamporto@gmail.com (M.P.G.); pu.yang991@gmail.com (P.Y.); ruimota16@gmail.com (R.M.)
[2] ICT—Instituto de Ciências da Terra, Departamento de Geociências, Universidade de Évora, Rua Romão Ramalho, 59, 7000-671 Évora, Portugal; aam@uevora.pt
[3] Centre for Nuclear Technologies, Technical University of Denmark, DTU Risø Campus, DK-4000 Roskilde, Denmark; jabu@dtu.dk
[4] Nordic Laboratory for Luminescence Dating, Aarhus University, DTU Risø Campus, DK-4000 Roskilde, Denmark; anmu@dtu.dk
[5] IDL-FCUL, Instituto Dom Luís, Faculdade de Ciências da Universidade de Lisboa, Campo Grande, 1749-016 Lisboa, Portugal; font_eric@hotmail.com
[6] Department of Earth Sciences, University of Coimbra, Rua Sílvio Lima, Univ. Coimbra—Pólo II, 3030-790 Coimbra, Portugal
[7] ICArEHB—Interdisciplinary Center for Archaeology and Evolution of Human Behaviour, Faculdade de Ciências Humanas e Sociais, Universidade do Algarve, Campus da Penha, 8005-139 Faro, Portugal; telmojrpereira@gmail.com
[8] Geosciences Center—U.C., Portuguese Center of Geo-History and Prehistory, Polytechnic Institute of Tomar, Quinta do Contador, Estrada da Serra, 2300-313 Tomar, Portugal; silverio.figueiredo@ipt.pt
[9] Geosciences Center—U.C., Portuguese Center of Geo-History and Prehistory, Earth and Memory Institute, 6120-721 Mação, Portugal; ferreira.cris.00@gmail.com
[10] Department of Geography, Durham University, South Road, Durham DH1 3LE, UK; d.r.bridgland@durham.ac.uk
[11] Universidade Federal de Mato Grosso do Sul/Camus, Avenida Ranulpho Marques Leal, 3484, Três Lagoas 76620-080, MS, Brazil; josecstevaux@gmail.com
*   Correspondence: pcunha@dct.uc.pt

Academic Editors: Jef Vandenberghe and Valentí Rull
Received: 11 October 2018; Accepted: 11 January 2019; Published: 18 January 2019

Abstract: Reconstruction of Pleistocene environments and processes in the sensitive geographical location of westernmost Iberia, facing the North Atlantic Ocean, is crucial for understanding impacts on early human communities. We provide a characterization of the lowest terrace (T6) of the Lower Tejo River, at Vila Velha de Ródão (eastern central Portugal). This terrace comprises a lower gravel bed and an upper division consisting of fine to very fine sands and coarse silts. We have used a multidisciplinary approach, combining geomorphology, optically stimulated luminescence (OSL) dating, grain-size analysis and rock magnetism measurement, in order to provide new insights into the environmental changes coincident with the activity of the last Neanderthals in this region. In addition, we conducted palynological analysis, X-ray diffraction measurement and scanning electron microscopy coupled with energy dispersive spectra of the clay fraction and carbonate concretions. We discuss these new findings in the context of previously published palaeontological and archeological data. The widespread occurrence of carbonate concretions and rizoliths in the T6 profile is evidence for episodic pedogenic evaporation, in agreement with the rare occurrence and poor preservation of phytoliths. We provide updated OSL ages for the lower two Tejo terraces, obtained by post infra-red stimulated luminescence: (i) T5 is c. 140 to 70 ka; (ii) T6 is c. 60 to 35 ka. The single archaeological and fossiliferous level located at the base of the T6 upper division, recording the last regional occurrence of megafauna (elephant and rhinoceros) and Mousterian artefacts, is now dated at 44 ± 3 ka. With reference to the arrival of Neanderthals in the region, probably by way of the Tejo valley (from central Iberia), new dating suggests a probable age of 200–170 ka for the earliest Mousterian industry located in the topmost deposits of T4.

Keywords: OSL dating; river terraces; Late Pleistocene; environmental change; western Iberia

1. Introduction

The Vila Velha do Ródão and Arneiro depressions are located in the furthest upstream reach of the Lower Tejo River, about 20 km from the Spanish border (this is Portuguese Reach I as defined by reference [1]) (Figure 1). From the sensitive location of the study area close to the North Atlantic, we can assume strong interaction between marine and terrestrial processes and environmental conditions. This highlights the relevance of this area for research on past climate and environmental change. Previous research in the study area has mainly focused on characterizing the geological setting, geomorphic genesis (e.g., of the terrace staircase), active tectonics, archaeological background and landscape evolution in general (e.g., see review in reference [2]). Climatic and environmental changes in the Lower Tejo Basin during the last glacial cycle and especially during Marine Isotope Stage (MIS) 3, remain poorly documented.

The Middle Palaeolithic of Iberia has received considerable attention in recent times in connection with the extinction of the Neanderthals [3–15], but is also relevant in connection with the early establishment of the Mousterian, the diversity and demise of early hominins and the widespread distribution of the first Neanderthals [16–21].

Since the 1970s, 37 archaeological contexts in the study area have been recorded, ranging from Acheulean to Mesolithic; amongst these, 17 are small surface assemblages of uncharacteristic chopper-like cores and flakes, broadly assigned to the Palaeolithic, and eight were obtained from excavation [22] (Foz do Enxarrique, Vilas Ruivas, Pegos do Tejo-2, Azinhal, Tapada do Montinho, Cobrinhos, Monte da Revelada and Alto da Revelada) (Figure 1).

In this contribution, we focus on the sedimentary record from the lowest terrace (T6) of the Tejo at Vila Velha de Rodão, which represents an important terrestrial archive of relevance to the possible relation between palaeoenvironmental conditions and early human occupation dynamics in the region. Herein, we reconstruct and discuss the climate and environmental conditions during the last glacial cycle, through integration of evidence from optically stimulated luminescence (OSL) dating, grain-size distribution, rock magnetic properties (low field magnetic susceptibility, frequency-dependent magnetic susceptibility and isothermal remanent magnetization curves), sediment mineralogy, phytolites and palynology, as well as reviewing published paleontological and archaeological data.

2. Geological and Geomorphological Setting

The Tejo is one of the largest systems of Western Europe and flows E–W across almost the whole of Iberia; it is an ancient river (c. 3.7 Ma) with an important sedimentary record [23,24]. In the uppermost Portuguese reach of the Lower Tejo, the river flows through two quartzite ridges by way of the Ródão gorge (named "Portas de Ródão"), which separates the Ródão (upstream) and Arneiro (downstream) depressions.

The oldest bedrock comprises the Neoproterozoic and lower Cambrian schists and metagreywackes of the Beiras Group and the Ordovician Armorican Quartzite Formation. The latter is dominated by resistant ridges that topographically dominate (by c. 150 m) the extensive adjacent planation surface developed on phylites and metagreywackes. The Cenozoic is represented by the Cabeço do Infante Formation, the Silveirinha dos Figos Formation and the Murracha Group. The first two are dominated by soft sandstones and gravels, while the Murracha Group consists of gravels interbedded with fine sediments [23,25].

In Lower Tejo Reach I, below a culminant sedimentary unit (the Falagueira Formation, at c. +260 m—above the river bed) corresponding to the ancestral Tejo River before drainage network entrenchment, the Pleistocene to Holocene record is summarized as follows [1,2,26] (Figure 1): (i) T1, with the surface at +111 m, without artefacts; (ii) T2, at +83 m, without artefacts; (iii) T3, at +61 m, without artefacts; (iv) T4, at +34 m, with Acheulean in the basal and middle levels and Mousterian in the uppermost levels; (v) T5, at +18 m, with Mousterian industries through the entire fluvial sequence; (vi) T6, at +10 m, with Mousterian industries at the lower deposits; (vii) Carregueira Formation (aeolian sands) (32 to 12 ka), with Upper Palaeolithic to Epi-Palaeolithic industries; (viii) Alluvial plain and a cover unit of aeolian sands (Holocene), with Mesolithic and more recent industries. Immediately upstream of the Ródão gorge, the modern river bed is at c. 72 m above sea level (a.s.l.). In this area, geomorphological evidence for late Cenozoic tectonics arises from interpretation of valley asymetry and drainage patterns, fault scarps, tectonic lineaments, fracture-controlled valleys, and vertical displacement of planation surfaces and terraces [26]. The sedimentary controls for the formation of the Lower Tejo terrace staircase (mainly glacio-eustasy and differential uplift) are different from those affecting the Middle and Upper Tejo, because of separation (between the Middle and Lower Tejo) by a lengthy knick zone through hard basement [27,28].

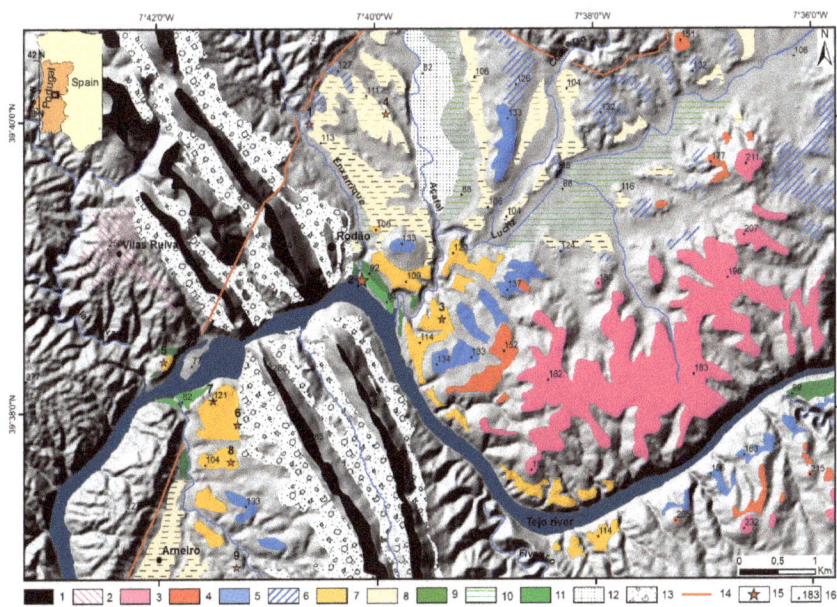

Figure 1. Geomorphological map of Lower Tejo Reach I (Vila Velha de Ródão and Arneiro depressions): 1—quartzite ridge; 2—erosion level (a strath without sedimentary deposits) correlative with T1; 3—T1; 4—T2; 5—T3; 6—erosion level correlative with T3; 7—T4; 8—erosion level correlative with T4; 9—T5; 10—erosion level correlative with T5; 11—T6; 12—alluvial plain; 13—colluvium; 14—Ponsul fault; 15—archaeological sites; 16—altitude (m). Palaeolithic sites: 1—Cobrinhos; 2—Foz do Enxarrique; 3—Monte do Famaco; 4—Monte da Revelada and Alto da Revelada; 5—Vilas Ruivas; 6—Tapada do Montinho; 7—Pegos do Tejo-2; 8—Arneiro; 9—Azinhal.

Previous dating of the Pleistocene litostratigraphic divisions was undertaken using Uranium-series, Thermoluminescence (TL), Optically stimulated luminescence on quartz (Quartz-OSL) and infra-red stimulated luminescence (IRSL) on K-feldspar. Quartz-OSL was used to date: the Carregueira Formation (32 to 12 ka; 3 ka) [2]; the alluvial deposits at the Azinhal archaeological site, linked to T6 (61 ± 2 ka: GLL code 050302); the topmost deposits of T4 at the Pegos do Tejo-2 archaeological site (minimum age of 135 ± 21 ka: GLL code 050301) [29]. In Reach I of the Tejo, IRSL dating provided a first temporal framework for the Lower Tejo terraces [26], but at that stage of research T5 and T6 were not yet separated as distinct terraces. Recently, T4 in Lower Tejo Reach IV was dated to c. 340 to 155 ka, using post infra-red stimulated luminescence (pIRIR) [28].

3. Methods

Geomorphological, stratigraphical, sedimentological and chronological data were obtained using standard methodology (e.g., [30]): (1) geomorphological study, complemented by local detailed investigations and the production of a detailed map using Geographic Information System (GIS), (2) field descriptions of the sedimentary units, (3) sedimentological characterization of the deposits and (4) luminescence dating.

3.1. Geomorphological Mapping

Geomorphological mapping was undertaken in three stages: (1) field mapping onto topographical (1/25,000) and geological (1/50,000) base maps, (2) analysis of 1/25,000 aerial photographs and of a digital elevation model (DEM) based upon a 1/25,000 and 1/10,000 topographic databases and (3) field ground truthing.

3.2. Field Work

The T6 deposits at Foz do Enxarrique were studied in detail in order to improve our understanding of the local stratigraphy and sedimentology. Exposures of T5 and T4 in the study area were also revisited. Fieldwork included stratigraphic logging and sedimentological characterization of the sedimentary deposits in order to obtain data on the depositional facies, including sediment colour, texture, maximum particle size, clast lithology, fossil content, bedding and depositional architecture.

At the studied stratigraphic section of T6 at Foz do Enxarrique, continuous sediment sampling was undertaken manually, every 1 cm, to a depth of 5.00 m. Samples were labelled as follows: e.g., for "T6FE0.21", T6 identifies the terrace code, FE the site and 0.21 the sample depth. We also collected a present-day sediment sample from the Foz do Enxarrique stream bed (FE-modern). Each sampled horizon (1 cm) was characterised according to its colour (using the Munsell system), texture and the relative abundance of carbonate concretions. Phytolith analyses were undertaken on carbonate concretions from five levels within the sequence (T6FE0.72, T6FE2.00, T6FE2.46–2.48, T6FE3.48 and T6FE5.00). An additional seven samples (spanning greater depth) were collected for clay mineralogy and palynological studies (T6FE0.70–0.74, T6FE1.13–1.17, T6FE1.98–2.02, T6FE2.88–2.92, T6FE3.46–3.50, T6FE4.10–4.14 and T6FE4.49–4.52). Palynological study also included samples collected from the T6 archaeological level with fossil bones: codes T6FE/15 0.25–0.30, T6FE/15 0.35–0.40 and T6FE/15 0.45–0.50 (here, the depth refers to the top of the archaeological level). The T6 upper unit at Foz do Enxarrique was sampled for optically stimulated luminescence (OSL) dating: sample 062201 collected at 0.89–0.93 m depth (below the terrace surface); sample 052202 collected at a 5.20–5.30 m depth; sample 052201 collected at a 5.40–5.50 m depth (top of the archaeological and fossiliferous layer).

At the Foz do Enxarrique site, a 1 m thick exposure reveals the lower part of T5. Two samples (T5FE0.15–0.23 and T5FE0.92–1.00) were collected here for clay mineralogy and palynological studies. At this site, two further samples for OSL dating (052204 and 052247) were collected at depths of 1.50 and 2.00 m below the T5 surface. T5 was also sampled for OSL dating at the Vilas Ruivas site (052207, 052231 and 052253).

From T4, samples for OSL dating were obtained from the Vilas Ruivas, Rodense Bolaria (Vila Velha de Ródão) and Pegos do Tejo-2 (Arneiro) sites (Figure 1).

3.3. Optically Stimulated Luminescence Dating

From Reach I of the Lower Tejo, samples for OSL dating were previously collected from the sedimentary successions of the T4, T5 and T6 terraces and dated by IRSL, with a correction for the anomalous fading effect [26], because the quartz-OSL signal was found to be in saturation. However, it was later documented that the fading correction used was inappropriate, leading to age underestimation. In order to improve the chronology of the terrace sequences and their associated lithic industries, we now use the pIRIR protocol. K-feldspar grains from the samples dated in 2008 (IRSL) were measured (Equivalent doses) in 2013 by pIRIR (the most up-to-date protocol); these new results are now presented. The storage of K-feldspar grains does not affect the luminescence properties or the resulting ages. In summary, several samples were selected for OSL dating (pIRIR protocol): three from T6 at Foz do Enxarrique (upper unit), five from T5 (collected at Foz do Enxarrique and Vilas Ruivas; Figure 1) and four from T4 (collected at Pegos do Tejo-2, Vilas Ruivas and Rodense Bolaria/V.V.Ródão; Figure 1).

OSL is an absolute dating technique that measures the time elapsed since sedimentary quartz or feldspar grains were last exposed to daylight [31]. Exposure to daylight during sediment transport removes the latent luminescence signal from those minerals. After burial, the luminescence signal (trapped charge) starts to accumulate in the mineral grains due to ionising radiation. The annual dose of a sediment sample is related to the decay of ^{238}U, ^{232}Th and ^{40}K present in the sediment itself, to cosmic ray bombardment and to the water content of the sediment. In the laboratory, the equivalent dose (D_e, assumed to be the dose absorbed since the last exposure to light, i.e., the burial dose, expressed in Grays —Gy) is determined by comparing the natural luminescence signal resulting from charge trapped during burial with that trapped during a laboratory irradiation. In this study, the radionuclide concentrations were measured by high-resolution gamma spectrometry [32]. These concentrations were then converted to environmental dose rates using the specified conversion factors [33]. For the calculation of the dose rate of sand-sized K-feldspar grains, an internal K content of 12.5 ± 0.5% was assumed [34]. Dividing the D_e by the environmental dose rate (in Gy/ka) gives the luminescence age of the sediment.

Sample preparation for luminescence analyses was carried out in darkroom conditions, at the Department of Earth Sciences of the University of Coimbra. Samples were wet-sieved to separate the 180–250 µm grain-size fraction, followed by HCl (10%) and H_2O_2 (10%) treatments to remove carbonates and organic matter, respectively. The K-feldspar-rich fraction was floated off using a heavy liquid solution of sodium polytungstate (ρ = 2.58 g/cm^3). The K-feldspar fraction was treated with 10% HF for 40 min to remove the outer alpha-irradiated layer and to clean the grains. After etching, the fraction was treated with HCl (10%) to dissolve any remaining fluorides.

At the Nordic Laboratory for Luminescence Dating (NLL), OSL were conducted using automated luminescence Risø TL/OSL-20 readers (Roskilde, Denmark), each containing a calibrated beta source. Small (2 mm) aliquots of K-feldspar were mounted in stainless steel cups. The K-feldspar equivalent doses (D_e) were measured with a pIRIR SAR protocol using a blue filter combination [35,36]. Preheating was at 320 °C for 60 s and the cut-heat 310 °C for 60 s. After preheating the aliquots were IR bleached at 50 °C for 200 s (IR$_{50}$ signal) and subsequently stimulated again with IR at 290 °C for 200 s (pIRIR$_{290}$ signal). It has been shown [36] that the post-IR IRSL signal measured at 290 °C can give accurate results without the need to correct for signal instability. For all IR$_{50}$ and pIRIR$_{290}$ calculations, the initial 2 s of the luminescence decay curve less a background derived from the last 50 s was used.

3.4. Grain-Size Measurements

Grain-size analyses of uncemented sediment samples was carried out using a Beckman Coulter LS230 laser granulometer (Brea, CA, USA), with a measurement range of 0.04 to 2000 µm and a relative error less than 2%. Visual inspection of grain-size distribution curves allowed the identification and interpretation of unimodal or multimodal subpopulations. The T6FE sediment samples of the 5.00–3.20 m depth interval were analysed at a 5 cm spacing; the 3.18–2.60 m and 2.50–0.30 m depth intervals were analysed at a 1 cm spacing in order to provide a better distinction between fluvial and aeolian deposition.

3.5. Mineral Composition

Analyses of sediment composition were based on binocular microscope observation and X-ray diffraction (Department of Earth Sciences—University of Coimbra), as well as Scanning Electron Microscopy (SEM) and Energy Dispersive Spectometry (EDS) of selected carbonate concretions (UNESP Laboratory, at the University of Rio Claro—Brazil). A Philips PW 3710 X-ray diffractometer (Virginia, USA), with a Cu tube, at 40 kV and 20 nA was used for mineralogical identification within carbonate concretions and for clay mineralogy. The mineralogical composition of the <2 μm fraction was obtained in oriented samples before and after ethylene glycol treatment and heating up to 550 °C. The percentages of the clay minerals in each sample were determined through the peak areas of the mineral present, with the use of specific correction parameters.

3.6. Rock Magnetism

For magnetic susceptibility measurement, samples were dried at 40 °C and transferred into plastic bags for subsequent analysis. Rock magnetic properties were measured in the Instituto Dom Luis, University of Lisboa and in the Department of Earth Sciences of the University of Coimbra, and consisted of low field mass specific magnetic susceptibility (χ in m^3/kg), frequency-dependent magnetic susceptibility (Kfd in %) and isothermal remnant magnetization (IRM). Magnetic susceptibility measures the ability of a material to be magnetized and includes contributions (in proportion to their abundance) from all diamagnetic (calcite), paramagnetic (clays), and ferromagnetic (magnetite) minerals present in the sediment. Low-field magnetic susceptibility was measured with a MFK1 (AGICO Inc, Brno, Czech Republic) apparatus operating with magnetic field intensity of 200 A/m and frequency of 978 Hz. Data were reported as mass-normalized values (m^3/kg). Frequency-dependent magnetic susceptibility is an indicator of the presence of superparamagnetic particles (SP), generally produced during pedogenic processes. Low (0.47 kHz) and high (4.7 kHZ) frequency-dependent magnetic susceptibility was measured with a Bartington Instruments magnetic susceptibility meter coupled to a MS2B sensor and reported in percentage as follows: Kfd (%) = 100*(Klf-Khf)/Klf. After cleaning by alternating field demagnetization up to 100 mT, samples were subsequently submitted to stepwise isothermal remanent magnetization (IRM) acquisition with an impulse magnetizer (model IM-10-30). We applied maximum fields of 1.2T following approximately 30 steps. Remanence was measured with a JR-6A (AGICO Inc, Brno, Czech Republic) magnetometer.

Data were analysed using a cumulative log-Gaussian (CLG) function with software developed for the purpose [37]. The S-ratio was calculated with the formula $-\text{IRM}_{-0.3T}/\text{IRM}_{1T}$.

3.7. Phytoliths

For phytolith analyses, samples with a volume of 1 cm^3 were placed in an Erlenmeyer flask and dissolved in 20 mL of HNO$_3$ and H$_2$SO$_4$ solution at 1:4. The material was heated for 3 h at 90 °C on a hot plate. After cooling at ~25 °C, 10 ml of H$_2$O$_2$ was added, before washing in distilled water, centrifuging (1500 rotations per minute up to neutralization (pH ~ 7.0), and washing with alcohol. For slide preparation, 50 μL of material was extracted by pipette, placed on slides and dried on hot plates. Coverslips were fixed using Entelan® resin (Hatfield, UK). Phytoliths were analyzed through optical microscopy (×160 and ×640), identified with reference to literature [38–40] and named according to the International Code of Phytolith Nomenclature [41].

3.8. Palynology

For palynological studies, thirteen sediment samples were selected: two from lower and middle levels within T5; three from the T6 archaeological level and eight samples from the upper division of T6 (silty very fine sands and sandy silts). These samples were subjected to a physical and chemical pollen concentration pre-treatment. The pollen residue was isolated with a standard palynological preparation methodology [42], with some modifications: omitting acetolysis and sieving, the latter in an attempt to increase pollen concentration. The pollen residue was assembled on thin glass slides to allow its identification and counting. It was embedded in glycerine and sealed with histolaque, to permit movement of the grains for more complete observation of the morphological features of pollen and non-pollen-palynomorphs. Grains were identified (based on references [43–46]) and counted using an optical transmitted-light microscope.

3.9. Geochemical Analyses

Geochemical analyses on sediment samples collected from the upper unit of T6 were performed, at the laboratory, with a X-ray fluorescence spectrometer (Niton XL3t Ultra Analyser—Thermo Fisher Scientific; Waltham, MA, USA).

4. Results

4.1. Geomorphology, Lithostratigraphy and Sedimentology

In the confluence area of the Enxarrique stream with the Tejo, several geomorphic units are represented: T1, at 179 m a.s.l.; T2, at 149 m a.s.l.; T3, at 132 m a.s.l.; T4, at 106 m a.s.l.; T5, at 89 m a.s.l.; T6, at 83 m a.s.l.; alluvial plain of the Açafal stream, at 78 m a.s.l. (Figure 2).

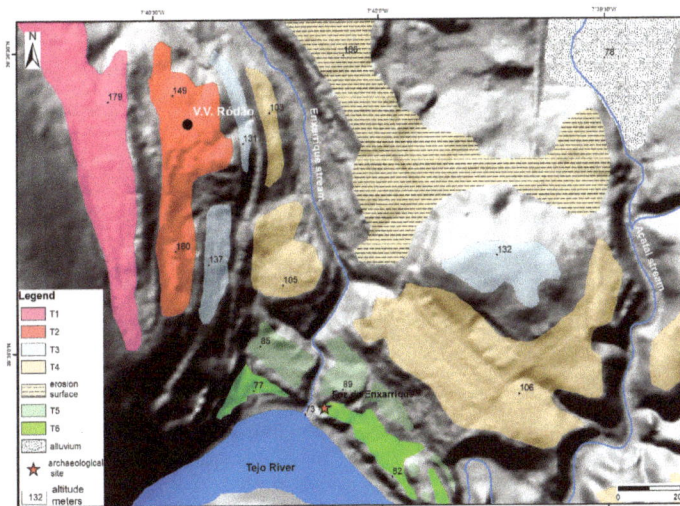

Figure 2. Geomorphological map of the neighbourhood of the Foz do Enxarrique site (Vila Velha de Ródão). This digital elevation model was performed with 2 m of equidistant contours and pixels of 5 × 5 m. The fluvial terraces, numbered from higher (T1) to lower (T6), are disposed in a staircase. The mapped erosion surface (a strath without sedimentary deposits) correlates with the T4 terrace. The location of the Foz do Enxarrique archaeological site is indicated.

In the Vila Velha de Ródão—Arneiro area, T4 comprises a lower boulder gravel (up to 2 m-thick) and an upper division of gravelly coarse sands (up to 2 m thick). The main exposures are at Pegos do

Tejo-2 and Vilas Ruivas, both located in the Arneiro depression (Figure 1). The massive clast-supported boulder gravel comprises clasts that are subrounded, with MPS (mean diameter of the 10 largest clasts) = 32 cm, of quartzite (75%), white quartz and rare slates/metagreywackes.

The T5 terrace usually has a basal gravelly pavement (c. 0.2 m thick) overlain by very fine sands (up to 2.3 m thick), with some very thin (<1 cm) levels containing calcium carbonate concretions. The main outcrops are provided by the sites at Foz do Enxarrique (up to c. 2.5 m thick) and Vilas Ruivas (0.5 m thick).

The sedimentary deposits of T6 are well exposed at the Foz do Enxarrique site (Figures 2–4) and were studied in detail during the work reported here. The sedimentary infill of T6 comprises a lower gravel, up to 0.40 m thick, and an upper unit, c. 5.60 m thick, dominated by very fine sands and coarse silts.

Figure 3. View of the T6 section at the Foz do Enxarrique archaeological site, with sampling underway. The surface of the T6 terrace is at 82 m a.s.l.; the exposures show the full thickness of the upper unit. The lower artificial pavement is placed approximately at the top of the bed containing artefacts and fossil bones (the archaeological level); the lower gravel bed is located just below this. A metal staircase and a platform has been built in front of the exposed section and provides access to the top as well some protection for the exposure. Note one of the several explanatory panels provided at the site. The Tejo River can be seen in the right background.

The upper unit of T6, generally lacking bedding but with rare lamination, can be divided into three layers: (i) from the surface to a depth of 4.55 m, an upper bed comprising sandy silts; (ii) from 4.55 to c. 5.40 m, a middle division consisting of micaceous very fine to fine sands with some interbedded thin gravel stringers; (iii) from c. 5.40–5.60 m, a bed comprising Mousterian artefacts and fossil bones in a matrix of micaceous fine sands. The predominant colour of the T6 upper unit deposits is yellowish brown to bright brown. The thin (1–3 cm thick) levels with calcium carbonate concretions and rhizoliths (Figure 5), 1–2 cm wide, are intercalated between thick intervals of uncemented sediment. At the depths of 4.55–4.57 and 4.73–4.82 m, quartzite and quartz clasts (<0.5 cm and <10 cm in size, respectively) and rolled concretions were observed dispersed in micaceous very fine to fine sands. A massive calcium carbonate level is present at a depth of 5.33–5.36 m, just above laminated fine sand. The levels containing calcium carbonate concretions do not show evidence of any erosive surface and they dip (up to 5 degrees) toward the Tejo River, progressively increasing in thickness.

The lower bed T6 consists of clast-supported boulder–pebble gravels, c. 0.4 m thick. The maximum clast diameter is 40 cm, with MPS = 31 cm. The gravel clast lithologies are quartzite (58%), milky quartz (27%) and metagreywakes/phylites (15%). The clasts are sub-rounded to angular. At about 76 m a.s.l., the T6 deposits overlie metamorphic basement, by an unconformity.

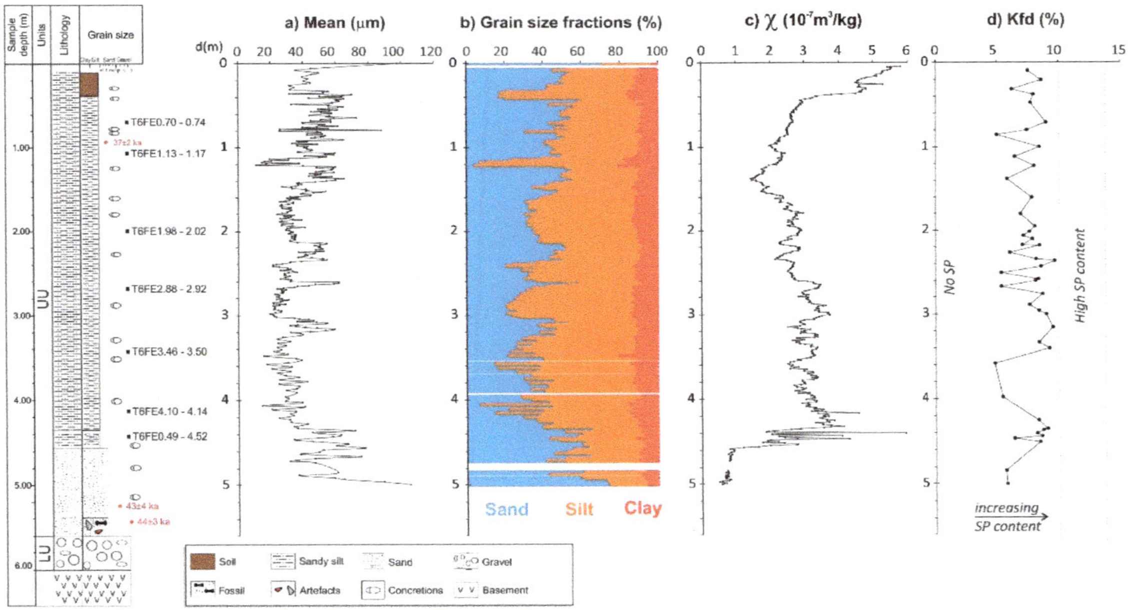

Figure 4. Stratigraphic log of the T6 terrace sequence at the Foz do Enxarrique archaeological site, also showing the (**a**) mean grain-size diameter, (**b**) main grain-size fractions in % (sand, 2000 to 63 μm; silt, 63 to 4 μm; clay, <4 μm), (**c**) mass specific magnetic susceptibility (χ) and (**d**) frequency-dependent magnetic susceptibility (Kfd in %), indicative of supermagnetic particles (SP). Black squares beside the stratigraphic log correspond to the sample intervals where palynological analysis and identification of clay minerals were conducted. Red circles indicate the stratigraphic location of optically stimulated luminescence (OSL) dated samples, with ages shown. LU—lower unit (gravels); UU—upper unit (dominated by very fine sands and coarse silts).

Figure 5. Examples of calcium carbonate concretions occurring in the coarse silts of T6: (**A**) three concretions collected, respectively, from depths of 3.56, 3.57 and 3.60 m; (**B**) a 2.5 cm thick concretion, obtained from a depth of 3.91–3.93 m; (**C**) two concretions and an agglomerate of two quartz pebbles, collected from a level at a depth of 4.73–4.83 m; (**D**) fragment of a quite continuous caliche level, 3 cm thick, that occurs at a depth of 5.33–5.36 m; (**E**) a 1 cm thick rhizolith, collected from a depth of 3.91–3.93 m.

4.2. Luminescence Dating Results

The luminescence dating results of the samples collected from T6, T5 and T4 are summarized in Tables 1 and 2. The geographical and stratigraphic locations of each sample are provided by Table 2. Four out of the twelve samples have D_e values lying above $2 \times D_0$ (corresponding to 86% of luminescence saturation). In view of the possibility of small systematic errors in the measurement of the luminescence signal, we feel it prudent to adopt the strategy of only presenting minimum doses for these samples (equivalent to $2 \times D_0$) and the derived minimum ages [47].

Table 1. Burial depth, Radionuclide activities (^{238}U, ^{226}Ra, ^{232}Th and ^{40}K) and water content used for dose-rate calculations of the luminescence dating samples.

NLL and Field Codes	X-Y Coordinates	U-238 (Bq kg^{-1})	Ra-226 (Bq kg^{-1})	Th-232 (Bq kg^{-1})	K-40 (Bq kg^{-1})	Water Content (%)
052201 PC1	39°58'59" N 7°40'13" W	108 ± 6	109.2 ± 1.4	191.4 ± 2.0	784 ± 19	10
052202 PC2	39°58'59" N 7°40'13" W	89 ± 6	77.4 ± 0.8	98.4 ± 0.9	727 ± 10	10
052204 PC4	39°58'59" N 7°40'13" W	52 ± 3	52.3 ± 0.5	75.5 ± 0.7	715 ± 9	10
052207 PC8	39°38'29" N 7°41'56" W	63 ± 11	70.3 ± 2.0	108.6 ± 2.1	621 ± 33	20
052208 PC9	39°39'19" N 7°40'07" W	19 ± 6	27.6 ± 0.6	29.6 ± 0.6	902 ± 17	25
052231 VRU4	39°38'29" N 7°41'56" W	78 ± 9	74.5 ± 1.0	114.3 ± 1.2	583 ± 12	20
052246 VRU5	39°38'29" N 7°41'56" W	27 ± 4	23.1 ± 0.4	28.0 ± 0.5	843 ± 12	10
052247 ENXAR1	39°58'59" N 7°40'13" W	44 ± 4	48.7 ± 0.5	75.1 ± 0.7	765 ± 8	10

Table 1. *Cont.*

NLL and Field Codes	X-Y Coordinates	U-238 (Bq kg^{-1})	Ra-226 (Bq kg^{-1})	Th-232 (Bq kg^{-1})	K-40 (Bq kg^{-1})	Water Content (%)
052253 VRU3	39°38'29" N 7°41'56" W	69 ± 7	61.3 ± 0.8	96.3 ± 1.1	536 ± 10	10
062201 ENXAR2	39°58'59" N 7°40'13" W	64 ± 7	68.4 ± 0.9	92.0 ± 1.1	764 ± 14	10
062202 PARN1	39°38'11" N 7°41'35" W	25 ± 5	27.9 ± 0.5	21.6 ± 0.5	843 ± 14	10
062203 PARN2	39°38'11" N 7°41'35" W	18 ± 5	24.9 ± 0.5	27.1 ± 0.5	995 ± 12	10

Table 2. Summary of the luminescence ages obtained from sediment samples from the study area. All ages were obtained by using a post IRIR$_{290}$ protocol (K-feldspar), in the Nordic Laboratory for Luminescence Dating (NLL). For samples having natural pIRIR$_{290}$ signals >86% of the saturation level of the dose response curves, a minimum age is given based on the 2*D$_0$ value. Previous IR$_{50}$ age estimates, including fading correction, are also shown [26].

NLL and Field Code	Sampled Site	Terrace Unit	Depth (cm)	IR$_{50}$, with Fading Corr. Age (ka)	pIRIR$_{290}$ Age (ka)	pIRIR$_{290}$ 2*Do (Gy)	pIRIR$_{290}$ D$_e$ (Gy)	pIRIR$_{290}$ Aliquots (n)	Dose Rate (Gy/ka)
052201 PC1	Foz de Enxarrique	T6, base middle part	550	38.5 ± 1.5	44 ± 3		350 ± 14	12	7.94 ± 0.32
052202 PC2	Foz de Enxarrique	T6, lower middle part	530	34.8 ± 1.3	43 ± 4		251 ± 21	9	5.83 ± 0.22
052204 PC4	Foz de Enxarrique	T5, middle part	150	136 ± 10	135 ± 9		683 ± 34	12	5.05 ± 0.18
052207 PC8	Vilas Ruivas	T5, middle part	400	105 ± 8	99 ± 7		507 ± 28	12	5.10 ± 0.19
052208 PC9	Rodense Bolaria	T4, lower part	200	277 ± 17	>220	>835 ± 51		6	3.95 ± 0.13
052231 VRU4	Vilas Ruivas	T5, upper part	400		71 ± 4		367 ± 17	9	5.18 ± 0.18
052246 VRU5	Vilas Ruivas	T4, lower part	300	277 ± 17	>220	>863 ± 13		6	4.19 ± 0.15
052247 ENXAR1	Foz de Enxarrique	T5, lower part	200	125 ± 7	145 ± 10		741 ± 40	12	5.11 ± 0.19
052253 VRU3	Vilas Ruivas	T5, middle part	400	113 ± 6	117 ± 7		587 ± 27	6	5.02 ± 0.19
062201 ENXAR2	Foz de Enxarrique	T6, topmost part	90	31.6 ± 1.3	37 ± 2		212 ± 6	6	5.73 ± 0.22
062202 PARN1	Pegos do Tejo-2 (Arneiro)	T4, lower middle part	400	209 ± 11	>190	>767 ± 27		9	4.14 ± 0.15
062203 PARN2	Pegos do Tejo-2 (Arneiro)	T4, lower top part	300	129 ± 11	>160	>743 ± 31		8	4.62 ± 0.17

Two of the four samples collected from T4 were collected from the lower gravel (052208 and 052246) and provided age estimates of c. 280 ka by IRSL and minimum ages of >220 ka by pIRIR. Samples 062202 and 062203 were collected from the base and middle of the T4 upper division which mainly consists of soft sandstones, and gave underestimated ages of 209 ± 11 ka and 129 ± 11 ka by IRSL and minimum ages of >190 ka and >160 ka by pIRIR, respectively. Sample 062203 was collected below the Middle Palaeolithic site of Pegos do Tejo-2, which is located c. 1 m bellow the T4 surface. As the uppermost deposits of T4 were recently dated at 155 ka ([28] (Cunha et al., 2017)) and sample 062203 provided a minimum age of 190 ka by pIRIR, it is very probable that the Middle Palaeolithic in the region had started by 200–170 ka.

From T5, which is onlaped by T6 at the Foz do Enxarrique site, ages of 145 ± 10 ka (at the terrace base) and 135 ± 9 ka (1.5 m below the terrace surface) were obtained by pIRIR dating. Samples collected from the T5 sequence at Vilas Ruivas were also dated, giving ages of 117 ± 7 ka, 99 ± 7 ka and 71 ± 4 ka by pIRIR.

From the T6 upper unit at Foz do Enxarrique, pIRIR dating provided the following three ages: 44 ± 3 ka, sample 052201 collected from the archaeological and fossiliferous bed (at a depth of 5.50–5.40 m); 43 ± 4 ka, sample 052202 collected from the micaceous fine sands located 20 cm higher (5.30–5.20 m); 37 ± 2 ka, sample 062201 collected c. 90 cm below the terrace surface (0.93–0.89) m. Previously, the archaeological bed was dated by Uranium series on equid and bovid teeth, providing an average age of 33.6 ± 5 ka ([48] (Raposo, 1995)), and by fading-corrected IRSL, giving an age of 38.5 ± 1.6 ka ([26] (Cunha et al., 2008)). It appears that both the U-series and conventional IRSL ages are underestimates, compared with the pIRIR ages. As previously noted, the IRSL dating used a correction for the anomalous fading effect which is not fully appropriate, leading to age underestimation. It should also be noted that U-series on teeth can easily be impaired as a result of the uptake of young uranium. The pIRIR protocol constitutes the best approach to date the type of material in consideration and should be considered to provide the best estimates for burial ages and for dating the associated lithic industries.

4.3. Sediment Characterization of T6

4.3.1. Mineral Composition

The sand and silt fractions of the T6 upper unit mainly comprise detrital grains of quartz (predominant), K-feldspar, plagioclase and muscovite, locally cemented by calcite. SEM and EDS analysis of 17 selected carbonate concretions also confirmed that these consist of calcite (micrite) developed in detrital silt–fine sand consisting of quartz (predominant), microcline, plagioclase and mica grains.

Exoscopy (SEM) (Figures 6 and 7) document well preserved microcline and quartz grains. The quartz grains are sub-angular and show conchoidal fractures, but with no evidences of abrasion or dissolution. Biotite grains show some dissolution features.

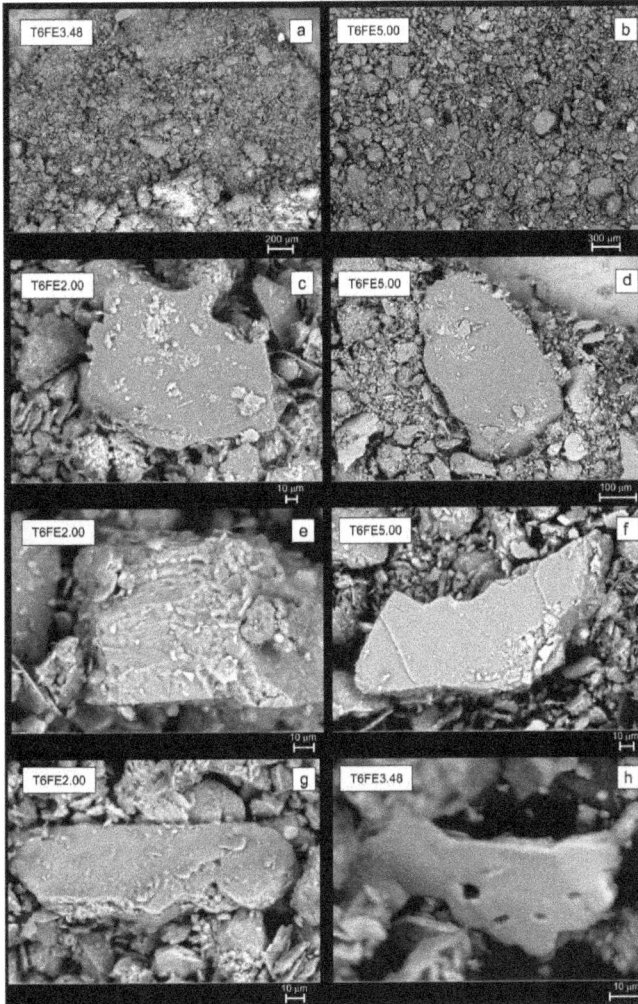

Figure 6. Back-scattered Scanning Electron Microscopy (SEM) photographs of samples from the T6 profile at Foz do Enxarrique. (**a**,**b**) poorly-sorted detrital material; (**c**,**d**) sub-angular quartz grain with conchoidal fractures and smoothed surfaces; (**e**,**f**) well preserved K-feldspar (microcline); and (**g**,**h**) mica grains, mainly biotite, with dissolution features.

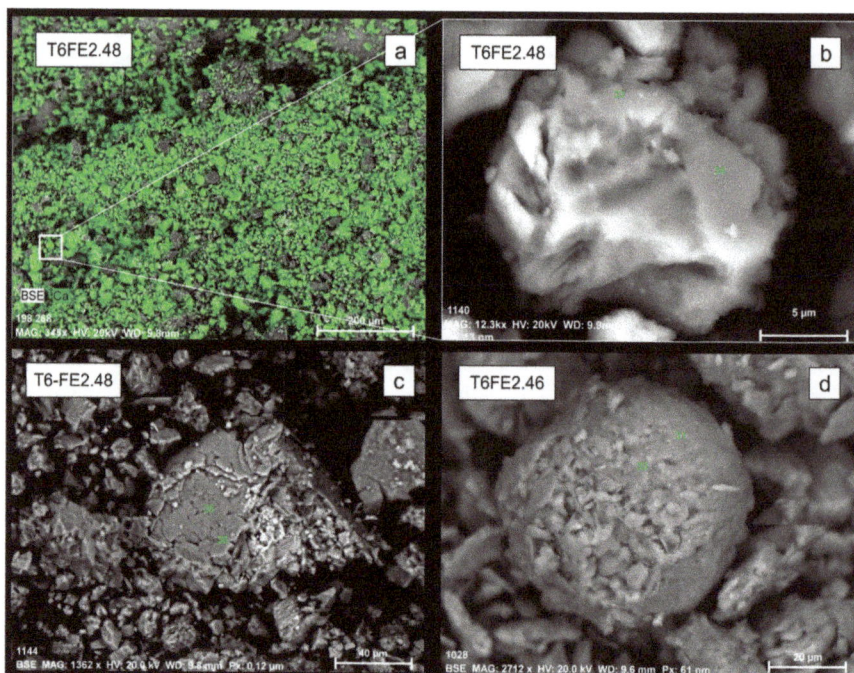

Figure 7. SEM photographs of samples T6FE2.46 and T6FE2.48. (**a**) compositional mapping showing the widespread distribution of calcium; (**b**,**c**) calcium particles and (**d**) nodules with size varying from 5 to 40 μm.

The mineralogical composition of the <2 μm fraction (Table 3) of 8 samples collected from the T6 upper unit does not indicate any vertical change and consists of smectite 35–52%, illite 22–35%, kaolinite 20–31% and vermiculite 17% (only present in sample T6FE 3.46–3.50). For comparison, two samples from T5 were also studied and contained smectite 53–54%, illite 21–29% and kaolinite 18–25%. Thus both terraces have a similar clay-mineral association, although T6 has less smectite. The modern sample also has vermiculite and chlorite.

Table 3. Mineralogical composition of the <2 μm fraction of samples collected from the T5, T6 upper unit and modern river bed of the Enxarrique stream.

Sample Code	Smectite (%)	Vermiculite (%)	Kaolinite (%)	Illite (%)	Chlorite (%)	Clay Mineral Association
T5FE0.15–0.23	54	0	25	21	0	S k i
T5FE0.92–1.00	53	0	18	29	0	S i k
T6FE0.70–0.74	48	0	26	26	0	S k i
T6FE1.13–1.17	52	0	23	25	0	S i k
T6FE1.97–2.02	41	0	27	32	0	S i k
T6FE2.88–2.92	47	0	31	22	0	S k i
T6FE3.46–3.50	35	16	25	24	0	S k i v
T6FE4.10–4.14	42	0	24	34	0	S i k
T6FE4.49–4.52	46	0	24	30	0	S i k
T6FE4.98–5.02	45	0	20	35	0	S i k
FE modern	17	14	22	35	12	I k s v c

4.3.2. Grain-Size Analysis

The sediment samples collected from the upper unit of T6 and analyzed by laser granulometry provided the results presented in Figure 4 and Supplementary Material. This unit is clearly dominated by coarse silt, but sampling at the 1 cm scale has documented significant grain-size oscillations as well as major grain size cycles.

From the 0.05–0.30 m depth, an organic dull yellowish to brown soil (10YR 5/3 and 10YR 4/6) has an average mean grain size of 41 μm. The samples are very poorly sorted (4.60) with very fine skewed (−0.72) distributions. In average, the sediment comprises 49% sand, 41% silt and 10% clay.

For the 0.31–5.00 m depth, the average grain-size data shows a mean grain size of 40.55 μm (very coarse silt), a standard deviation of 5.57 (very poorly sorted), fine skewed (−1.09), leptokurtic and bimodal distributions in the silt fraction (6–20 μm and 40–60 μm) (Figure 8). In average, the sediment comprises 44% sand, 45% silt and 9% clay. The range of mean grain size is 11–110 μm; all main fractions change significantly: 6–72% sand; 25–78% silt; 3–20% clay. The following intervals can be differentiated to a depth of 5.00 m:

- For the 0.31–4.32 m depth, the samples have a yellowish to bright yellowish brown color, (10YR 5/3, 10YR 5/6, 10YR 5/8 and 10YR 6/8). The mean grain-size is c. 37 μm and is usually very coarse-grained silt. The average grain-size fractions consist of 40% sand, 49% silt and 10% clay. Some thin levels have calcium carbonate concretions reaching 1.5 cm of diameter (e.g., T6FE0.88; T6FE 0.95). There are some layers (e.g., at 1.15–1.20 m and 2.39–2.58 m depth) with much more silt (e.g., 76%; 68%) than sand.

- For the 4.33–5.00 m depth, the sediment is increasingly coarser, with a mean grain-size between 32 and 110 μm. In average, the sediment consists of 57% sand, 36% silt and 7% clay. Between depths of 4.73 and 4.82 m, rolled calcium carbonate concretions occur.

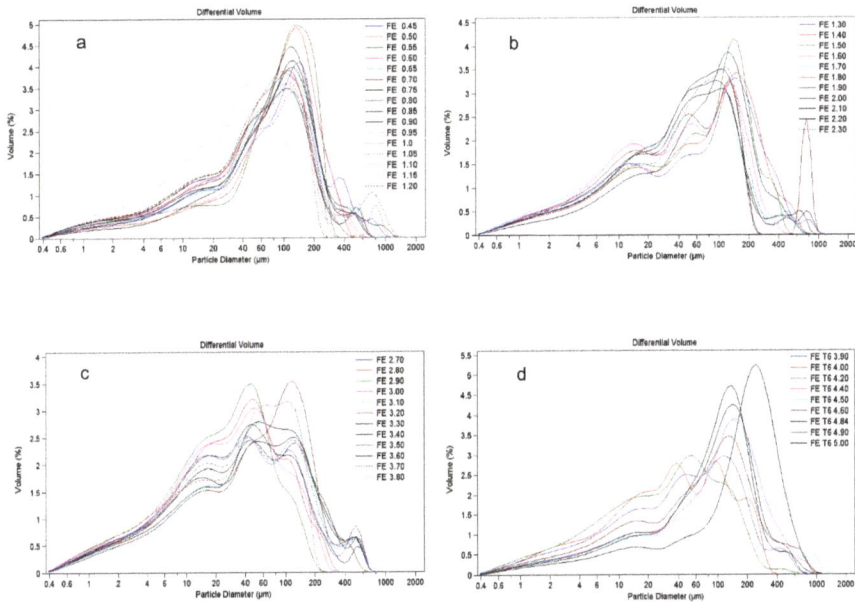

Figure 8. Grain size distributions of selected samples: (**a**) 0.45–1.20 m depth range; (**b**) 1.30–2.30 m depth range; (**c**) 2.70–3.80 depth range; (**d**) 3.90–5.00 m depth range.

4.3.3. Rock Magnetism

Magnetic Susceptibility

Mass specific magnetic susceptibility of samples from the Foz do Enxarrique T6 profile varies from c. 8×10^{-8} m^3/kg to c. 6×10^{-7} m^3/kg (Table S1—Supplementary Material; Figures 4, 9 and 10). These values are comparable with those of suspended sediments from other fluvial systems, such as the Jackmoor Brook catchment in South West England [49]. The lowest values of magnetic susceptibility (c. 7×10^{-8} m^3/kg) are found at the base of the profile, from c. 5.00 up to 4.58 m, and corresponding to sand-rich sediment (Figure 4). From 4.57 up to 4.32 m, there are some dispersed quartzite pebbles; magnetic susceptibility increases and exhibits saw-tooth oscillations from c. 1.2×10^{-8} to 4.1×10^{-8} m^3/kg. Between 4.32 and 0.40 m, magnetic susceptibility shows short-term and discrete cyclical oscillations from 1.4×10^{-7} to c. 3.7×10^{-7} m^3/kg and gradually increases up to c. 5.5–8×10^{-8} m^3/kg in the uppermost 0.40 m.

Magnetic susceptibility is primary controlled by lithological changes, but can also reflect pedogenic and/or post-depositional weathering driven by environmental and climatic forcing in the case of continental sediments [50]. Quartz is diamagnetic (negative and very low magnetic susceptibility), explaining the very low values of magnetic susceptibility observed at the base of the profile (Figure 4). Phyllosilicates, including clays, are paramagnetic (positive and weak magnetic susceptibility), giving significant higher values of magnetic susceptibility in the silt interval from 4.32 m to 0.40 m. The short-term cyclical variations observed in this part of the profile do not seem to be controlled by the lithology, which is very homogenous in the field, but may reflect changes in the magnetic mineralogy (ex. proportion of magnetite and hematite), including magnetic enhancement by weathering and/or pedogenic processes. The saw-tooth oscillation observed from 4.57 m to 4.32 m probably reflects the mixture of gravel with different origin and composition. In contrast, the increase of magnetic susceptibility in the topsoil horizon, which corresponds in the field to a black sediment, rich in organic material, probably reflects magnetic enhancement resulting from burning, pedogenic processes and/or anthropogenic pollution [51–53].

Frequency-Dependence Magnetic Susceptibility

Frequency-dependence magnetic susceptibility varies between 4 and 10 along the entire profile, indicating a significant contribution from superparamagnetic particles. These superparamagnetic particles are generally ultra-fine magnetite and/maghemite produced during pedogenic processes, as a by-product of the metabolism of iron-reducing bacteria [54,55]. The presence of pedogenic magnetic particles in the T6 profile agrees with the ubiquitous occurrence of rhizoliths (root moulds and concretions) observed in the field. However, these pedogenic particles can also have a detrital origin, from the incorporation of soil sediments or clays.

Isothermal Remanent Magnetization Curves

Isothermal Remanent Magnetization (IRM) measurements were conducted on eight samples from the T6 profile in order to investigate the nature and origin of the ferromagnetic particles present in the sediment and their contribution in the short-term cyclical oscillations observed in the magnetic susceptibility curve. In addition, a sample of modern sediments was analyzed for comparison. Results are shown in Table 3 and Figure 9. After treatment of the raw data by the cumulative Log-Gaussian function [37], two main components are obtained in all samples. Component 1 has $B_{1/2}$ values of 21–24 mT and DP (dispersion parameter) of 0.25–0.30, corresponding to the coercivity range of detrital magnetite [56]. Magnetite contributes to 77–83% of the total remanence. Component 2 has $B_{1/2}$ values of 200–234 mT and DP of 0.50–0.55, typical of hematite [56]. Hematite contributes to 17–22% of the total remanence. Goethite has not been identified although the data are noisy above imparted fields of 1 T. The S-ratio varies between 0.86 and 0.92 and reflects the dominance of magnetite in the total remanence.

Except for the slight variations of the S-ratio, no significant changes in the magnetic mineralogy (coercivity and grain-size) was noted within the eight analyzed samples, nor with modern sediments (Figure 9). However, there is a clear correlation between magnetic susceptibility and saturation isothermal remnant magnetization (SIRM), suggesting that magnetic susceptibility is mostly controlled by the concentration of ferromagnetic iron oxides (magnetite and hematite) (Figure 10). In addition, samples having the lowest magnetic susceptibility values also have lower S-ratio (Table 4), suggesting that the short-term cyclical variations observed in the magnetic susceptibility curve reflect variation in the relative proportion of magnetite and hematite.

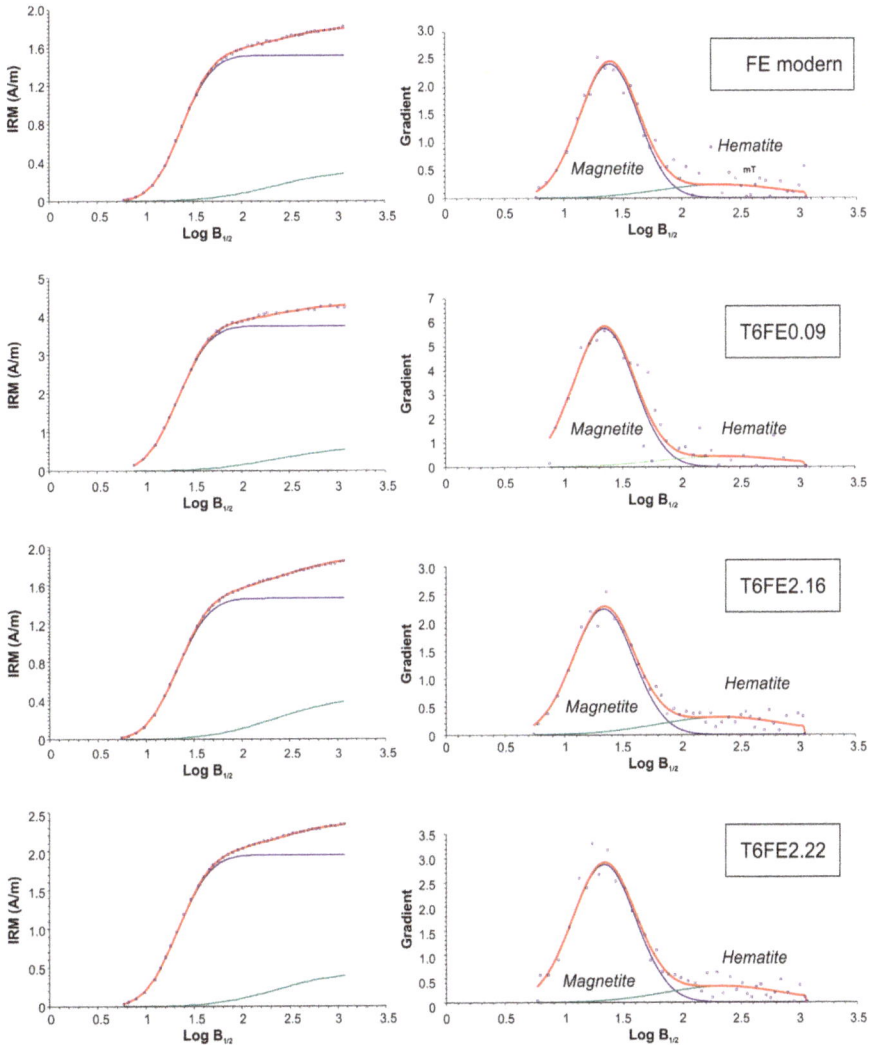

Figure 9. Isothermal Remnant Magnetization (IRM) curves treated by the log-Gaussian function, of representative samples from the upper unit of T6 (the sample depth is indicated by the sample reference) and present-day sediment from the Enxarrique stream (sample FE modern), all collected at the Foz do Enxarrique site.

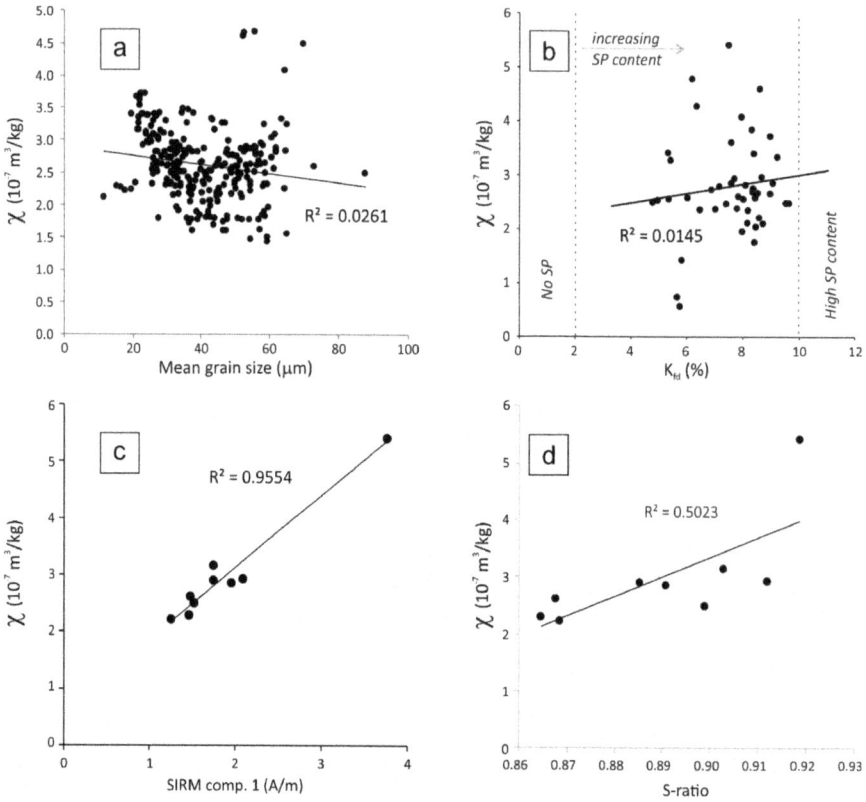

Figure 10. Correlations between mass specific magnetic susceptibility and (**a**) grain size, (**b**) frequency-dependent magnetic susceptibility and (**c**) Saturation Isothermal Remanent Magnetization of component 1 (i.e., magnetite content) and (**d**) S-ratio (relative proportion of magnetite and hematite). R^2 is the determinant factor.

Table 4. Rock magnetic parameters calculated from the analysis of the IRM curves (see Figure 9). The % column represents the percentage of the contribution of the magnetic phase; SIRM is the IRM at saturation is proportional to the concentration of the magnetic phase; B1/2 corresponds to the coercivity and DP is the dispersion parameter of the Gaussian curves.

Sample Code	Component 1 (Magnetite)				Component 2 (Hematite)				
	%	SIRM	B1/2	DP	%	SIRM	B1/2	DP	S-Ratio
FE modern	83.1	1.52	24.0	0.25	16.9	0.31	218.8	0.52	0.90
T6FE0.09	86.2	3.75	21.9	0.26	13.8	0.60	223.9	0.55	0.92
T6FE0.62	77.8	1.47	21.9	0.26	22.2	0.42	223.9	0.55	0.87
T6FE2.00	80.6	1.75	21.9	0.3	19.4	0.42	223.9	0.55	0.89
T6FE2.16	77.2	1.46	21.4	0.26	22.8	0.43	223.9	0.55	0.86
T6FE2.22	82.3	1.95	21.4	0.27	17.7	0.42	223.9	0.50	0.89
T6FE2.33	76.7	1.25	21.4	0.26	23.3	0.38	199.5	0.53	0.87
T6FE3.74	84.3	2.09	22.4	0.25	15.7	0.39	199.5	0.53	0.91
T6FE3.84	79.5	1.75	22.4	0.25	20.5	0.45	158.5	0.50	0.90

4.3.4. Phytolith Analysis

Phytoliths were identified in those carbonate concretions corresponding to samples with codes T6FE0.72, T6FE2.00, T6FE2.90, T6FE3.48 and T6FE4.12, but constituted fewer than 10 units per observed concretion, which limits their potential for environmental interpretation. The identified morphotypes are predominantly larger than 20 µm, of Bulliform and Cuneiform types (Figure 11) and rare Elongates and Trapeziforms. The Bulliform and Cuneiform types are usually produced by grasses undergoing hydric stress. The poor preservation of the studied phytoliths is probably due to the alkalinity of the concretions, although there is some evidence of aeolian abrasion.

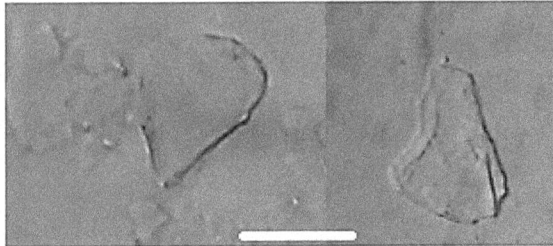

Figure 11. Phytoliths of Bulliform and Cuneifom types, identified in carbonate concretions. Scale bar: 20 µm.

4.3.5. Palynology

From a general point of view, the prepared residues from both the T6 and T5 terraces are very poor in organic matter and pollen. A small number of highly degraded pollen grains were observed, of which some are indeterminable because of their damaged state. The small number of identified pollen grains and, in some cases, the sterility of the samples, precludes the construction of pollen diagrams. The results are presented in Table 5, organized by sample; taxa for each sample are divided into arboreal types, followed by shrubs and herbaceous and, finally, indeterminate pollen grains.

Table 5. Palynology of the T6 and T5 terrace deposits, from the stratigraphic sections at Foz do Enxarrique site.

Sample Code	Taxa	Count
Basal layer (archaeological strata) of the T6 upper unit		
FE/15 0.25–0.30	Indeterminate	1
FE/15 0.35–0.40	Chenopodiaceae	2
	Indeterminate	4
FE/15 0.45–0.50	Indeterminate	2
Upper part of the T6 upper unit (silty sands and sandy silts)		
T6FE0.70–0.74	*Pinus* sp.	4
	Ericaceae	2
	Poaceae	2
	Artemisia	1
	Chenopodiaceae	2
	Indeterminable	9

Table 5. *Cont.*

Sample Code	Taxa	Count
Upper part of the T6 upper unit (silty sands and sandy silts)		
T6FE1.13–1.17	*Pinus* sp.	1
	Poaceae	2
	Indeterminate	10
T6FE1.97–2.02	*Pinus* sp.	1
	Asteraceae tubuliflora	1
	Indeterminate	6
T6FE2.88–2.92	Indeterminate	1
T6FE3.46–3.50	Indeterminate	3
T6FE4.10–4.14	Indeterminate	1
T6FE4.49–4.52	Indeterminate	1
T6FE4.98–5.02	*Pinus* sp.	1
	Indeterminate	1
T5 lower part (sandy silts)		
T5FE0.15–0.23	Indeterminate e	1
T5FE0.92–1.00	*Pinus* sp.	1
	Indeterminate	2

The two samples from the lower part of the T5 sequence from outcrops at the Foz do Enxarrique site (T5FE015–0.23 and T5FE0.92–1.00) only allowed the identification of a pollen grain of *Pinus* sp. For the three samples collected from the basal layer (archaeological strata) of the T6 upper member only Chenopodiacea could be identified. In the samples collected in the upper part of T6, the few taxa identified indicate the presence of *Pinus* sp., the only tree species, Ericaceae (shrubs) and, in the herbaceous group, Poaceae, *Artimisia*, Chenopodiaceae and *Asteraceae tubuliflora*. In conclusion, the results obtained from the palynological study of the T6 and T5, from oxidized sediments, revealed very few palynomorphs and do not allow any palaeoenvironmental interpretation.

4.3.6. Geochemical Analyses

The results of geochemical analyses performed with the Thermal Scientific Analyzer Nikon XL3t spectrometer, on samples collected from the upper unit of T6, indicate that Ca has high values on the calcium carbonate concretions/rizoliths and very low values on uncemented siliciclastic sediment (coarse silt to fine sand) (Appendix A, Figure A1).

4.4. Palaeolithic Sites and Fauna Associated

No evidence has been found to date of industries prior to the Acheulean in this area. However, considering the presence of Oldowan material at several sites in Iberia, some dating back to c. 1.4 Ma, this absence must be seen as a problem of archaeological visibility [57].

The oldest archaeological evidence in this region appears in the basal deposits of T4, which are dated as older than 280 ka and perhaps as old as c. 340 ka [26,28]. A large Acheulean quartzite assemblage, with handaxes, cleavers, very large flakes and respective cores, was recovered from Monte do Famaco (Vila Velha de Ródão), in a colluvium resulting from the erosion of basal T4 deposits [58]. Test pits were excavated at this site and a large assemblage was also collected from the surface, but no thorough excavation was undertaken, nor has the assemblage been published in detail. Other evidence of Acheulean (e.g., identification of single or small numbers of handaxes and cleavers, sometimes along with quartzite flakes) has been reported from T4 deposits at Vilas Ruivas (Arneiro), but also at the surface of Pegos do Tejo-1, Monte das Nove Oliveiras, Monte do Arneiro and Monte da Cabeça Gorda; all these sites at Arneiro are still to be excavated [59,60].

The earliest Mousterian (Middle Paleolithic) industries occur on both sides of the Tejo in the uppermost deposits of T4. At Pegos do Tejo-2 (probably dating from c. 200–170 ka), the identification of a ring of cobbles was interpreted as a possible hearth (fire pit). These occur along with Levallois, discoidal, opportunistic and bipolar cores, flakes (often thick), denticulates, notches and sidescrapers [61,62]. A similar age (c. 165–155 ka) was attributed to Cobrinhos (Vila Velha de Ródão), a dense lithic assemblage spread over 1300 m² and dominated by Levallois, discoidal, radial and opportunistic cores and Levallois and *Kombewa* flakes (and also some points and blades), pseudo-Levallois points, sidescrapers, denticulates and notches [22].

At Monte da Revelada (Vila Velha de Ródão), also at the top of T4, the lowest archaeological context has discoidal and Levallois cores, flakes with dihedral and faceted platforms, along with radial dorsal patterns, including some Levallois flakes and pseudo-Levallois points that can be associated with a Mousterian occupation. These artefacts occur along with oval stone features that that again may correspond to hearths. At the nearby site of Alto da Revelada, Mousterian artefacts occur scattered on the surface, on the Cabeço do Infante Formation (Paleogene), in the western area where it has not been covered by aeolian sands [63].

In the T5 deposits at Vilas Ruivas (Arneiro), now dated to c. 117–71 ka, it was again possible to recognize hearths, windshields and post-holes [59,60], associated with a small but well-preserved Mousterian assemblage. This was composed of Levallois, discoidal and centripetal cores, Levallois flakes, pseudo-Levallois points, some blades and points, denticulates, notches and sidescrapers, mostly quartzite, but without faunal remains.

At Azinhal (Arneiro), archaeological evidence located at an alluvial deposit dated to c. 61 ka and connected with T6 terrace, corresponds with a lithic assemblage dominated by quartzite (62%), quartz (36%) and some flint (2%), with discoidal, centripetal and opportunistic cores, Levallois flakes and blades, choppers, perforators, burins, denticulates and handaxes, suggesting some mixture with both Acheulean and Upper Paleolithic [61,62].

At Tapada do Montinho (Arneiro), a site in an alluvial fan unit dated to c. 51 ka and connected with T6, the lithic assemblage is composed of quartzite (84%), quartz (10%) and flint (5%) and comprises *Kombewa*, Levallois, and discoidal cores and Levallois flakes along with notches [61,62].

At the Foz do Enxarrique site, several archaeological campaigns were undertaken between 1982 and 2001. A single archaeological level was identified, at the base of the T6 upper unit, with a diverse Pleistocene faunal assemblage associated with a rich Mousterian industry represented by more than 10,000 artefacts [48,64]. Among them are Levallois, discoidal and centripetal cores, Levallois flakes and pseudo-Levallois points, plus blades and points, denticulates, notches and sidescrapers, most of them produced from quartzite (c. 66%), quartz (c. 25%) and flint (c. 9%). A complete sequence of debitage is present. The explotation of raw materials was interpreted as resulting from local and opportunistic procurement [48].

The Foz do Enxarrique faunal assemblage from the campaigns carried out between 1982 and 1999, a total of 785 bone specimens, was studied by Brugal and consists of red deer (*Cervus elaphus*—54.8%), horse (*Equus caballus*—36.5%), aurochs (*Bos primigenius*—2.2%), elephant (1.2%), rhinoceros (0.5%), rabbit (4.2%), and a very few carnivores such as fox (0.5%), hyena (0.2%) and lynx (0.2%) [65–68]. The fauna collected during later campaigns (2000 and 2001) has the same characteristics: a clear predominance of deer, horse and aurochs; the predominance of these herbivores and the weak occurrence of carnivores points to an anthropogenic accumulation [69]. This seems to be reinforced by the distribution of the anatomical units of the skeleton identified from the most commonly represented species (*Cervus* and *Equus*). The cranial specimen is the most represented (47.3% for the deer and 75.3% for equidae), the axial and appendicular skeleton representing only 52.7% and 24.7%, respectively [66]. This suggested the idea that the site served as a hunting zone, because the edible parts (associated to the appendicular and axial skeleton) have been removed. Such interpretation is more difficult for the megafauna assemblage as, for instance, in the case of the elephant remains (classified as *Elephas antiquus* (=*Paleoloxodon antiquus*)) that are represented by a lamella of a superior molar and four bone fragments [66–68]. Worthy of note is the occurrence of birds, fish and mollusc remains [66]; however, in a subsequent study no bird bones have been identified [69]. Amongst the studied remains, just over 50% are small fragments of unidentified bones; there are 42% of identified remains, which represents a high rate compared to other Paleolithic sites, where the fauna remains appear very fragmented (10 to 20% of total remains: TNR) [66]. About 13 specimens were identified with marks from the action of carnivores and 15 with marks from human butchery (nine with cutmarks and six with evidence of burning) [66]. Results from use–wear study on the lithic assemblage performed on 110 quartzite artefacts showed a dominance of wear traces associated to butchering, cutting of soft-animal tissue and scraping on bone activities [70]. Based on this, the authors conclude that the site was primarily a hunting zone, in which the animals were killed, quartered and then mainly taken to other places.

The Carregueira Formation (the cover unit of aeolian sans) contains Late Palaeolithic to Neolithic industries. At Monte da Revelada, the upper bed is composed of aeolian sands, disturbed by ploughing, and contains a mixture of Epipaleolithic and Neolithic material. A similar context, but here with backed bladelets, was found within the aeolian sands covering the eastern area of Alto da Revelada. At Vilas Ruivas, the aeolian unit, again disturbed by ploughing, has a Late Upper Palaeolithic or Epipaleolithic industry, in which flint occurs in greater quantity and the lithic assemblage has prismatic cores, blades and bladelets, small flakes, burins and endscrapers. Therefore, it is possible that the MIS 2 aeolian sands of the Carregueira Formation may preserve the first modern human occupations in the region.

5. Discussion

Regarding the environmental interpretation of the T6 record in the study area, the lower boulder–pebble gravel, 0.4 m thick and overlying a strath cut in metamorphic basement, corresponds to the coarse river-bed sedimentation near the margin of the energetic Tejo palaeochannel, probably during the interval c. 60–45 ka (MIS 3). No lithic artefacts or fossils, which could have helped in the interpretation, were found in this bed.

The T6 upper unit, mainly consisting of fine to very fine sands, grading upwards to coarse silt, is attributed overbank sedimentation. The detailed environmental interpretation of the various stratigraphic subunits is discussed in the following paragraphs.

The c. 20 cm lower bed (at 5.60–5.40 m depth), comprising Mousterian artefacts and fossil bones in a matrix of gravelly micaceous fine sands, is interpreted as overbank deposits close to the channel margin that also record hominin activities of hunting and butchery [48,64,66,70]. During this period, the Tejo channel moved laterally towards the west, preserving this record of human occupation. Thus, this confined place at the confluence of the Enxarrique and Açafal streams was used by animals for drinking and they were easily hunted. As this thin layer, now dated to 44 ± 3 ka, represents the last regional evidence of Mousterian industries and the megafauna, it may correspond to a cold and dry period that negatively impacted the animals and the Neanderthals. It is possible the Neanderthals relied heavily on some specific biotic resource that may have been reduced during cold climatic conditions, so that they faced difficulties in adapting. The remaining Neanderthals could have been induced to move toward the better climate conditions of SW Iberia and were later absorbed into modern human populations. Also relevant to this discussion is the fact that the ages of the stratigraphic levels with the youngest Mousterian industries in westernmost Iberia are progressively younger toward the SSW: the Cantabrian region, c. 48–45 ka [71]; central Portugal, c. 44–34 ka (e.g., Foz do Enxarrique, c. 44 ka; Almonda, c. 32 ka; Mira Nascente, c. 42–40 ka; Gruta da Figueira Brava, c. 34 ka) [58,72–75]; Murcia, 37 ka [11]; Gibraltar, between c. 33 and 24 ka [9].

The sediments of the 5.40–4.55 m depth interval within T6, consisting of micaceous fine sands with some interbedded thin gravel stringers, are also attributed to fluvial overbank deposition close to the channel margin. No fossils or artefacts were found here or in the upper deposits of T6. The onset of a new sedimentation phase (overbank fine-grained sediments) without artefacts is not necessarily evidence for cultural breakdown. However, in other areas of the Lower Tejo where T6 is preserved, no Mousterian artefacts were found in younger stratigraphic levels.

Regarding the T6 upper unit, the coarse silts from 4.55 cm depth to the surface are attributed to overbank sedimentation, but some characteristics point to the possibility of short-distance transport by wind, namely the lack of lamination and of erosion surfaces, the absence of dispersed coarser grains, the low clay content, a mean grain size in the coarse silt range, the fine skewed distributions and evidence of aeolian abrasion provided by phytolith analysis. The literature shows grain-size-distribution curves of loess deposits to be very similar to these silty T6FE deposits (e.g., [76–80]. Possible short-distance aeolian transport of exposed overbank fines could have been promoted by strong winds coming from the west and penetrating through the Ródão gorge (Figure 1). However, if there had there been significant aeolian transport, the resultant sediments should also cover higher terrace levels, and this is not evidenced by field observation.

Sediment magnetic properties have been widely applied to fluvial sediments and loess and may provide useful information about fluvial activity, climate and environmental changes, as well as pedogenesis [50,51,81–83]. In fluvial sediments from Beijing, for example, high magnetic susceptibility values generally reflect warm-climate conditions, whereas lower values match colder periods [84]. In wind-blown sediments and buried soils from southern Siberia [81], colder high-wind periods that are associated with an absence of soil formation show low values of frequency-dependence of magnetic susceptibility, whereas higher values are observed in episodes with less wind. In the classic loess–palaeosol sequence of the central Chinese Loess Plateau, there is a striking correlation between magnetic susceptibility and grain size [85], which are good indicators of summer and winter monsoon intensity respectively [86,87]. In general, ferromagnetic crystals in soils derive from both primary (detrital) and secondary (enhanced) iron minerals. The latter are most often of stable single-domain size or less and associated with the clay fraction, whereas the former are usually associated with sand and coarse silt-size fractions [49]. Regarding the studied T6 sediments, no significant correlation ($R^2 = 0.0261$) is observed between sedimentary grain size and magnetic susceptibility, suggesting that the latter is not controlled by mineralogy (ex. paramagnetic clay and phyllosilicates versus diamagnetic quartz) (Figure 5). In contrast, a striking correlation between SIRM and magnetic susceptibility ($R^2 = 0.9554$ for magnetite; $R^2 = 0.769$ for hematite) indicates that magnetic susceptibility is dominantly controlled by the iron-oxide content, in the form of magnetite and hematite (Figure 10), probably of a detrital origin. The proportion and contribution of magnetite and hematite in the bulk remanence is mostly similar in all samples, as illustrated by narrow S-ratio values of 0.86–0.92. In particular, the IRM curves of the T6 samples are very similar to those of modern sediments (Figure 9), suggesting a common source for the entire sedimentary profile during the last millennium. Superparamagnetic particles, generally interpreted as a product of pedogenic processes, are present in all samples and may have been formed in situ or transported from the surrounding soils. The presence of numerous rhizoliths observed in the field suggests that pedogenic magnetic particles may have precipitated in-situ, during soil development. However, the poor correlation ($R^2 = 0.0145$) between mass specific magnetic susceptibility and frequency-dependent magnetic susceptibility indicates that pedogenic processes alone cannot explain the short-term cyclical variations observed in the magnetic susceptibility curve of the T6 profile (Figure 4). Conversely, a slight but significant correlation ($R^2 = 0.5023$) between the S-ratio, i.e., the relative proportion of magnetite versus hematite, and magnetic susceptibility imply that the short-term cycles may correspond to changes in weathering regime and climate. More exactly, warmer/drier periods would enhance the oxidation of magnetite (or maghemite) and promote precipitation of hematite, and the reverse. Because the magnetic susceptibility of hematite is lower than magnetite, this provides a potential explanation for the cyclical oscillations observed in the magnetic susceptibility curve.

The 0.60–5.40 m depth interval contains c. 30 thin levels of calcium carbonate concretions and rizoliths intercalated in uncemented coarse silt. The characteristics of these levels, which have no evidence of any erosive surface and dip toward the palaeochannel, progressively increasing in thickness, point to a secondary origin for the carbonate concretions. A relatively stable surface and a certain amount of rainfall during the represented periods are indicated. The most probable source of calcium carbonate for these pedogenetic concretions is the dissolution of dolomitic and calcium carbonates that occur at the base of the Cabeço do Infante Formation (Paleogene), which crops out at a short distance from the site. Upstream sources of calcium carbonates are located at least at 200 km away (in the Madrid Cenozoic basin).

The aggradation of T5 (c. 140–70 ka) correlates with the very high sea levels of MIS 5, whereas the following period of river down-cutting (c. 70–60 ka) indicates to have been mainly determined by the low-sea-level conditions during MIS 4 and the aggradation of T6 (c. 60–32 ka) seems to correlate with the higher sea levels and high sediment supply coeval with MIS 3 [2,27,28]. During this interval (c. 140–32 ka) Iberia was influenced by several climatic (e.g., [88–92]) and oceanographic changes [93,94], registered in the North Atlantic region and in records from Greenland Ice Cores (e.g., [95,96]).

The results obtained from palynological study of the T6 and T5 deposits do not allow palaeoenvironmental interpretation. Regarding the interval represented by the T6 terrace, the palynological record of the MD95-2039 ocean core points to an open landscape with steppe vegetation and low values of tree pollen, suggesting a severely cold and dry climate, during 61–59 ka; during the interval 57–31 ka, there are fluctuations in the expansion and contraction of arboreal pollen and Ericaceae related to the alternation of warmer and humid conditions during the interstadials and the cold and dry stadial minima of the last glacial cycle [92]. In the Puente Pino sequence (Toledo, Spain), for the 42–34 ka interval, declining woodland and the increasing herbaceous pollen taxa are observed, related to adverse climate conditions of cold and dry character [97]. The palaeoenvironmental data provided by these two adjacent contexts support the possibility of an open landscape in the region of the study area and could explain the several levels with pedogenic calcium carbonate rizoliths and concretions in the T6 upper unit.

The mineralogical composition of the <2 μm fraction of the samples collected from T6 and T5 are similar, consisting of smectite, illite and kaolinite, although T6 has less smectite. Even if a significant part of the clay minerals could have been sourced from erosion of the local Paleogene and Miocene formations, very rich in smectite [23,25], this clay mineral association seems compatible with the regional climatic conditions during MIS 5–3 [91,92].

By 32 ka, the climate had changed to cold and dry conditions and aeolian deposition dominated the valley landscape, preserving Upper Palaeolithic industries.

6. Conclusions

Updated ages from the three Lower Tejo terrace sequences, containing Mousterian industries, were obtained by pIRIR, as follows:

(i) OSL dating of the oldest Mousterian industry, stratigraphically situated in the uppermost T4 deposits, suggests a probable age of c. 200–170 ka for the arrival of the Neanderthals in this region, probably by way of the Tejo River valley from central Iberia;

(ii) T5 dates from c. 140 ka at the base and 70 ka at the top;

(iii) T6 dates from c. 60 ka at the base and c. 35 ka near the top;

(iv) The new date of 44 ± 3 ka for a level located at the base of the T6 upper unit records the last regional occurrence of a Mousterian industry and of the megafauna.

The T6 upper unit (so far without Mousterian or Early Upper Palaeoleolithic industries), consists of fine sands to coarse silts interpreted as overbank sediments. It has a large number of intercalated thin levels of carbonate concretions and rizoliths, suggesting episodic evaporation and development of paleosols in a seasonal dry period, in agreement with the occurrence of phytoliths.

Supplementary Materials: The following is available online at http://www.mdpi.com/2571-550X/2/1/3/s1. Table S1—Grain-size statistical parameters and mass specific magnetic susceptibility of samples collected from of the T6 upper unit at Foz do Enxarrique site.

Author Contributions: Conceptualization, P.P.P.; Methodology, P.P.C. and E.F.; Validation, P.P.C., J.-P.B., M.P.G., E.F., P.Y. and R.M.; Formal Analysis, J.-P.B., A.S.M, M.P.G., E.F., C.F., P.Y., J.C.S. and R.M.; Investigation, P.P.C, A.A.M., J.-P.B, A.S.M, M.P.G., E.F., T.P., S.F.,C.F., P.Y., J.C.S. and R.M.; Writing-Original Draft Preparation, P.P.C., A.M., M.P.G., E.F., T.P., S.F., C.F., D.R.B.; Writing-Review & Editing, P.P.C., A.A.M, J.-P.B., E.F., T.P., S.F., C.F., D.R.B.; Visualization, A.A.M., P.P.C., M.P.G., E.F., P.Y., R.M.; Supervision, P.P.C.

Funding: This research was part funded by the Fundação para a Ciência e a Tecnologia, through projects, UID/MAR/04292/2013—MARE (PPC, MPG, YP and RM), UID/GEO/04683/2013—ICT (SF and CF), UID/GEO/04683/2013—ICT (AAM) and IF/01075/2013 (TP). MPG has the FCT PhD grant SFRH/BD/116038/2016.

Acknowledgments: Thanks to Ana Bárbara Costa, Johannes Remhof, José Erbolato Filho, Alexandre Cabral and Paulo Pedrosa for assistance during fieldwork and at the Sedimentology Laboratory. Lídia Catarino is thanked for assistance during the use of the Thermal Scientific Analyzer Nikon XL3t spectrometer. Luis Raposo and Nelson Almeida provided very constructive comments regarding the Palaeolithic of the area. Thanks to the reviewers and editor for manuscript improvements.

Conflicts of Interest: The authors declare no conflict of interest. The funders had no role in the design of the study; in the collection, analyses, or interpretation of data; in the writing of the manuscript, or in the decision to publish the results.

Appendix A

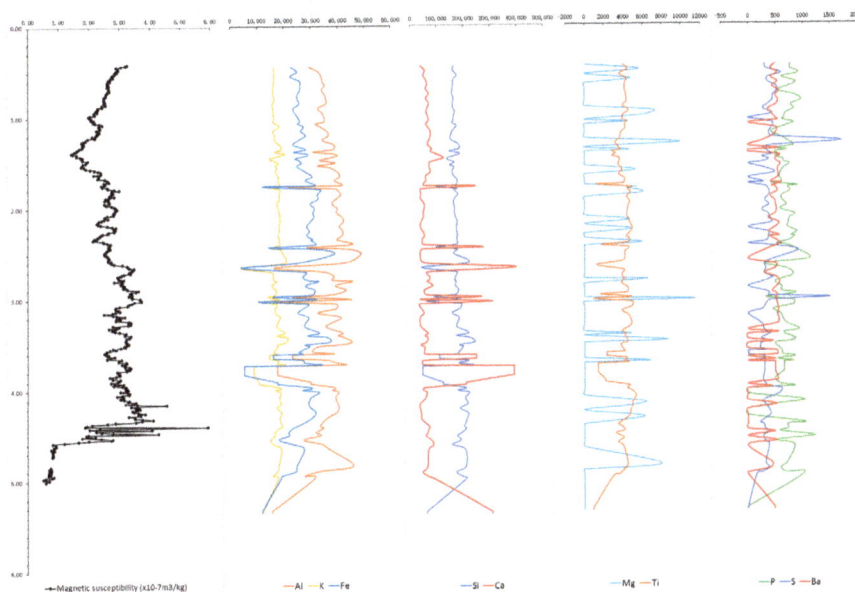

Figure A1. Comparison between the mass specific magnetic susceptibility results and the results of geochemical analyses performed with a X-ray fluorescence spectrometer (Al, K, Fe; Si, Ca; P, S, Ba, Mg and Ti), on samples collected from the upper unit of T6 at Foz do Enxarrique (sample depth in meters).

References

1. Cunha, P.P.; Martins, A.A.; Daveau, S.; Friend, P.F. Tectonic control of the Tejo river fluvial incision during the late Cenozoic, in Ródão—Central Portugal (Atlantic Iberian border). *Geomorphology* **2005**, *64*, 271–298. [CrossRef]

2. Cunha, P.P.; Almeida, N.A.C.; Aubry, T.; Martins, A.A.; Murray, A.S.; Buylaert, J.-P.; Sohbati, R.; Raposo, L.; Rocha, L. Records of human occupation from Pleistocene river terrace and aeolian sediments in the Arneiro depression (Lower Tejo River, central eastern Portugal). *Geomorphology* **2012**, *165–166*, 78–90. [CrossRef]

3. Zilhão, J. Le passage du Paléolithique moyen/Paléolithique supérieur dans le Portugal. In *El Origin Del Hombre Moderno En El Suroeste de Europ*; UNED: Madrid, Spain, 1993; pp. 127–146.

4. Zilhão, J. The Ebro Frontier: A Model for the Late Extinction of Iberian Neanderthals. In *Neanderthals on the Edge: 150th Anniversary Conference of the Forbes' Quarry Discovery, Gibraltar*; Oxbow Books Limited: Oxford, UK, 2000; pp. 111–121.

5. Zilhão, J. Chronostratigraphy of the Middle-to-Upper Paleolithic Transition in the Iberian Peninsula. *Pyrenae* **2006**, *37*, 7–84. [CrossRef]

6. Zilhão, J. The Ebro frontier revisited. In *The Mediterranean from 50,000 to 25,000 BP: Turning Points and New Directions*; Camps, M., Szmidt, C., Eds.; Oxbow Books: Oxford, UK, 2009; pp. 293–311.

7. Jöris, O.; Álvarez Fernández, E.; Weninger, B. Radiocarbon evidence of the Middle to Upper Palaeolithic transition in Southwestern Europe. *Trab. Prehist.* **2003**, *60*, 15–38. [CrossRef]

8. Finlayson, C.; Giles Pacheco, F.; Rodríguez-Vidal, J.; Fa, D.A.; María Gutierrez López, J.; Santiago Pérez, A.; Finlayson, G.; Allue, E.; Baena Preysler, J.; Cáceres, I.; et al. Late survival of Neanderthals at the southernmost extreme of Europe. *Nature* **2006**, *443*, 850–853. [CrossRef] [PubMed]

9. Finlayson, C.; Fa, D.A.; Jiménez Espejo, F.; Carrión, J.S.; Finlayson, G.; Giles Pacheco, F.; Rodríguez Vidal, J.; Stringer, C.; Martínez Ruiz, F. Gorham's Cave, Gibraltar—The persistence of a Neanderthal population. *Quat. Int.* **2008**, *181*, 64–71. [CrossRef]

10. Zilhão, J.; Davis, S.J.M.; Duarte, C.; Soares, A.M.M.; Steier, P.; Wild, E. Pego do Diabo (Loures, Portugal): Dating the emergence of anatomical modernity in westernmost Eurasia. *PLoS ONE* **2010**, *5*, e8880. [CrossRef]

11. Zilhão, J.; Anesin, D.; Aubry, T.; Badal, E.; Cabanes, D.; Kehl, M.; Klasen, N.; Lucena, A.; Martín-Lerma, I.; Martínez, S.; et al. Precise dating of the Middle-to-Upper Paleolithic transition in Murcia (Spain) supports late Neandertal persistence in Iberia. *Heliyon* **2017**, *3*, e00435. [CrossRef]

12. Bradtmöller, M.; Pastoors, A.; Weninger, B.; Weniger, G.-C.C. The repeated replacement model—Rapid climate change and population dynamics in Late Pleistocene Europe. *Quat. Int.* **2012**, *247*, 38–49. [CrossRef]

13. De la Peña, P. The beginning of the Upper Paleolithic in the Baetic Mountain area (Spain). *Quat. Int.* **2013**, *318*, 69–89. [CrossRef]

14. Galván, B.; Hernández, C.M.; Mallol, C.; Mercier, N.; Sistiaga, A.; Soler, V. New evidence of early Neanderthal disappearance in the Iberian Peninsula. *J. Hum. Evol.* **2014**, *75*, 16–27. [CrossRef] [PubMed]

15. Bicho, N.; Marreiros, J.; Cascalheira, J.; Pereira, T.; Haws, J. Bayesian modeling and the chronology of the Portuguese Gravettian. *Quat. Int.* **2015**, *359–360*, 499–509. [CrossRef]

16. Santonja, M.; Pérez-González, A. La industria lítica del miembro estratigráfico medio de Ambrona (Soria, España) en el contexto del Paleolítico Antiguo de la Península Ibérica. *Zephyrvs* **2006**, *59*, 7–20.

17. Terradillos-Bernal, M.; Díez-Fernández-Lomana, J.C. La transition entre les Modes 2 et 3 en Europe: Le rapport sur les gisements du Plateau Nord (Péninsule Ibérique). *Anthropologie* **2012**, *116*, 348–363. [CrossRef]

18. Ollé, A.; Mosquera, M.; Rodríguez, X.P.; de Lombera-Hermida, A.; García-Antón, M.D.; García-Medrano, P.; Peña, L.; Menéndez, L.; Navazo, M.; Terradillos, M.; et al. The Early and Middle Pleistocene technological record from Sierra de Atapuerca (Burgos, Spain). *Quat. Int.* **2013**, *295*, 138–167. [CrossRef]

19. Álvarez-Alonso, D. First Neanderthal settlements in northern Iberia: The acheulean and the emergence of mousterian technology in the Cantabrian region. *Quat. Int.* **2014**, *326–327*, 288–306. [CrossRef]

20. Santonja, M.; Pérez-González, A.; Panera, J.; Rubio-Jara, S.; Méndez-Quintas, E. The coexistence of Acheulean and Ancient Middle Palaeolithic techno-complexes in the Middle Pleistocene of the Iberian Peninsula. *Quat. Int.* **2016**, *411*, 367–377. [CrossRef]

21. Daura, J.; Sanz, M.; Arsuaga, J.L.; Hoffmann, D.L.; Quam, R.M.; Ortega, M.C.; Santos, E.; Gómez, S.; Rubio, Á.; Villaescusa, L.; et al. New Middle Pleistocene hominin cranium from Gruta da Aroeira (Portugal). *Proc. Natl. Acad. Sci. USA* **2017**, *114*, 3397–3402. [CrossRef]

22. Pereira, T.; Cunha, P.P.; Martins, A.A.; Nora, D.; Paixão, E.; Figueiredo, O.; Raposo, L.; Henriques, F.; Caninas, J.; Moura, D.; et al. Geoarchaeology of the Cobrinhos site (Vila Velha de Ródão, Portugal)—A record of the earliest Mousterian in western Iberia. *J. Archaeol. Sci. Rep.* **2019**, in press.

23. Cunha, P.P. Estratigrafia e Sedimentologia dos Depósitos do Cretácico Superior e Terciário de Portugal Central, a Leste de Coimbra/Stratigraphy and Sedimentology of the Upper Cretaceous and Tertiary of Central Portugal, East of Coimbra. Ph.D. Thesis, University of Coimbra, Coimbra, Portugal, July 1992.

24. Cunha, P.P.; Barbosa, B.P.; Pena dos Reis, R. Synthesis of the Piacenzian onshore record, between the Aveiro and Setúbal parallels (Western Portuguese margin). *Ciências da Terra* **1993**, *12*, 35–43.

25. Cunha, P.P. Unidades litostratigráficas do Terciário da Beira Baixa (Portugal). *Comum. Inst. Geol. Min.* **1996**, *82*, 87–130.

26. Cunha, P.P.; Martins, A.A.; Huot, S.; Murray, A.S.; Raposo, L. Dating the Tejo River lower terraces in the Ródão area (Portugal) to assess the role of tectonics and uplift. *Geomorphology* **2008**, *102*, 43–54. [CrossRef]

27. Cunha, P.P.; Martins, A.A.; Gouveia, M.P. As escadarias de terraços do Ródão à Chamusca (Baixo Tejo)—Caracterização e interpretação de dados sedimentares, tectónicos, climáticos e do Paleolítico/The terrace staircases of the Lower Tagus River (Ródão to Chamusca)—Characterization and interpretation of the sedimentary, tectonic, climatic and Palaeolithic data. *Estudos do Quaternário* **2016**, *14*, 1–24. [CrossRef]

28. Cunha, P.P.; Martins, A.A.; Buylaert, J.P.; Murray, A.S.; Raposo, L.; Mozzi, P.; Stokes, M. New data on the chronology of the Vale do Forno sedimentary sequence (Lower Tejo River terrace staircase) and its relevance as a fluvial archive of the Middle Pleistocene in western Iberia. *Quat. Sci. Rev.* **2017**, *166*, 204–226. [CrossRef]

29. Almeida, N.; Deprez, S.; De Dapper, M. The Palaeolithic occupation of the Northeastern of Alen Tagus (Portugal): A geoarchaeological approach. In *Graphical Markers and Megalith Builders in the International Tagus, Iberian Peninsula—British Archaeological Reports International Series*; Bueno-Ramirez, P., Barroso-Bermejo, R., Balbín Berhmann, R., Eds.; Archaeopress: Oxford, UK, 2008; Volume 1765, pp. 19–26. ISBN 9781407302546.

30. Stokes, M.; Cunha, P.P.; Martins, A.A. Techniques for analysing Late Cenozoic river terrace sequences. *Geomorphology* **2012**, *165–166*, 1–6. [CrossRef]

31. Duller, G.A.T. Luminescence dating of Quaternary sediments: Recent advances. *J. Quat. Sci.* **2004**, *19*, 183–192. [CrossRef]

32. Murray, A.; Marten, R.; Johnston, A.; Martin, P. Analysis for naturally occurring radionuclides at environmental concentrations by gamma spectrometry. *J. Radioanal. Nucl. Chem.* **1987**, *115*, 263–288. [CrossRef]

33. Olley, J.M.; Murray, A.S.; Roberts, R.G. The effects of disequilibria in the uranium and thorium decay chains on burial dose rates in fluvial sediments. *Quat. Sci. Rev.* **1996**, *15*, 751–760. [CrossRef]

34. Huntley, D.J.; Baril, M.R. The K content of the K-feldspars being measured in optical dating or in thermoluminescence dating. *Ancient TL* **1997**, *15*, 11–13.

35. Thomsen, K.J.; Murray, A.S.; Jain, M.; Bøtter-Jensen, L. Laboratory fading rates of various luminescence signals from feldspar-rich sediment extracts. *Radiat. Meas.* **2008**, *43*, 1474–1486. [CrossRef]

36. Buylaert, J.-P.; Jain, M.; Murray, A.S.; Thomsen, K.J.; Thiel, C.; Sohbati, R. A robust feldspar luminescence dating method for Middle and Late Pleistocene sediments. *Boreas* **2012**, *41*, 435–451. [CrossRef]

37. Kruiver, P.P.; Dekkers, M.J.; Heslop, D. Quantification of magnetic coercivity components by the analysis of acquisition curves of isothermal remanent magnetisation. *Earth Planet. Sci. Lett.* **2001**, *189*, 269–276. [CrossRef]

38. Costa, F.G.C.M.; Bove, C.P.; Arruda, R.; Philbrick, C.T. Silica bodies and their systematic implications at the subfamily level in Podostemaceae. *Rodriguesia* **2011**, *62*, 937–942. [CrossRef]

39. Piperno, D.R. *Phytoliths: A Comprehensive Guide for Archaeologists and Paleoecologists*; Rowman Altamira Press: Oxford, UK, 2006; p. 238. ISBN 9780759103856.

40. Twiss, P.C.; Suess, E.; Smith, R.M. Morphological classification of grass phytoliths. *Soil Sci. Soc. Am. Proc.* **1969**, *33*, 109–115. [CrossRef]

41. Madella, M.; Alexandre, A.; Ball, T. International code for phytolith nomenclature 1.0. *Ann. Bot.* **2005**, *96*, 253–260. [CrossRef] [PubMed]

42. Faegri, K.; Iversen, J. *Textbook of Pollen Analysis*, 4th ed.; Faegri, K., Kaland, P.E., Krzywinski, K., Eds.; John Wiley and Sons: Chichester, UK, 1989; ISBN 0 471 92178 5.

43. Moore, P.D.; Webb, J.A.; Collinson, M.E. *Pollen Analysis*; Blackwell: Oxford, UK, 1991; 216p, ISBN 0632021764.

44. Reille, M. *Pollen et Spores d'Europe et d' Afrique du Nord*; Laboratoire de Botanique Historique et Palynologie: Marseille, France, 1992; 543p, ISBN 2-9507175-0-0.

45. Reille, M. *Pollen et Spores d'Europe et d' Afrique du Nord*; Supplement 1; Laboratoire de Botanique Historique et Palynologie: Marseille, France, 1995; ISBN 2950717519.

46. Reille, M. *Pollen et Spores d´Europe et d´Afrique du Nord*; Laboratoire de Botanique Historique et Palynologie: Marseille, France, 1999; ISBN-10: 2950717535.
47. Wintle, A.G.; Murray, A.S. A review of quartz optically stimulated luminescence characteristics and their relevance in single-aliquot regeneration dating protocols. *Radiat. Meas.* **2006**, *41*, 369–391. [CrossRef]
48. Raposo, L. Ambientes, Territorios y Subsistencia en el Paleolítico Medio de Portugal. *Complutum* **1995**, *6*, 57–78.
49. Thompson, R.; Oldfield, F. *Environmental Magnetism*; Allen and Unwin: London, UK, 1986; p. 227.
50. Evans, M.E.; Heller, F. *Environmental Magnetism: Principles and Applications of Enviromagnetics*; Academic: San Diego, CA, USA, 2003; ISBN 9780080505787.
51. Grygar, T.; Svetlik, I.; Lisa, L.; Koptikova, L.; Bajer, A.; Wray, D.S.; Ettler, V.; Mihaljevic, M.; Novakova, T.; Koubova, M.; et al. Geochemical tools for the stratigraphic correlation of floodplain deposits of the Morava River in Straznicke Pomoravi, Czech Republic from the last millennium. *Catena* **2010**, *80*, 106–121. [CrossRef]
52. Desenfant, F.; Petrovsky, E.; Rochette, P. Magnetic signature of industrial pollution of stream sediments and correlation with heavy metals: Case study from South France. *Water Air Soil Pollut.* **2004**, *152*, 297–312. [CrossRef]
53. Zhang, C.X.; Appel, E.; Qiao, Q.Q. Heavy metal pollution in farmland irrigated with river water near a steel plant-magnetic and geochemical signature. *Geophys. J. Int.* **2013**, *192*, 963–974. [CrossRef]
54. Maher, B.A.; Thompson, R. Pedogenesis and Paleoclimate—Interpretation of the Magnetic-Susceptibility Record of Chinese Loess-Paleosol Sequences—Comment. *Geology* **1994**, *22*, 857–858. [CrossRef]
55. Maher, B.A.; Alekseev, A.; Alekseeva, T. Magnetic mineralogy of soils across the Russian Steppe: Climatic dependence of pedogenic magnetite formation. *Palaeogeogr. Palaeoclimatol. Palaeoecol.* **2003**, *201*, 321–341. [CrossRef]
56. Egli, R. Characterization of Individual Rock Magnetic Components by Analysis of Remanence Curves, 1. Unmixing Natural Sediments. *Studia Geophysica et Geodaetica* **2004**, *48*, 391–446. [CrossRef]
57. Cunha, P.P.; Cura, S.; Cunha Ribeiro, J.P.; Figueiredo, S.; Martins, A.A.; Raposo, L.; Pereira, T.; Almeida, N. As indústrias do Paleolítico Inferior e Médio associadas ao Terraço T4 do Baixo Tejo (Portugal central)—Arquivos da mais antiga ocupação humana no oeste da Ibéria, com ca. 340 ka a 155 ka/The Lower and Middle Palaeolithic industries associated with the T4 Terrace of the Lower Tejo River—Archives of the human occupation during ca. 335 ka to 155 ka ago. *J. Lithic Stud.* **2017**, *4*, 27–56. [CrossRef]
58. Raposo, L. Os mais antigos vestígios de ocupação humana paleolítica na região de Ródão. In *Da Pré-História, Homenagem a O. Veiga Ferreira*; Editorial Delta: Madrid, Spain, 1987; pp. 153–178.
59. Silva, A.; Pimenta, C.; Lemos, F.; Zilhão, J.; Mateus, J.; Raposo, L.; Coutinho, M. Vilas Ruivas: Um acampamento do Paleolítico Médio. *História e Sociedade* **1980**, *7*, 29–33.
60. Raposo, L.; Silva, A.C. A estação Paleolítica de Vilas Ruivas (Ródão). Campanha de 1979. *O Arqueólogo Português* **1993**, *IV*, 15–38.
61. Almeida, N.A.C. Estruturas de habitat do final do Plistocénico médio em Portugal: O caso dos Pegos do Tejo 2, Portas de Ródão, Nisa. In *Arqueologia Em Portugal—150 Anos*; Associação dos Arqueólogos Portugueses: Lisboa, Portugal, 2013; pp. 243–250.
62. Almeida, N.A.C. O Paleolítico Médio das Portas de Ródão, a Margem Esquerda (Nisa, Portugal). Contributo para a sua Caracterização Cronoestratigráfica. Ph.D Thesis, Universidade de Évora, Évora, Portugal, January 2014.
63. Monteiro, M.; Henriques, F.; Clélia, S.; Évora, M.; Nora, D.; Alves, C.; Mendes, C.; Carvalho, E.; Anacleto, C.; Silva, D.; et al. Monte da Revelada 2—Resultados preliminares. *AÇAFA On-line* **2017**, *11*, 18–27.
64. Raposo, L. Campanha de Escavações Arqueológicas no Sítio da Foz do Enxarrique. Alto Tejo. *Boletim Informativo do Núcleo Regional de Investigação Arqueológica* **1991**, *9*, 1–2.
65. Cardoso, J.L. *Contribuição para o conhecimento dos grandes mamíferos do Plistocénico Superior de Portugal*; Câmara Municipal de Oeiras: Oeiras, Portugal, 1993.
66. Brugal, J.-P.; Raposo, L. Foz do Enxarrique (Ródão—Portugal): Preliminary results of the analysis of a bone assemblage from A Midle Palaeolithic open site. In *The Role of Early Humans in the Accumulation of European Lower and Middle Palaeolithic Bone Assemblages*; Römisch Germanisches Zentralmuseum Mainz: Mainz, Germany, 1999; pp. 367–379. ISBN 9783884670446.
67. Sousa, M.F.; Figueiredo, S.D. The Pleistocene Elephants of Portugal. In Proceedings of the Congresso La Terra degli Elefanti (Actas), Roma, Italy, 16–20 October 2001; pp. 611–616.

68. Figueiredo, S.D.; Sousa, M.F. Os Elefantes Pleistocénicos de Portugal. *Revista Evolução* **2003**, *1*, 3–32.

69. Figueiredo, S.D.; Raposo, L. As Aves Como Recurso Alimentar do Homem do Paleolítico Médio: Interpretação tafonómica das acumulações faunísticas da Gruta Nova da Columbeira e da Foz do Enxarrique. *Boletim do Centro Português de Geo-História e Pré-História* **2018**, *4*, 69–82.

70. Berruti, G.L.F.; Rosina, P.; Raposo, L. The use-wear analysis of the quartzite lithic assemblage from the middle Palaeolithic site of Foz do Enxarrique (Ródão, Portugal). *Mediterr. Archaeol. Archaeom.* **2016**, *16*, 107–126. [CrossRef]

71. Marín-Arroyo, A.B.; Rios-Garaizar, J.; Straus, L.G.; Jones, J.R.; de la Rasilla, M.; González Morales, M.R.; Richards, M.; Altuna, J.; Mariezkurrena, K.; Ocio, D. Chronological reassessment of the Middle to Upper Paleolithic transition and Early Upper Paleolithic cultures in Cantabrian Spain. *PLoS ONE* **2018**, *13*, e0194708. [CrossRef]

72. Zilhão, J. Middle Paleolithic settlement patterns in Portugal. In *Settlement Dynamics of the Middle Paleolithic and Middle Stone Age*; Conard, N., Ed.; Kerns Verlag: Tübingen, Germany, 2001; pp. 597–608.

73. Cardoso, J.L. The Mousterian complex in Portugal. *Zephyrus* **2006**, *59*, 21–50.

74. Benedetti, M.M.; Haws, J.A.; Funk, C.L.; Daniels, J.M.; Hesp, P.A.; Bicho, N.F.; Minckley, T.A.; Ellwood, B.B.; Forman, S.L. Late Pleistocene raised beaches of coastal Estremadura, central Portugal. *Quat. Sci. Rev.* **2009**, *28*, 3428–3447. [CrossRef]

75. Pereira, T.; Bicho, N.; Haws, J. O Paleolítico Médio em Território Português. *Mainake* **2012**, *33*, 11–30.

76. Martignier, L.; Nussbaumer, M.; Adatte, T.; Gobat, J.M.; Verrecchia, E.P. Assessment of a locally—Sourced loess system in Europe: The Swiss Jura Mountains. *Aeolian Res.* **2015**, *18*, 11–21. [CrossRef]

77. Újvári, G.; Kok, J.F.; Varga, G.; Kovács, J. The physics of wind-blown loess: Implications for grain size proxy interpretations in Quaternary paleoclimate studies. *Earth-Sci. Rev.* **2016**, *154*, 247–278. [CrossRef]

78. Vandenberghe, J.; Sun, Y.; Wang, X.; Abels, H.A.; Liu, X. Grain-size characterization of reworked fine-grained aeolian deposits. *Earth-Sci. Rev.* **2018**, *177*, 43–52. [CrossRef]

79. Varga, G.; Újvári, G.; Kovács, J. Interpretation of sedimentary (sub)populations extracted from grain size distributions of Central European loess-paleosol series. *Quat. Int.* **2017**. [CrossRef]

80. Wolf, D.; Ryborz, K.; Kolb, T.; Calvo Zapata, R.; Sanchez Vizcaino, J.; Zöller, L.; Faust, D. Origins and genesis of loess deposits in central Spain, as indicated by heavy mineral compositions and grain-size variability. *Sedimentology* **2018**, in press. [CrossRef]

81. Kravchinsky, V.A.; Langereis, C.G.; Walker, S.D.; Dlusskiy, K.G.; White, D. Discovery of Holocene millennial climate cycles in the Asian continental interior: Has the sun been governing the continental climate? *Glob. Planet. Chang.* **2013**, *110*, 386–396. [CrossRef]

82. Liu, Q.S.; Deng, C.L.; Yu, Y.J.; Torrent, J.; Jackson, M.J.; Banerjee, S.K.; Zhu, R.X. Temperature dependence of magnetic susceptibility in an argon environment: Implications for pedogenesis of Chinese loess/palaeosols. *Geophys. J. Int.* **2005**, *161*, 102–112. [CrossRef]

83. Maher, B.A. The magnetic properties of Quaternary aeolian dusts and sediments, and their palaeoclimatic significance. *Aeolian Res.* **2011**, *3*, 87–144. [CrossRef]

84. Shi, L.F.; Yang, Z.Y.; Zheng, L.D.; Jia, S.M.; Tong, Y.B.; Zhang, S.Q.; Xu, D.Y.; Guo, G.X. Environmental magnetic record of the fluvial sediments from the Tianzhu borehole in Beijing for the last 800 ka. *Earth Planets Space* **2010**, *62*, 631–645. [CrossRef]

85. Lu, H.Y.; Van Huissteden, K.O.; An, Z.S.; Nugteren, G.; Vandenberghe, J. East Asia winter monsoon variations on a millennial time-scale before the last glacial-interglacial cycle. *J. Quat. Sci.* **1999**, *14*, 101–110. [CrossRef]

86. Maher, B.A.; Thompson, R. Paleoclimatic significance of the mineral magnetic record of the Chinese loess and paleosols. *Quat. Res.* **1992**, *37*, 155–170. [CrossRef]

87. Maher, B.A. Environmental magnetism and climate change. *Contemp. Phys.* **2007**, *48*, 247–274. [CrossRef]

88. Sánchez Goñi, M.F.; Eynaud, F.; Turon, J.L.; Shackleton, N.J. High resolution palynological record off the Iberian Margin: Direct land-sea correlation for the Last Interglacial complex. *Earth Planet. Sci. Lett.* **1999**, *171*, 123–137. [CrossRef]

89. Sánchez Goñi, M.F.; Turon, J.-L.; Eynaud, F.; Gendreau, S. European climatic response to millennial-scale changes in the atmosphere-ocean system during the last glacial period. *Quat. Res.* **2000**, *54*, 394–403. [CrossRef]

90. Sánchez Goñi, M.F.; Cacho, I.; Turon, J.-L.; Guiot, J.; Sierro, F.J.; Peypouquet, J.-P.; Grimalt, J.O.; Shackleton, N.J. Synchroneity between marine and terrestrial responses to millennial scale climatic variability during the last glacial period in the Mediterranean region. *Climate Dyn.* **2002**, *19*, 95–105. [CrossRef]

91. Roucoux, K.H.; Shackleton, N.J.; de Abreu, L.; Schönfeld, J.; Tzedakis, P.C. Combined marine proxy and pollen analyses reveal rapid Iberian vegetation response to North Atlantic millennial-scale climate oscillations. *Quat. Res.* **2001**, *56*, 128–132. [CrossRef]

92. Roucoux, K.H.; de Abreu, L.; Shackleton, N.J.; Tzedakis, P.C. The response of NW Iberian vegetation to North Atlantic climate oscillations during the last 65 kyr. *Quat. Sci. Rev.* **2005**, *24*, 1637–1653. [CrossRef]

93. De Abreu, L.; Shackleton, N.J.; Schönfeld, J.; Hall, M.; Chapman, M.R. Millennial-scale oceanic climate variability off the Western Iberian margin during the last two glacial periods. *Mar. Geol.* **2003**, *196*, 1–20. [CrossRef]

94. Schönfeld, J.; Zahn, R.; de Abreu, L. Surface to deep water response to rapid climate changes at the western Iberian Margin. *Glob. Planet. Chang.* **2003**, *36*, 237–264. [CrossRef]

95. Dansgaard, W.; Johnsen, S.J.; Clauser, H.B.; Dahl, J.; Gundestrup, N.S.; Hommer, C.U.; Huidberg, C.S.; Steffensen, J.P.; Svernbjornsdottir, A.E.; Jouzel, J.; et al. Evidence for general instability of past climate from a 250 kyr ice-core record. *Nature* **1993**, *364*, 218–220. [CrossRef]

96. Grootes, P.M.; Stuiver, M.; White, J.W.C.; Johnsen, S.; Jouzel, J. Comparison of oxygen isotope records from the GISP2 and GRIP Greenland ice cores. *Nature* **1993**, *366*, 552–554. [CrossRef]

97. Ruiz Zapata, M.B.; Rodríguez de Tembleque, J.M.; Pérez González, A.; Gil García, M.J. Paleovegetación en el yacimiento achelense de Puente Pino y su entorno (Toledo, España). *Cuaternario y Geomorfología* **2009**, *23*, 113–126.

quaternary

MDPI

Article

Anatomy, Age and Origin of an Intramontane Top Basin Surface (Sorbas Basin, Betic Cordillera, SE Spain)

Martin Stokes [1,*], **Anne E. Mather** [1], **Ángel Rodés** [2], **Samantha H. Kearsey** [1,†] **and Shaun Lewin** [1]

[1]   School of Geography, Earth and Environmental Sciences, University of Plymouth, Drake Circus, Plymouth, Devon PL4 8AA, UK; amather@plymouth.ac.uk (A.E.M.); samantha.kearsey1@gmail.com (S.H.K.); shaun.lewin@plymouth.ac.uk (S.L.)
[2]   Natural Environment Research Council (NERC) Cosmogenic Isotope Analysis Facility, Scottish Universities Environmental Research Centre, Rankine Avenue, Scottish Enterprise Technology Park, East Kilbride G75 0QF, UK; Angel.Rodes@glasgow.ac.uk
*   Correspondence: mstokes@plymouth.ac.uk; Tel.: +44-1752-584772
†   Née Samantha H. Ilott.

Academic Editors: Xianyan Wang, Jef Vandenberghe and David Bridgland
Received: 29 June 2018; Accepted: 22 August 2018; Published: 24 August 2018

Abstract: Collisional mountain belts commonly develop intramontane basins from mechanical and isostatic subsidence during orogenic development. These frequently display a relict top surface, evidencing a change interval from basin infilling to erosion often via capture or overspill. Such surfaces provide markers that inform on orogenic growth patterns via climate and base level interplay. Here, we describe the top surface from the Sorbas Basin, a key intramontane basin within the Betic Cordillera (SE Spain). The surface is fragmentary comprising high elevation hilltops and discontinuous ridges developed onto the variably deformed final basin infill outcrop (Gochar Formation). We reconstruct surface configuration using DEM interpolation and apply ^{10}Be/^{26}Al cosmonuclides to assess surface formation timing. The surface is a degraded Early Pleistocene erosional pediment developed via autogenic switching of alluvial fan streams under stable dryland climate and base level conditions. Base-level lowering since the Middle Pleistocene focused headwards incision up interfan drainages, culminating in fan head capture and fan morphological preservation within the abandoned surface. Post abandonment erosion has lowered the basin surface by 31 m (average) and removed ~5.95 km^3 of fill. Regional basin comparisons reveal a phase of Early Pleistocene surface formation, marking landscape stability following the most recent Pliocene-Early Pleistocene mountain building. Post-surface erosion rate quantification is low and in accordance with ^{10}Be denudation rates typical of the low uplift Betic Cordillera.

Keywords: intramontane basin; pediment; glacis; alluvial fan; river terrace; DEM; interpolation; cosmonuclide; base level

1. Introduction

Intramontane basins are areas of fault and fold-related subsidence that develop within an evolving collisional mountain belt [1]. The tectonically dynamic nature of such settings means that intramontane basins can cyclically form, fill and erode over geological timescales [2,3]. The basins can be internally drained, dominated by alluvial fan and lacustrine settings, but can then switch to externally drained systems via lake overspill or river capture processes [4,5]. Studies of intramontane basins are either (1) geological, focusing on the sedimentary infill record for stratigraphic, palaeoenvironmental and tectonic purposes [6] or (2) geomorphological, using inset river-fan-lake terrace levels to reconstruct the basin incisional history linked to tectonic-climatic-capture-related changes in sediment supply and base level [7]. A key, but often overlooked stratigraphic unit is the surface that caps the final stage of intramontane basin infill. This surface can be (1) depositional, with a morphology reflecting the final depositional environment(s) (alluvial fan/lake) or (2) erosional, formed by regional subaerial processes. Such 'epigene' land surfaces (sensu Watchman and Twidale [8]) are scientifically important because they mark the point at which the basin has switched from erosion to deposition [9]. Furthermore, they can act as a regional marker, providing insight into patterns and drivers of the onset and subsequent basin incision [10] or as a marker for surface deformation assessments [11]. However, these surfaces can be problematic to study due to poor preservation, post depositional modification and dating challenges meaning the surfaces often only attract peripheral attention as the respective end or start context points of geological and geomorphological research. For example, surface remnants are often highly fragmentary and can be degraded by erosion or deformation causing across basin or between basin correlation problems [12,13]. Once abandoned, the surface can become modified due to cementation by pedogenic or groundwater processes [14]. Surface dating can be a significant challenge due to technique limitations or material suitability issues collectively related to surface composition, degradation because of surface antiquity (i.e., surface is beyond the technique age range limit) and post depositional degradation and modification also linked to antiquity [15,16]. To explore and overcome some of these challenges and to highlight the importance of intramontane top basin surfaces for understanding sedimentary basin evolution and longer-term Quaternary landscape development we examine the Sorbas Basin in SE Spain (Figure 1). The Sorbas Basin is a medium sized (30 × 20 km) Neogene sedimentary basin that has developed as part of the ongoing fault and fold related uplift of the Betic Cordillera, a major Alpine mountain range, formed because of the ongoing Africa-Europe collision [2]. The basin fill is dominated by marine Miocene sedimentation [17,18], with continental sedimentation forming the final stages of basin infill (Gochar Formation [19,20]).

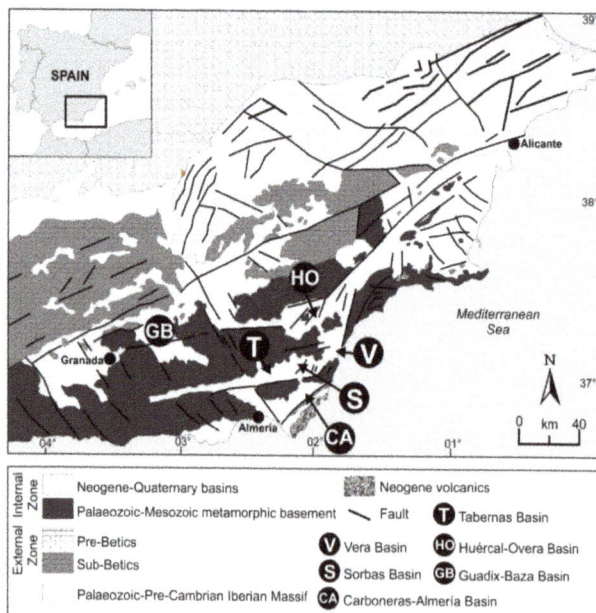

Figure 1. Tectonic zonation of the Betic Cordillera and key intramontane basins referred to within the text (modified from [21–23]).

A surface is developed onto the final stage of basin fill, commonly referred to as the "Gochar Surface" by studies examining long-term drainage evolution [10,24]. The purpose of this paper is to: (1) describe the relict morphology of the basin surface, (2) to digitally reconstruct the surface using interpolation of surface remnants, (3) to provide age estimates for surface development using cosmonuclide dating; (4) to use the interpolated and dated surface to quantify spatial and temporal patterns of basin erosion and (5) to consider the development of the surface as a Quaternary landscape feature in the context of the ongoing cyclic development of an intramontane basin.

2. Geological and Geomorphological Background

The Sorbas Basin (Figures 1 and 2) is one of a series of Neogene intramontane sedimentary basins within the Betic Cordillera [2]. It is defined to the north and south by mountain ranges of metamorphic basement (Figure 2) that are organized into km-scale regional antiformal fold structures formed in consequence of Miocene-Recent collision-related tectonic denudation [25,26]. The Sierra de los Filabres to north peaks at 1304 m (Ermita de la Virgen de la Cabeza) and comprises an embayed non-faulted mountain front with a relief of up to 700 m. To the south, the Sierra Alhamilla is characterised by a linear faulted mountain front [27], peaking at 1004 m (Cerrón de Lucainena) and with a relief of ~400 m. The intervening basin is infilled with a sequence of Miocene to Quaternary marine and continental sediments that are folded into an open E-W orientated syncline structure (Figure 2). The basin narrows to the west and east, joining the adjacent Tabernas and Vera Basins, delimited by poorly defined topographic highs developed into the sedimentary infill.

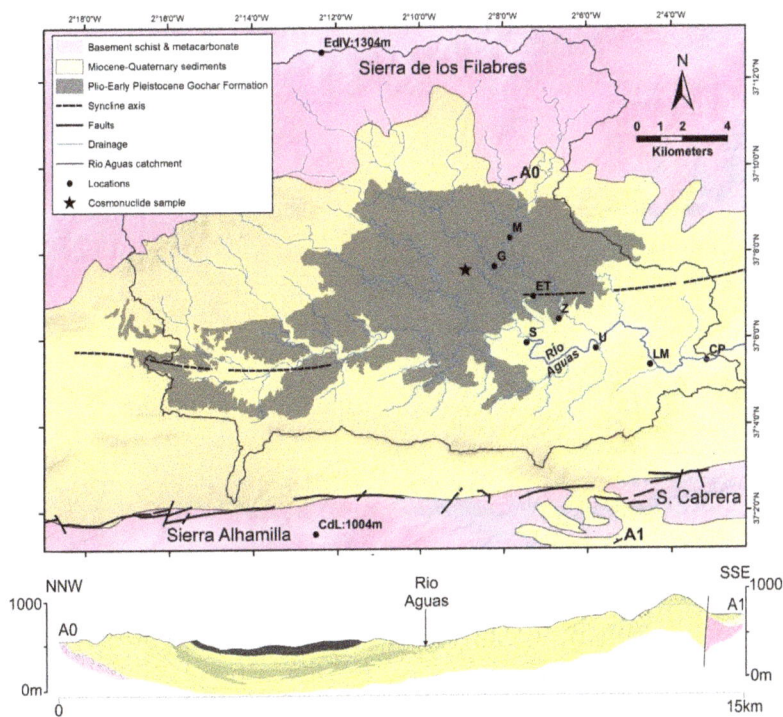

Figure 2. Simplified Sorbas Basin geology map and cross section (modified from [28,29]). EdlV: Ermita de la Virgen; M: Moras; G: Gochar; ET: El Tieso; S: Sorbas; Z: Zorreras; U: Urra; LM: Los Molinos; CP: capture point; CdL: Cerron de Lucainena; A0-A1: line of section.

Miocene marine sediments dominate the Sorbas Basin sedimentary infill (Figure 3), becoming progressively continental during the late Miocene represented by coastal plain sediments (Zorreras Member) and basin margin alluvial fan sequences (Moras Member) [19].

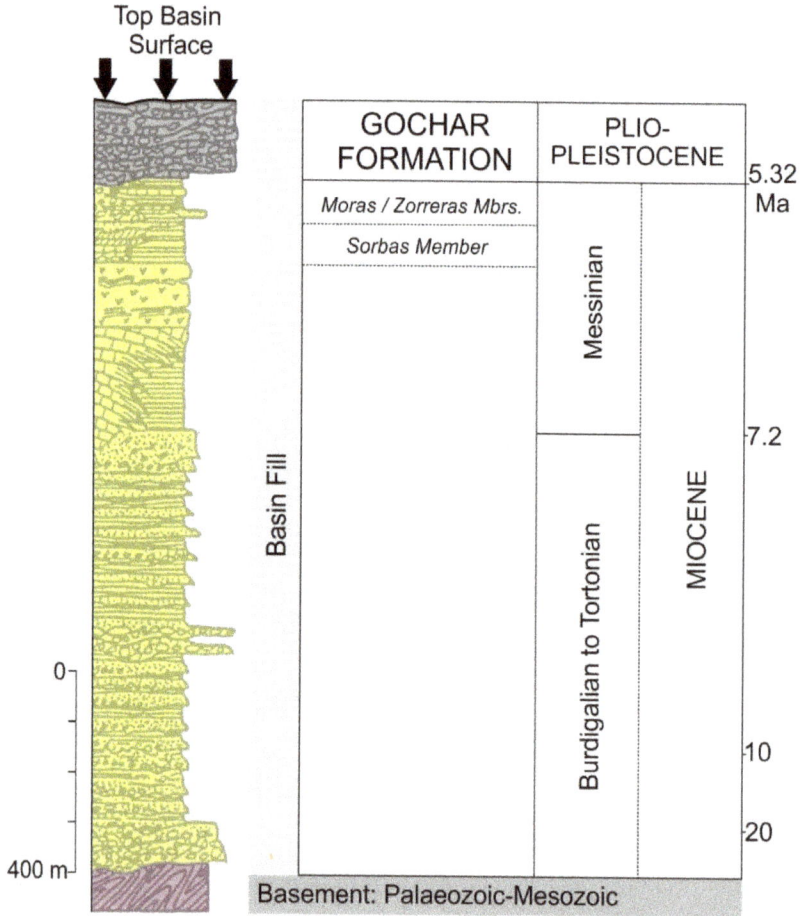

Figure 3. Simplified composite graphic log of the Sorbas Basin sedimentary infill (modified from [19]) illustrating key stratigraphic units referred to within the text and other figures.

The end Zorreras Member is stratigraphically important, being constrained to the Mio-Pliocene boundary from magnetostratigraphic and biostratigraphic studies [30,31]. Furthermore, the Zorreras Member lacustrine-marine bands have been used as marker horizons to demonstrate spatially variable Plio-Quaternary uplift patterns, ranging from 0.08 to 0.16 mm^{a-1} from the basin centre to the southern margin [19].

The overlying Gochar Formation (Figure 3) represents the final infilling stage of the Sorbas Basin, forming an outcrop of ~80 km^2 (Figure 2). It comprises a 40–200 m thick conglomerate and sandstone sequence deposited by alluvial fans and braided rivers [19,20,32] with spatially and temporally variable degrees of syn- and post-depositional deformation [19]. The fan and river systems are organised into four distinct drainage systems based on variations in sedimentology, provenance and palaeocurrent directions [19,20,32]. These drainage systems are important for the morphological development of the top basin surface, providing a relict topography onto which surface erosion occurred. The timing of the Gochar Formation is unclear as it lacks any direct age control, with a broad assignment to the Plio-Quaternary based upon stratigraphic bracketing with the Miocene basin fill (Zorreras Member) and Pleistocene river terraces.

Post Gochar Formation the Sorbas Basin has undergone incision, reflected in the development of an inset Pleistocene river terrace sequence [24] with coeval landslide, karst and badland development [33,34]. The river terraces (Figure 4) are configured into up to 5 inset levels (labelled A to E, where A: highest and oldest, and E: lowest and youngest), comprising up to 20 m thick aggradations of undeformed conglomerate capped by varying degrees of calcrete and soil reddening dependent on relative age [35].

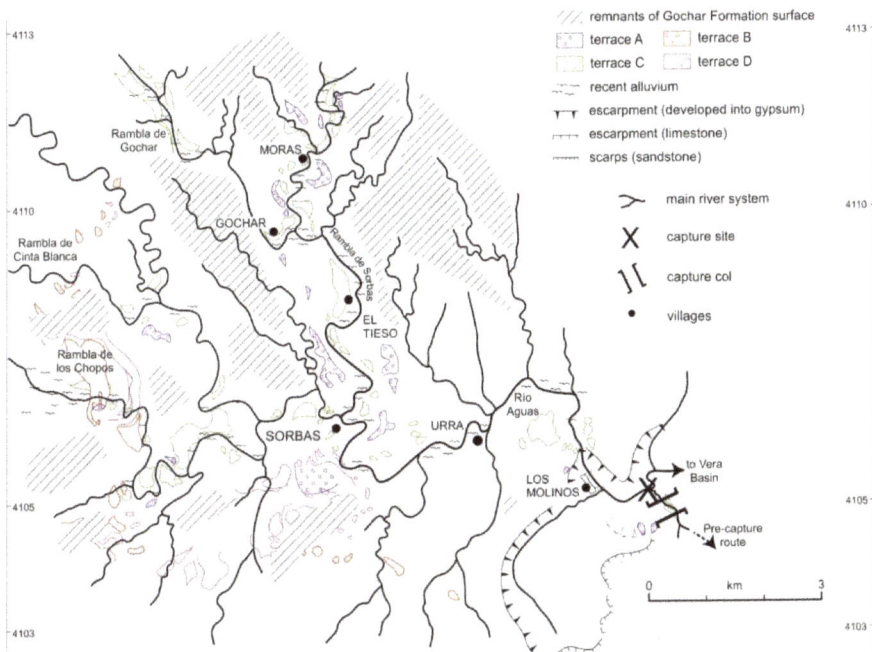

Figure 4. Sorbas Basin Middle-Late Pleistocene river terrace map (modified from [10]). See Figure 2 for catchment-scale overview.

Terrace level A can be inset by up to 20 m into the Gochar Formation sediments (Figure 5), with the entire terrace sequence recording between 40 m to 160 m of incision between upstream (Moras) and downstream (Los Molinos) regions [10]. These incision patterns are linked to spatially variable base-level lowering driven by combinations of regional uplift variability and river capture [12,24]. Terrace ages span the Middle-Late Pleistocene based on a range of radiometric and luminescence techniques [24,36–38]. The terraces are developed along the valleys of the trunk drainage (Río Aguas) and its major tributaries (Ramblas de Gochar, Moras, Cinta Blanca, los Chopos, etc.) (Figure 4). Terraces have formed within a catchment area of ~285 km^2 upstream of the Aguas-Feos capture point (Figures 2, 4 and 5), the site of a major basin-scale capture that occurred ~100 ka, beheading and re-routing the former southwards flowing drainage (Rambla de los Feos) to the east into the Vera Basin [24].

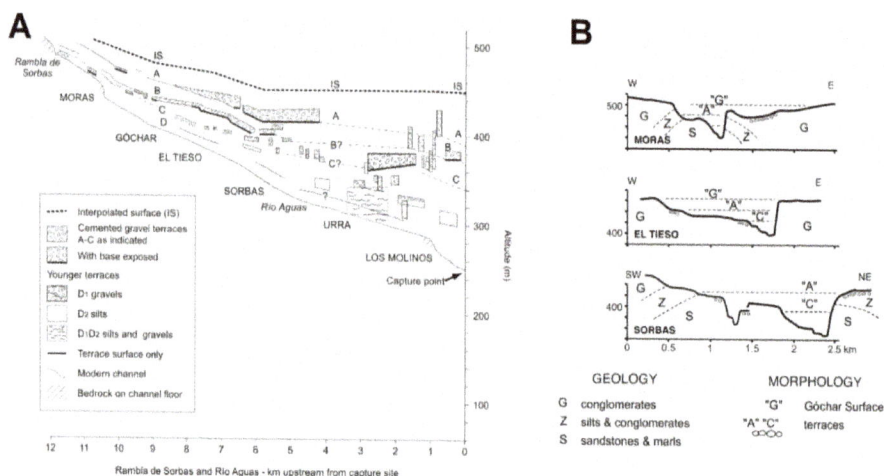

Figure 5. (**A**) River long profile and terraces of the (upstream) Rambla de Sorbas and (downstream) Río Aguas (modified from [10]). Marked steps in the Gochar surface and terraces A-C profiles around Sorbas relate to rock strength variations and a change in drainage orientation. (**B**) Cross valley profiles to illustrate the top basin surface (Gochar surface) and its relationship to key inset river terrace levels (modified from [10]).

The surface studied here is stratigraphically positioned between the Gochar Formation and Level A of the Pleistocene river terrace sequence (Figures 3–5) and is likely to be of Quaternary age based on relative dating. Similar high elevation surfaces occur in adjacent intramontane basins (Huércal-Overa, Vera, Tabernas: Figure 1) where they cap the basin fill and mark the onset of basin incision [16,39,40]. Similar surfaces with varying degrees of expression and quality of preservation are noted throughout the Betic Cordillera Internal Zone region where they are considered as an indicator of the most recent phase of relief generation within the Betics [41]. In the Sorbas Basin, the surface is fragmentary but appears to be a single and spatially extensive feature, comprising a series of rounded ridge crests and hilltops, developed primarily onto the Gochar Formation. Here, we focus on the most extensive surface remnants associated with the Gochar Formation outcrop.

3. Methods

3.1. Surface Morphology

We describe the top basin surface morphology using a combination of field and remote sensing approaches. The general surface configuration is imaged from different basin margin perspectives using elevated view points and oblique aerial drone imagery. Remote sensing of the surface used digital datasets, interrogated within Arc Map 10.5.1 (Esri, Redlands, CA, USA). The basin-scale outcrop of the Gochar Formation used digitized 1:50,000 geological maps [25,26]. The broader basin geomorphology used 5 m DEM data sourced online [42] with checks against other commonly used datasets (e.g., SRTM) to ensure visualization and analysis quality [43].

The top basin surface is an erosional feature that lacks any sedimentary deposits. As such, the surface remnants are preserved in the rounded ridge crests and hilltops within the highest elevation areas of the Gochar Formation outcrop (Figure 6).

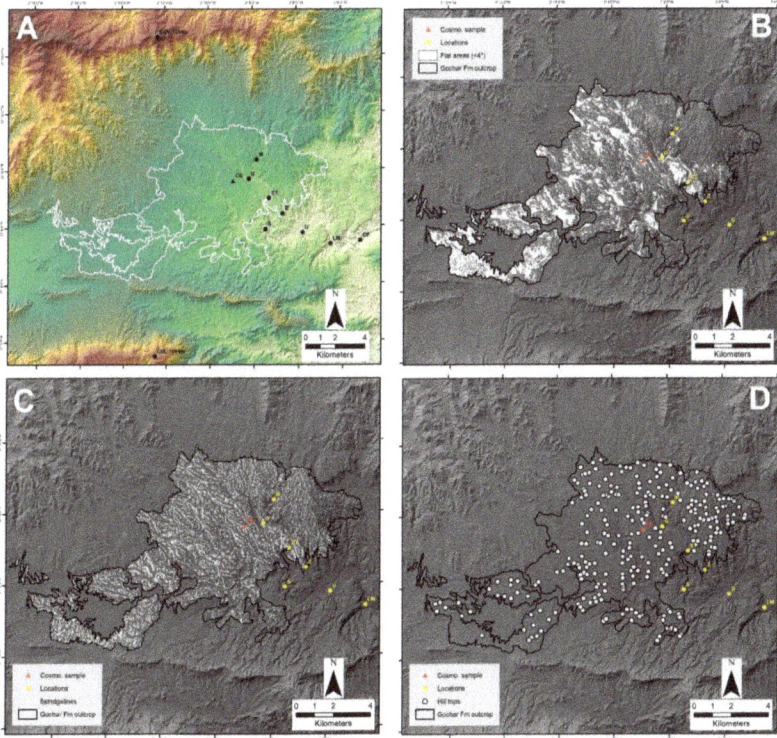

Figure 6. (**A**) DEM and hillshade showing Gochar Formation outcrop and key locations. (**B**) Slope map of areas of <4°. (**C**) Ridge lines within the Gochar Formation outcrop. (**D**) Final dataset of the highest elevation hilltops used for surface interpolation.

To map these areas, hilltop locations and elevations were combined with flat ridge crest regions. The assumption is that these highest-flattest ridges are the most representative surface remnants, since steeper dipping and lower elevation ridges will have been formed by incision into the top basin surface. Hilltops were extracted from spot heights using scanned 1:25,000 topographic maps [42] in combination with the 5 m DEM. Hilltops were removed from the dataset if (1) the spot height coincided with lower level inset river terrace locations (cross-referenced by using a combination of published terrace maps [10,35], terrace capping red soil regions identified from satellite imagery, and cross-valley profiles); (2) had no proximity relationship to the high elevation flat ridge areas (see below); (3) were anomalously low/high elevation occurrences compared to adjacent spot heights and (4) where the difference between the spot height and DEM elevation value was >5 m. Ridge crests were obtained from the DEM using an inverse stream extraction approach [44]. A reclassified slope map was then used to capture the flattest ridges (i.e., ridges coinciding with slopes of <5°). Hilltops that coincided with the flat ridges were then used as interpolation points from which to reconstruct the top basin fill surface.

3.2. Surface Reconstruction and Erosion Quantification

Digital surface reconstruction is a common geomorphological method for analysis of erosional landscapes at a range of spatial and temporal scales [45–48]. In this study we used the variable Inverse Distance Weighting (IDW var) approach [49] due to similarities of basin scale, landscape morphology and higher quality of method statistical performance. Digital points from the cleaned hilltop dataset (see above) were used for the interpolation. The IDW var interpolates between known points giving greater weights to points closest to the prediction location, with weights diminishing with distance away from the known points. The interpolation was extrapolated outside of the Gochar outcrop into the basin margin mountain reliefs to explore the wider configuration of the surface, noting that interpolation accuracy would have diminished due to the nature of IDW var method. The resultant interpolated top basin surface was combined with the modern landscape DEM to allow analysis of areas above and below the interpolated surface (a DEM of Difference). We consider the original top surface to dip towards the basin centre and to have an undulating morphology based on erosion due to lithological and tectonic substrate heterogeneities onto which the surface was developed. Surface hilltops (N = 278) within the Gochar Formation outcrop range from 582 m to 442 m with a mean elevation of 511 m and average distance between hilltops of 273 m. Elevations between groups of adjacent hilltops is typically <10 m. In areas adjacent to the river valleys the hilltop elevations (i.e., the surface remnants) range from 10–20 m above terrace level A (Figure 5). Thus, a buffer value of ±10 m was used to reclassify the DEM of Difference to model the extent of the top surface that is preserved within the modern landscape.

The interpolated top basin surface was used to assess the amount of erosion that has taken place since surface formation. Erosion was calculated by subtracting the interpolated surface from the modern landscape DEM. Since surface formation, the Sorbas Basin catchment area has been modified by capture-related drainage network re-organization [24] and we therefore use the Aguas-Feos capture site as the downstream limit for the erosion calculation (Figure 2, Figure 4, and Figure 5).

3.3. Surface Dating

Dating of the top basin surface was undertaken using a [10]Be-[26]Al cosmonuclide depth-profile originally sampled and analysed by Ilott [38] as part of a broader chronological investigation of the timing of Quaternary fluvial landscape development within the Sorbas Basin. The paired isotope and depth-profile approach allowed for surface exposure and burial age quantification [50]. The surface exposure technique measures the concentration of cosmonuclides at the surface [51], with concentrations affected by the time of exposure to cosmic radiation, cosmonuclide loss due to erosion, sediment density variability (affects cosmic ray attenuation) and cosmonuclide production variations [15,52]. Burial dating uses known radioactive decay rates of cosmonuclides and requires analysis of samples shielded (deep burial) from cosmic radiation after exposure [53], but with potential problems concerning cosmonuclide inheritance issues related to complex exposure-burial histories prior to deposition [54,55].

Sampling was undertaken on a road cutting (37.12692 −2.148214) that passed through one of the higher elevation flat ridges (~495 m) developed into Gochar Formation conglomerates in a north-central basin location (Figure 2). The section comprises ~2.5 m of massive and variably cemented gravel-cobble conglomerate capped by a 0.4 m soil unit, comprising a 0.1 m laminar calcrete and overlying 0.3 m red soil (Munsell = 7.5YR/4R). Sampling was undertaken up the section face at 0.5 m intervals from 2 m depth to the surface with >30 quartz clasts of >5 cm length sampled for each interval. The section location, aspect, angle of section repose, angle to highest topographic feature and surface altitude were quantified for data modelling inputs. The samples were crushed and milled, etched with HF for cleaning followed by dissolution, chemical separation (anion exchange and hydroxide precipitation) and a final metal mixing before AMS measurement.

The original age modelling [38] was undertaken using the CRONUS calculator [56] within Matlab (MathWorks, Natick, MA, USA). The concentration results revealed no hiatus within the profile so a simple exposure history was explored. This involved using a Chi square minimization method that was applied to the raw nuclide concentration data to allow fitting to the accumulation model equations of Lal [57] with variable inheritance, density and erosion data input values [15,50].

For the purpose of this study we remodelled the concentration data using the updated CRONUS 2.3 calculator [58]. New surface erosion estimates of 10 m and 4 m were inputted to represent the relationship of the cosmogenic sample site to the interpolated surface (see results). A value of 10 m was used to reflect the general elevation range between adjacent hilltop heights used for surface interpolation. A value of 4 m was also used as this is the height of the sample site below the interpolated surface. An average upstream altitude of 689 m was derived from the 5 m DEM as a modelling data input to improve the maximum burial age value.

Maximum and minimum exposure and burial ages were calculated. These values were considered alongside other published age data for the region to inform on the timing of surface formation. Combination of the remodelled ages with surface incision data enabled amounts and rates of basin erosion to be calculated.

4. Results

4.1. Surface Morphology and Erosion

The field expression of the surface is shown from a range of basin margin perspectives in Figure 7. The surface comprises high elevation isolated hilltops and gently dipping but discontinuous ridge crests, with numerous intervening topographic lows along the ridge lengths and between adjacent hilltops. The hilltops and ridges are further accentuated by incision of the modern drainage network and its tributaries. Despite the erosion, the various landscape panoramic perspectives and along ridge slope profiles (Figure 7) clearly demonstrates a visual correlation and reconstruction of a single surface in a downslope basin centre direction.

Reconstruction of the surface using IDW var interpolation of the hilltop dataset within the Gochar Formation outcrop shows that the top basin surface is contained almost entirely within the broader sedimentary infill of the Sorbas Basin (Figure 8). The surface is particularly prevalent in northern, central and western regions, with low preservation in the south (Figure 8). Areas eroded below the surface coincide with the modern drainage network, concentrated along the major tributary valleys and becoming widespread towards the east along the Río Aguas as it routes into the Vera Basin (Figure 9). Other extensive areas below the surface occur in the headwaters of the Tabernas Basin (west) and the Carboneras-Almería Basin (south). Areas above the surface are mainly concentrated in the mountains of metamorphic basement that border the Sorbas Basin, but there are notable areas where Miocene basin fill sediments form topographic highs within the west and south of the basin. When compared to the modern Río Aguas catchment upstream of the capture site (285 km²), the maximum extent of the interpolated surface covers 144 km², some 50% of the modern catchment. The amounts of incision below the interpolated surface increase downstream to a maximum of −254 m (Figure 9) with a mean basin surface lowering of ~31 m. This incision is concentrated along the lower reaches of tributaries draining to the basin centre and downstream along the main Rio Aguas valley, especially between Sorbas and the capture point east of Los Molinos. The volume of sediment removed by the erosion is 5.95 km³.

The areas of better surface preservation are associated within the confines of the Gochar Formation outcrop. Within this region, the interpolated surface comprises an area of 35 km², some 44% of the Gochar Formation outcrop. The hilltops, ridges of the interpolated surface and the incised drainage pick out a series of relict fan-shaped bodies (Figure 10) that broadly correspond to the dip slopes of the synclinal fold configuration of the Sorbas Basin (Figure 2). These are most evident along the northern basin margin, comprising at least two fans of 5–6 km length that backfill into the embayed Sierra de los Filabres mountain front (Figure 10). The clearest of the fans, the eastern 'Cariatiz Fan' (Figure 10B), was used by Mather et al. [59] as part of a regional morphometric study of modern and older Plio-Quaternary fans in SE Spain to illustrate the importance of capture-related re-organizations of fan source areas.

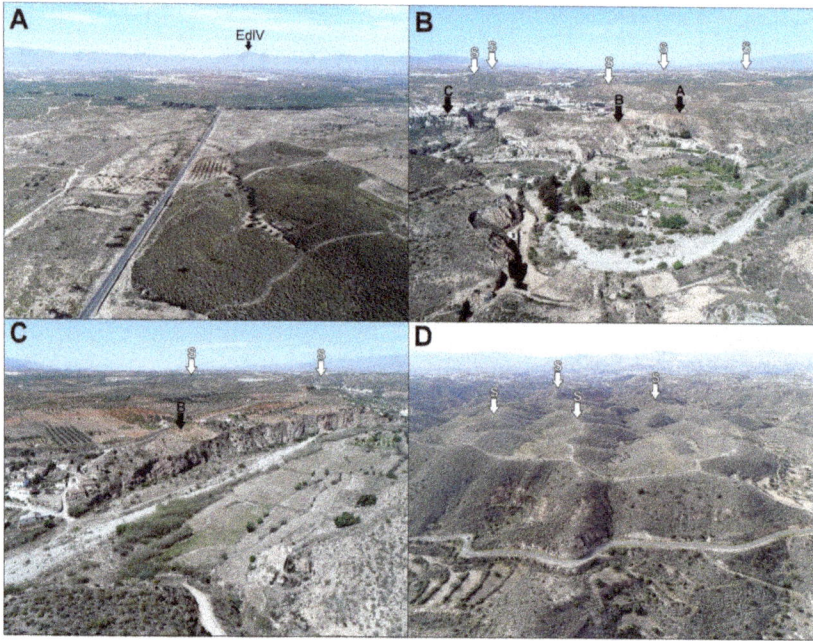

Figure 7. Field imaging of relict surface. (**A**) View from south western basin margin (37.06899 −2.19864) looking north across the basin surface with little dissection. EdIV = Ermita de la Virgen 1304 m. (**B**) View from southeastern basin margin (37.10498 −2.11419) looking west across the Rambla de Sorbas inset terrace sequence (A, B, C) in the Sorbas town region. Surface remnants (S) visible in distance. (**C**) View from eastern basin margin (37.12254 −2.11848) looking northwest across the El Tieso 'B' terrace with extensive surface remnants visible in far ground (S). (**D**) View south-southwest from the northeastern basin margin (37.145648 −2.099306) along ridgelines of the relict surface (S).

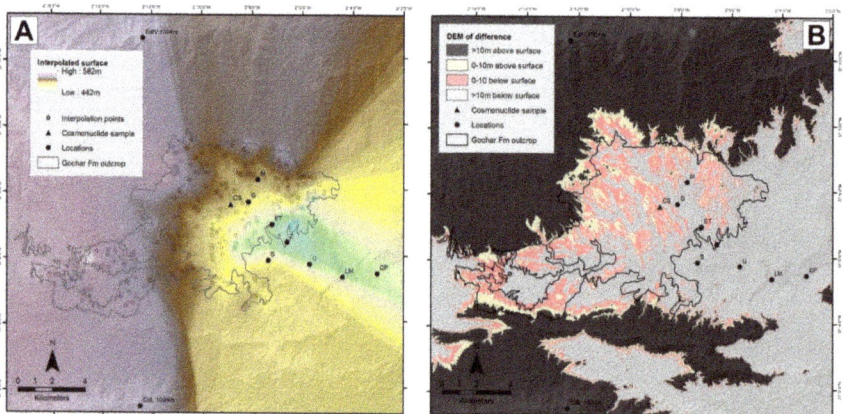

Figure 8. (**A**) Interpolated surface results. (**B**) Comparison of the interpolated surface with the modern landscape highlighting areas that are 10 m above and below the interpolated surface.

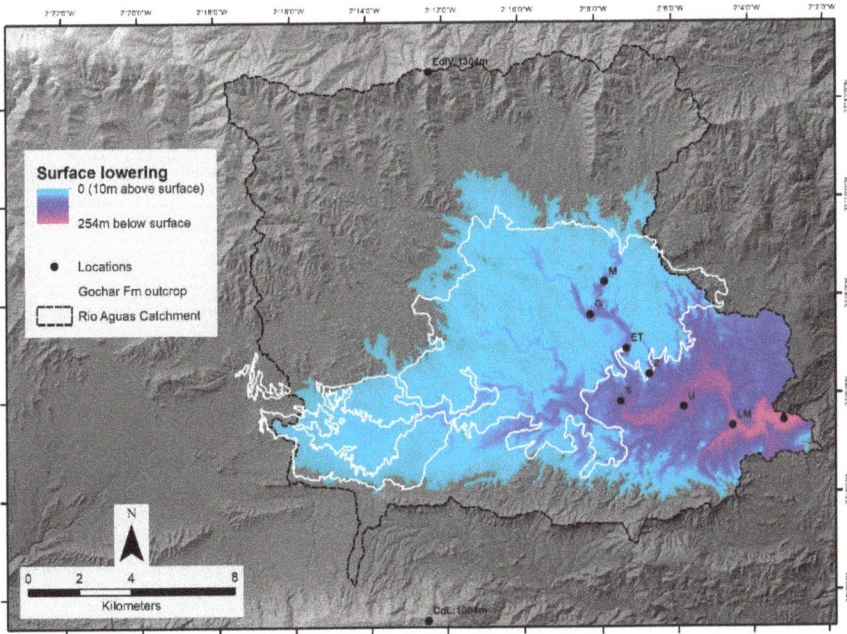

Figure 9. Surface lowering map showing concentrated erosion in the east and upstream along tributary channels.

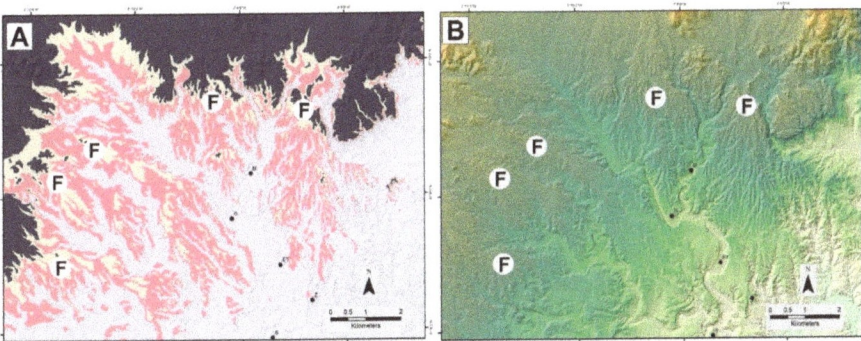

Figure 10. Fan-shaped geometries (F) enhanced by surface interpolation (**A**). Modern landscape DEM (**B**) for comparison. Cariatiz Fan = NE fan.

A series of 4–7 km long fans are also evident along the western and northwestern basin margins, but their morphology is less clear. The surface interpolation (Figure 10) accentuates these fan features suggesting that the formation of the surface erosion and its subsequent incision is accentuating and exploiting the Gochar Formation palaeogeography and its drainage morphology of the Marchalico and Gochar systems [20]. Fan morphologies are not evident in the surface remnants along the southern basin margin, possibly reflecting a more fragmentary surface record or that the higher uplift rate and greater degree of deformation along the southern basin margin [60] has destroyed any Gochar Formation drainage morphology in that area.

4.2. Surface Age and Erosion

The cosmonuclide sample site location examined by Ilott [38] is located on a gently dipping NW-SE orientated ridge with rounded edges that slope into an adjacent incised drainage network that visually appears to be part of the relict surface. Within the broader landscape, the sampled ridge is slightly inset when compared to adjacent ridge hilltop elevations (Figure 11). The interpolation modelling confirms the inset configuration (Figure 11), with the site occurring at −4 m below the interpolated surface and within the broad −10 m buffer zone (see Section 3). As such, the sample site does not provide the best representation of the 'true' surface but instead relates to the onset of incision into it. However, this incision amount is too small for the sampled ridge to be part of terrace Level A, which is typically positioned at 20 m below the interpolated surface (Figure 5). A benefit of knowing surface and terrace elevation variability is that the values provide erosion data inputs for modelling the cosmonuclide exposure and burial ages (see Section 3).

The remodelled cosmonuclide data are presented in the Supplementary Materials and summary results in Table 1. Using the higher 10 m erosion value provides exposure ages of 1990 ka (maximum) and 169 ka (minimum) and burial ages of 1056 ka (maximum) and 679 ka (minimum). In contrast, using a 4 m erosion value provides exposure ages of 798 ka (maximum) and 169 ka (minimum) and burial ages of 1048 ka (maximum) and 679 ka (minimum). These ages span the Early-Middle Pleistocene (maximum exposure-burial ages) and Middle-Late Pleistocene (minimum exposure-burial ages). Stratigraphic convention should mean that the sediment (burial) age should be older than that of the surface (exposure) age. However, the age inconsistencies are explainable as they reinforce the surface origin as an erosional form as opposed to a depositional top basin fill surface. Furthermore, despite the age variability, the results provide some insight into the broad timing of surface formation. The minimum 679 ka burial ages suggest that surface is older than 679 ka and probably more in keeping with the Early Pleistocene. Indeed, the more realistic surface age scenarios are probably closer to the maximum burial age range 1056–798 ka for both the erosion amount scenarios. An Early Pleistocene surface age is also supported by the chronologies of the inset river terrace sequence where U-Series dating of pedogenic terrace capping calcretes show that terraces A and B are Middle Pleistocene landforms [36,37].

Table 1. Remodelled cosmonuclide exposure and burial age results. See Supplementary Materials for detail.

Surface erosion (m)	Min. exposure age (ka)	Max. exposure age (ka)	Min. sediment burial age (Ma)	Max. sediment burial age (Ma)	Min. surface erosion rate (m/Ma)	Max. surface erosion rate (m/Ma)	Min. upstream basin erosion rate (m/Ma)	Max. upstream basin erosion rate (m/Ma)	Reduced chi-square	Min. depositional age (ka)	Max. depositional age (ka)
10	169	1990	0.679	1.056	0.04	5.98	6.8	9.3	2.8	191	1056
4	169	798	0.679	1.048	0.05	5.72	6.6	8.7	2.9	191	798

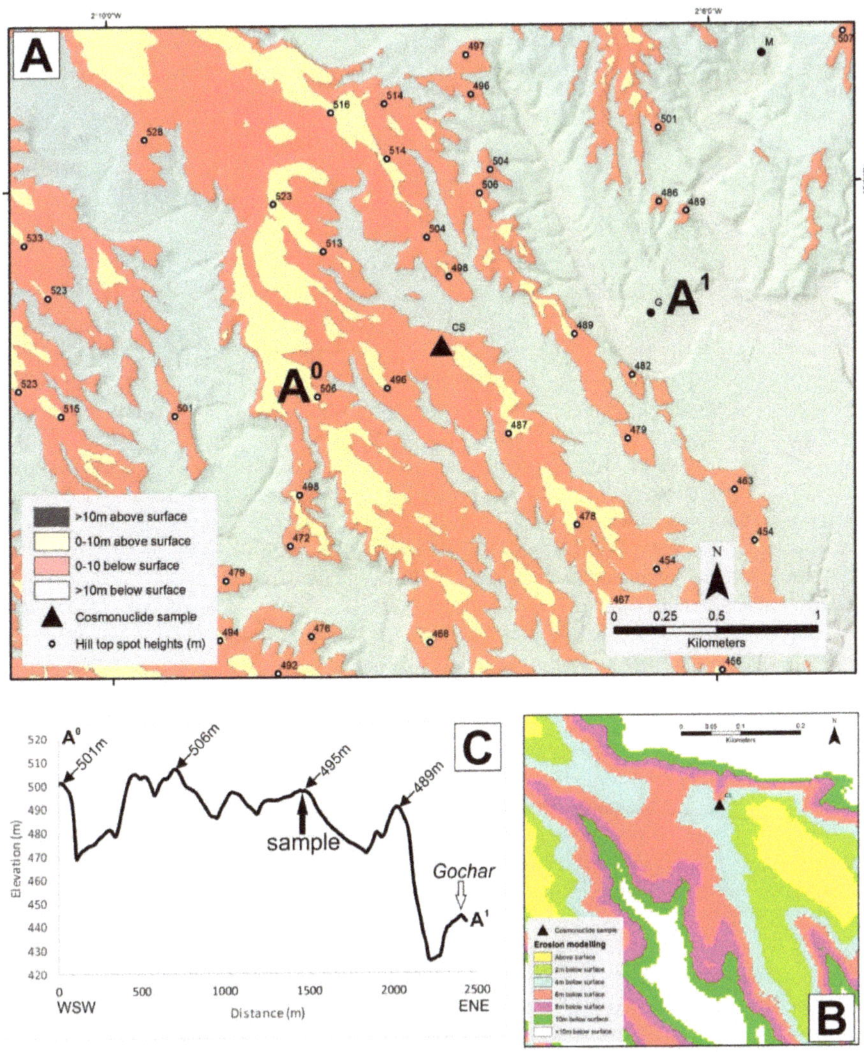

Figure 11. Visualization of the interpolated surface at 10 m (**A**) and 2 m (**B**) intervals, showing that the cosmonuclide sample site is located ~4 m below the interpolated surface. (**C**) Topographic profile further illustrating the inset nature of sample site.

The AMS measurements collectively revealed high concentrations of inherited [10]Be and [26]Al (Supplementary Materials) and this begins to inform on the transport history and relative landscape stability of the end Gochar Formation period prior to surface formation. It suggests that sediments were generated under low basin erosion rate conditions, implying a relatively stable landscape with recycling of the basin fill most likely from the Gochar Formation sediments into which the surface has developed [19].

The cosmonuclide age data can be combined with the interpolated surface to provide insights into rates of basin erosion. Because the surface is most likely Early Pleistocene (see above discussion) we use the maximum and minimum burial ages in conjunction with the surface lowering (~31 m) and volume (5.95 km^3) data to calculate the surface lowering and volume erosion rates. Surface lowering rates range from 46 mm/ka (minimum burial age: 679 ka) to 29 mm/ka (maximum burial age—10 m: 1056 ka). Volume rates range from 0.001 km^3/ka (minimum burial age: 679 ka) to 0.004 km^3/ka (maximum burial age—10 m: 1056 ka).

5. Discussion

5.1. Controls on Surface Formation

Despite the fragmentary nature of high elevation hilltops and ridges within the Sorbas Basin, they link together to form a single surface developed across the basin fill. Its crosscutting relationship with the underlying Gochar Formation suggests it represents a key basin wide erosional event that marks the onset of basin incision. The erosion has cut across deformed Gochar Formation sediments, meaning that surface construction post-dated a basin wide deformation event. Although surface remnants form a single surface that grades from the basin margins to the basin centre there are local elevation differences between adjacent surface remnants. These differences may relate to variations in strength, stratigraphy and localised deformation of the basin fill or a passive exploitation of the basin fill palaeogeography and its relict morphology of the depositional environment. For example, surfaces developed into flat lying and fine grained lacustrine dominated intramontane basin infills (e.g., Guadix-Baza [11]) are more likely to be well developed and spatially extensive than those developed into dipping and coarse-grained alluvial intramontane basin fills (this study).

Surfaces are evident throughout Betic Cordillera intramontane basins (Figure 12), occupying mountain fronts where surface remnants dip towards the basin centre [13]. These surfaces are either (1) degraded forms, lacking in sediment cover and developed onto the Plio-Pleistocene continental alluvial basin infill (e.g., Sorbas Basin) or (2) are well preserved, with a <20 m thick cover of coarse-grained alluvial conglomerates, that unconformably overlie Neogene marine basin infill (e.g., Tabernas and eastern Vera Basin) (Figure 12). The well-preserved surfaces often comprise a pedogenic calcrete cap, with groundwater calcretes sometimes developed along the basal unconformity contact [14,61]. Although surfaces may have origins associated with alluvial fan environments [13], they are more typical of pediments (sensu King [62]) that have been observed worldwide, with examples throughout SE Spain often referred to using the French term 'glacis' [63]. The degraded surface considered here could be a highly eroded pediment remnant, most likely a bedrock pediment or the remnants of the bedrock base of a pediment due to absence of calcrete and alluvial cover. Studies of pediment formation [64] suggest they form at mountain fronts where bedrock weathers to sediment; in climates with a soil hydrology, vegetation cover and weathering style that suppresses fluvial incision and deep bedrock weathering; and a balanced mountain front sediment flux and base level position. If the top basin surface follows these criteria for autogenic formation, then the surface informs indirectly on Quaternary climate and tectonics. The climatic criteria are fulfilled due to a persistence of seasonally variable cool/warm dryland climatic conditions throughout the Quaternary [65–68]. However, the base level configuration has changed, particularly with respect to the top basin surface as it marks a key point at which the basin switches from sedimentation to erosion, after which there is a sustained base-level lowering linked to tectonic uplift and capture [13,69]. For the top surface to form as a basin wide feature means that dryland conditions must have coincided with a stable and sustained basin level position during a time of relative tectonic quiescence and a time when the drainage network configuration was not conducive to capture. Uplift rate quantifications for the Sorbas Basin are time averaged from the lower Pliocene (70–160 m Ma^{-1}: [19,69] and thus lack temporal clarity to inform on the restricted pediment formation timescale. However, direct evidence for deformation is restricted to the Gochar Formation sediments into which the surface is developed, implying a marked reduction in tectonic activity at the time of surface formation and thus base level stabilization. Tectonics would have also played a passive role in surface formation, with the overall basin syncline configuration forming fold limb dip slope drainages routed to a basin centre axial drainage coincident with the basin syncline axis. Subsequent fluvial incision appears to have concentrated along the synclinal axis, dissipating upstream along the fold limb configured streams (Figures 8 and 9). The passive influence of fold structures on drainage pattern configuration and development is a commonly reported feature in collisional mountain belt settings [70].

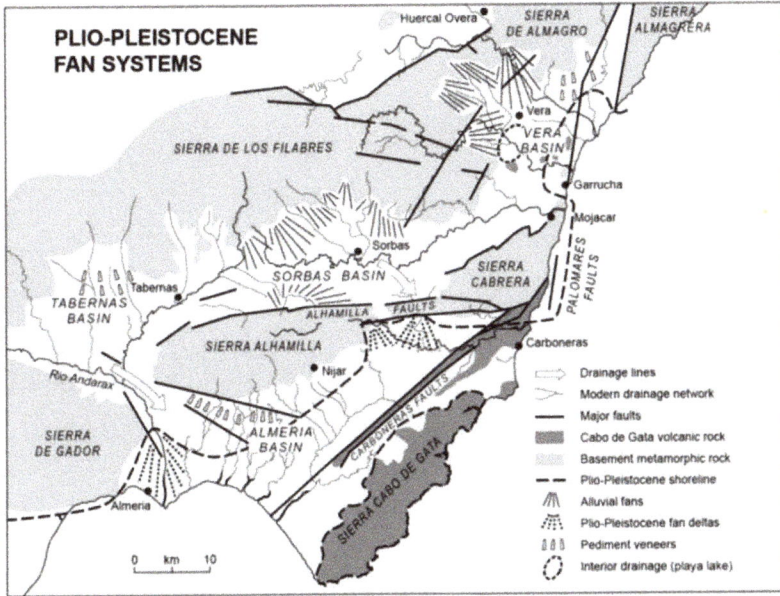

Figure 12. Regional occurrence of Plio-Pleistocene alluvial fan/pediment systems within the east-central Betic Cordillera [13]. Degraded surfaces are developed onto Plio-Pleistocene continental (alluvial fan) sediments, whilst well preserved surfaces (pediment veneers) are developed onto Neogene marine sediments.

The very nature of the surface as a continuous basin wide feature implies the absence of an incised drainage network for it to form by autogenic processes, e.g., [64]. Drainage routing throughout the Plio/Quaternary has recorded a persistent pattern of basin margin streams feeding an axial drainage [12,20,24]. Because the surface has formed as an interval in-between the final basin infilling and pre-basin incision, it too is likely to have formed by the same basin convergent drainage pattern (Figure 12). If the basin was undissected then radiating streams with collective fan-shaped forms would have dominated the palaeogeography (in-keeping with the Gochar Formation), with autogenic lateral shifting of the radiating streams being responsible for creating the pediment like surface, noting that any pediment cover sediments are not preserved due to the eroded/degraded surface form. The surface remnants and interpolation mapping (Figure 10) provides strong evidence for large fan-shaped bodies along the northern and western basin margins. These morphologies, particularly along the northern margin, are accentuated because of progressive surface incision and localised captures.

Headwards erosion by the axial drainage has exploited the inter-fan drainage areas (Figure 13). It is common for alluvial fans to develop an incised drainage along their axial feeder channel due to a connectivity interplay between fan head and fan toe base-level variations [71]. Because incision and headwards erosion has been concentrated along the interfan areas it suggests that the fans responsible for autogenically creating the surface were undissected with insufficient axial drainage to be exploited. As headwards erosion has proceeded up the interfan areas it has captured the fan feeders, resulting in fan abandonment [72]. The fans forming the pediment surface would have also possessed an overall convex morphology with topographic lows present within interfan areas. This convex morphology may have also played a role in passively influencing interfan drainage exploitation.

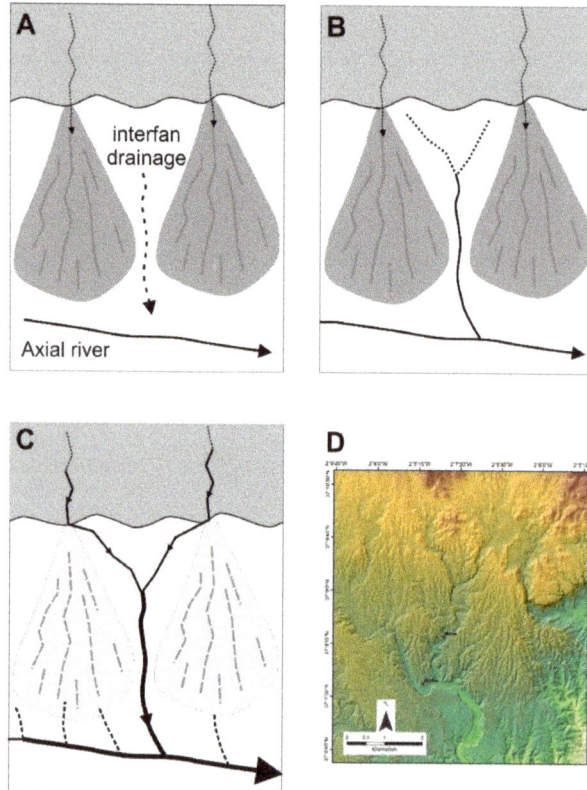

Figure 13. Fan/pediment abandonment model based on Sorbas Basin northern margin. (**A–C**) Interfan development and capture of mountain front fan feeder streams. (**D**) Relict fan morphology with former interfan drainage now forming a key component of the current drainage network.

5.2. Timing of Surface Formation

The remodelled cosmonuclide data suggest that the surface is an Early Pleistocene feature, with the max-min burial ages (1056–679 ka: Table 1) providing the most coherent age range indicators for surface development. This means that the underlying Gochar Formation into which the surface is developed spans the Pliocene and probably the earliest Pleistocene based upon bracketing between a basal Mio-Pliocene boundary age [30] and a top Early Pleistocene age (this study). From a geological perspective the Early Pleistocene surface age presented here is significant for understanding the Late Miocene geological history of the Sorbas Basin which has received considerable attention for its role in documenting the Mediterranean Messinian Salinity Crisis. Clauzon et al. [31] describe the same surface studied here (see Figure 8G in [31], and Figure 7C of this study) as a fan-delta abandonment feature assigning a Mio-Pliocene (~5.3 Ma) boundary age to the surface through downslope extrapolation to a biostratigraphically dated Zorreras Member type location section, the Zorreras Hill (Figure 2). This 450 m elevation hilltop is capped by Gochar Formation conglomerates and fits within our interpolated surface dataset. However, its Early Pleistocene cosmonuclide age bears no relationship to the immediate post Messinian Salinity Crisis recovery of the Sorbas basin as implied by Clauzon et al. [31].

The regional significance of the style and timing of Sorbas Basin surface formation within the Betic Cordillera can be further explored through comparison with adjacent intramontane basins (Figure 12). To the east, the Vera Basin is like Sorbas, comprising a deformed continental basin infill (Salmerón Formation) that grades into a high elevation pediment surface and an inset fan pediment-river terrace sequence [73–75]. Electron Spin Resonance (ESR) dating brackets the Salmerón Formation and its pediment to the Early Pleistocene (~2.4–1.3 Ma) [76,77]. The timing appears co-eval with the latter stages of the Gochar Formation, attributed to regional uplift timing and amount variability between the Sorbas (earlier and greater uplift) and Vera Basins [40]. The inset Vera Basin pediment-river terrace sequence spans the Middle to Late Pleistocene based upon ESR and OSL chronologies [78,79]. This timing is in-keeping with the U-Series dated Middle-Late Pleistocene Sorbas Basin river terrace sequence [36,37]. Other adjacent basins (Huércal-Overa, Tabernas, Carboneras-Almería) show varying degrees of geological-geomorphological similarity: (1) Pliocene-Early Pleistocene basin fill, (2) Early Pleistocene deformation and (3) Middle-Late Pleistocene pediment-river terrace sequence formation [16,39,50]. Farines et al. [41] attributes the Early Pleistocene to the most recent phase of Betic Cordillera relief generation, highlighting a poorly understood interplay between mechanical and isostatic relief building processes, with ductile crustal flow cited as a key Plio/Quaternary uplift mechanism. Of note, is the Guadix-Baza Basin, the largest and most intensively studied intramontane basins in the region. This basin occupies a central-interior location within the Betics and differs in timing to Sorbas and its adjacent basins. The Guadix-Baza Basin is characterised by a continuous Miocene-Late Pleistocene continental sedimentary infill [68], capped by a single Late Pleistocene pediment into which extensive basin wide erosion has occurred following capture by the Río Guadalquivir sometime between 350 to 68 ka [4,47]. This difference in timing and pattern of basin geological-geomorphological development reflects variations and connectivity of regional base-levels. Sorbas and adjacent basins occupy marginal mountain belt locations with better connectivity to the Mediterranean coastlines, thus responding more effectively to regional base-level change. In contrast, the Guadix-Baza Basin has an interior mountain belt location with an internal drainage disconnected from regional base-level variability, until captured very recently geologically speaking.

High elevation Early Pleistocene pediment surfaces are also present within intraplate basins as part of the largest drainage systems in Iberia such as the Duero and Tajo [80]. These surfaces have alluvial fan origins and show development within wide-shallow valleys that form the beginnings of river terrace staircases that record hundreds of metres of incision [80]. Thus, the Early Pleistocene is an important interval for surface development and a key marker for subsequent fluvial landscape incision, both within the Betic Cordillera (this study) and within Iberia [81]. Climate and base level (tectonic and capture) variability are widely cited controlling mechanisms for Early Pleistocene Iberian landscape development [48,80–82]. Surface formation within the Sorbas Basin clearly demonstrates interplay of these factors, but the surface itself probably reflects a sustained period of climate stability and base level position to allow the surface to form autogenically at a basin scale. Marked changes to the global climate [83] and regional base levels [12,41] are then driving the surface abandonment and incision.

5.3. Basin Erosion

The interpolation derived basin erosion rates quantified in this study (Figure 9) can be compared with erosion of the Sierra de los Filabres using ^{10}Be [84]. Rates of 52 ± 6 mm/ka were derived from basement schist dominated catchments of tributaries to the Río Jauto along the northeastern margin of the Sorbas Basin [84]. These catchments were formerly part of the main Sorbas Basin drainage before being captured and routed to the southern Vera Basin sometime during the Middle-Late Pleistocene [12]. The average basin surface lowering rates calculated in this study using the dated interpolated surface cover a lower range at 29–46 mm/ka. This could be due to rock strength differences between variably cemented conglomerate basin infill (this study) vs easily weathered basement schist [84]. However, the low value from the Sorbas surface is still broadly in keeping with Betic Cordillera mean (64 $\pm$ 54 mm/ka^{-1}), reflecting low tectonic uplift and possibly a steady state topography where denudation balances uplift [84].

6. Conclusions

- Despite a fragmentary nature, the top Sorbas Basin surface can be reconstructed using GIS interpolation (IDW var) where a sufficiently high-resolution DEM is available;
- The surface is an erosional pediment (glacis) form and not the depositional surface of the Gochar Formation;
- The surface is an Early Pleistocene feature, developed onto deformed basin fill;
- The surface reconstruction approach used here could be used to inform on sampling strategy for dating or could help clarify local surface erosion for age modelling purposes;
- The basin wide configuration of the surface suggests surface formation by autogenic processes that are operating within a stable landscape characterized by a sustained dryland climate and fixed base-level;
- The relict fan-morphology picked out by the surface remnants suggests the surface was autogenically eroded by undissected radiating mountain front streams that formed fan-shaped bodies;
- The Early Pleistocene surface age helps stratigraphically bracket the underlying Gochar Formation to the Pliocene. This clarifies the degraded pediment surface as a Quaternary landscape feature and not a Mio-Pliocene fan delta abandonment surface linked to the post Messinian salinity crisis recovery;
- Surface abandonment took place during the Middle Pleistocene with preferential incision along interfan drainage lines, resulting in capture to preserve the relict fan morphologies;
- Early Pleistocene surfaces are evident throughout Betic Cordillera intramontane basins as either (1) well developed pediments, developed onto Neogene marine basin fill sediments (e.g., Tabernas, Vera Basins) or (2) degraded pediments developed onto Plio-Pleistocene continental alluvial basin fill sediments (Sorbas Basin). Collectively these pediments are regionally and temporally significant, with formation occurring during a stable phase that post-dates deformation of the Plio-Pleistocene continental sediments that form the final basin infill. The deformation and subsequent surface formation probably correspond to the most recent major uplift and relief building phase of the Betic Cordillera;
- Surface form reflects differences in substrate lithology, passive basin tectonic configuration and depositional setting (e.g., lake vs fan);
- Regional variations in surface preservation and differences in formation timing relates to base-level connectivity with the Mediterranean coastal margins of the Betic Cordillera;
- Surface lowering and erosion amounts, and rates are low, comparing well with other denudation techniques (e.g., ^{10}Be) and are in keeping with the Betic Cordillera as a low uplift rate mountain range. The base-level lowering since surface formation is probably an ongoing response to the low uplift rates and basin scale capture events.

Supplementary Materials: The following are available online at http://www.mdpi.com/2571-550X/1/2/15/s1, Figures S1–S4: cosmonuclide results graphs for 4 m erosion scenario, Figures S5–S8: cosmonuclide results graphs for 10 m erosion scenario, Tables S1–S7: cosmonuclide datasets for 4 m erosion scenario; Tables S8–S14: cosmonuclide datasets for 10 m erosion scenario.

Author Contributions: Conceptualization, M.S. and A.E.M.; Methodology, M.S., A.E.M., Á.R., S.L.; Software, Á.R., S.L.; Validation, M.S., Á.R., S.L.; Formal Analysis, M.S., Á.R., S.L.; Investigation, M.S., A.E.M., Á.R., S.H.K., S.L.; Resources, Á.R., S.L.; Data Curation, M.S., Á.R.; Writing—Original Draft Preparation, M.S.; Writing—Review

& Editing, A.E.M., Á.R., S.H.K., S.L.; Visualization, M.S.; Supervision, M.S.; Project Administration, M.S.; Funding Acquisition, M.S.

Funding: This research was part funded by NERC grants CIAF 9039-1007 and NE/F00642X/1.

Acknowledgments: Thanks to Lindy Walsh and Paco Contreras of the Cortijo Urra Field Centre (Sorbas) for fieldwork support. Thanks to the reviewers and editor for manuscript improvements.

Conflicts of Interest: The authors declare no conflict of interest.

References

1. Kingston, D.R.; Dishroon, C.P.; Williams, P.A. Global basin classification system. *AAPG Bull.* **1983**, *67*, 2175–2193.

2. Sanz De Galdeano, C.; Vera, J.A. Stratigraphic record and palaeogeographical context of the Neogene basins in the Betic Cordillera, Spain. *Basin Res.* **1992**, *4*, 21–35. [CrossRef]

3. Sobel, E.R.; Hilley, G.E.; Strecker, M.R. Formation of internally drained contractional basins by aridity-limited bedrock incision. *J. Geophys. Res.* **2003**, *108*. [CrossRef]

4. Calvache, M.L.; Viseras, C. Long-term control mechanisms of stream piracy processes in southeast Spain. *Earth Surf. Process. Landf.* **1997**, *22*, 93–105. [CrossRef]

5. Craddock, W.H.; Kirby, E.; Harkins, N.W.; Zhang, H.; Shi, X.; Liu, J. Rapid fluvial incision along the Yellow River during headward basin integration. *Nat. Geosci.* **2010**, *3*, 209–213. [CrossRef]

6. Soria, J.M.; Fernández, J.; Viseras, C. Late Miocene stratigraphy and palaeogeographic evolution of the intramontane Guadix Basin (Central Betic Cordillera, Spain): Implications for an Atlantic–Mediterranean connection. *Palaeogeogr. Palaeoclimatol. Palaeoecol.* **1999**, *151*, 255–266. [CrossRef]

7. Benvenuti, M.; Bonini, M.; Moroni, A. Tectonic control on the Late Quaternary hydrography of the Upper Tiber Basin (Northern Apennines, Italy). *Geomorphology* **2016**, *269*, 85–103. [CrossRef]

8. Watchman, A.L.; Twidale, C.R. Relative and 'absolute'dating of land surfaces. *Earth-Sci. Rev.* **2002**, *58*, 1–49. [CrossRef]

9. Viseras, C.; Fernández, J. Sedimentary basin destruction inferred from the evolution of drainage systems in the Betic Cordillera, southern Spain. *J. Geol. Soc.* **1992**, *149*, 1021–1029. [CrossRef]

10. Stokes, M.; Mather, A.E.; Harvey, A.M. Quantification of river-capture-induced base-level changes and landscape development, Sorbas Basin, SE Spain. *Geol. Soc. Lond. Spec. Publ.* **2002**, *191*, 23–35. [CrossRef]

11. García-Tortosa, F.J.; Alfaro, P.; de Galdeano, C.S.; Galindo-Zaldívar, J. Glacis geometry as a geomorphic marker of recent tectonics: The Guadix–Baza basin (South Spain). *Geomorphology* **2011**, *125*, 517–529. [CrossRef]

12. Harvey, A.M.; Whitfield, E.; Stokes, M.; Mather, A. The Late Neogene to Quaternary drainage evolution of the uplifted Neogene sedimentary Basins of Almería, Betic Chain. In *Landscapes and Landforms of Spain*; Gutiérrez, F., Gutiérrez, M., Eds.; Springer: Dordrecht, The Netherlands, 2014; pp. 37–61, ISBN 978-94-017-8628-7.

13. Harvey, A.M.; Stokes, M.; Mather, A.; Whitfield, E. Spatial characteristics of the Pliocene to modern alluvial fan successions in the uplifted sedimentary basins of Almería, SE Spain: Review and regional synthesis. *Geol. Soc. Lond. Spec. Publ.* **2018**, *440*, SP440-5. [CrossRef]

14. Stokes, M.; Nash, D.J.; Harvey, A.M. Calcrete 'fossilisation' of alluvial fans in SE Spain: The roles of groundwater, pedogenic processes and fan dynamics in calcrete development. *Geomorphology* **2007**, *85*, 63–84. [CrossRef]

15. Rodés, Á.; Pallàs, R.; Braucher, R.; Moreno, X.; Masana, E.; Bourlés, D.L. Effect of density uncertainties in cosmogenic ^{10}Be depth-profiles: Dating a cemented Pleistocene alluvial fan (Carboneras Fault, SE Iberia). *Quat. Geochronol.* **2011**, *6*, 186–194. [CrossRef]

16. Geach, M.R.; Thomsen, K.J.; Buylaert, J.P.; Murray, A.S.; Mather, A.E.; Telfer, M.W.; Stokes, M. Single-grain and multi-grain OSL dating of river terrace sediments in the Tabernas Basin, SE Spain. *Quat. Geochronol.* **2015**, *30*, 213–218. [CrossRef]

17. Martín, J.; Braga, J.C. Messinian events in the Sorbas Basin in southeastern Spain and their implications in the recent history of the Mediterranean. *Sediment. Geol.* **1994**, *90*, 257–268. [CrossRef]

18. Haughton, P.D. Deposits of deflected and ponded turbidity currents, Sorbas Basin, southeast Spain. *J. Sediment. Res.* **1994**, *64*, 233–246. [CrossRef]

19. Mather, A.E. Cenozoic Drainage Evolution of the Sorbas Basin SE Spain. Ph.D. Thesis, University of Liverpool, Liverpool, UK, 1991.
20. Mather, A.E.; Harvey, A.M. Controls on drainage evolution in the Sorbas basin, southeast Spain. In *Mediterranean Quaternary River Environments*; Lewin, J., Macklin, M.G., Woodward, J.C., Eds.; Balkema: Rotterdam, The Netherlands, 1995; pp. 65–75, ISBN 9054101911.
21. IGME. *Mapa Geológico de España, 1:200 000. Almería-Garrucha, 84–85*, 2nd ed.; IGME: Madrid, Spain, 1980.
22. IGME. *Mapa Geológico de España, 1:200 000. Baza, 78*, 2nd ed.; IGME: Madrid, Spain, 1983.
23. IGME. *Mapa Geológico de España, 1:200 000. Murcia, 78*, 2nd ed.; IGME: Madrid, Spain, 1983.
24. Harvey, A.M.; Wells, S.G. Response of Quaternary fluvial systems to differential epeirogenic uplift: Aguas and Feos river systems, southeast Spain. *Geology* **1987**, *15*, 689–693. [CrossRef]
25. Vázquez, M.; Jabaloy, A.; Barbero, L.; Stuart, F.M. Deciphering tectonic-and erosion-driven exhumation of the Nevado–Filábride Complex (Betic Cordillera, Southern Spain) by low temperature thermochronology. *Terra Nova* **2011**, *23*, 257–263. [CrossRef]
26. Platt, J.P.; Kelley, S.P.; Carter, A.; Orozco, M. Timing of tectonic events in the Alpujárride Complex, Betic Cordillera, southern Spain. *J. Geol. Soc. Lond.* **2005**, *162*, 451–462. [CrossRef]
27. Giaconia, F.; Booth-Rea, G.; Martínez-Martínez, J.M.; Azañón, J.M.; Pérez-Peña, J.V.; Pérez-Romero, J.; Villegas, I. Geomorphic evidence of active tectonics in the Sierra Alhamilla (eastern Betics, SE Spain). *Geomorphology* **2012**, *145*, 90–106. [CrossRef]
28. IGME. *Mapa Geológico de España, 1:50 000. Sorbas, 1031, 24–42*; IGME: Madrid, Spain, 1973.
29. IGME. *Mapa Geológico de España, 1:50 000. Tabernas, 1030, 23–42*; IGME: Madrid, Spain, 1973.
30. Martín-Suárez, E.; Freudenthal, M.; Krijgsman, W.; Fortuin, A.R. On the age of the continental deposits of the Zorreras Member (Sorbas Basin, SE Spain). *Geobios* **2000**, *33*, 505–512. [CrossRef]
31. Clauzon, G.; Suc, J.P.; Do Couto, D.; Jouannic, G.; Melinte-Dobrinescu, M.C.; Jolivet, L.; Quillévéré, F.; Lebret, N.; Mocochain, L.; Popescu, S.M.; et al. New insights on the Sorbas Basin (SE Spain): The onshore reference of the Messinian Salinity Crisis. *Mar. Pet. Geol.* **2015**, *66*, 71–100. [CrossRef]
32. Mather, A.E. Basin inversion: Some consequences for drainage evolution and alluvial architecture. *Sedimentology* **1993**, *40*, 1069–1089. [CrossRef]
33. Griffiths, J.S.; Hart, A.B.; Mather, A.E.; Stokes, M. Assessment of some spatial and temporal issues in landslide initiation within the Río Aguas Catchment, South–East Spain. *Landslides* **2005**, *2*, 183–192. [CrossRef]
34. Mather, A.E.; Stokes, M.; Griffiths, J.S. Quaternary landscape evolution: A framework for understanding contemporary erosion, southeast Spain. *Land Degrad. Dev.* **2002**, *13*, 89–109. [CrossRef]
35. Harvey, A.M.; Miller, S.Y.; Wells, S.G. Quaternary soil and river terrace sequences in the Aguas/Feos river systems: Sorbas basin, southeast Spain. In *Mediterranean Quaternary River Environments*; Lewin, J., Macklin, M.G., Woodward, J.C., Eds.; Balkema: Rotterdam, The Netherlands, 1995; pp. 263–281, ISBN 9054101911.
36. Kelly, M.; Black, S.; Rowan, J.S. A calcrete-based U/Th chronology for landform evolution in the Sorbas basin, southeast Spain. *Quat. Sci. Rev.* **2000**, *19*, 995–1010. [CrossRef]
37. Candy, I.; Black, S.; Sellwood, B. U-series isochron dating of immature and mature calcretes as a basis for constructing Quaternary landform chronologies for the Sorbas basin, southeast Spain. *Quat. Res.* **2005**, *64*, 100–111. [CrossRef]
38. Ilott, S.H. Cosmogenic Dating of Fluvial Terraces in the Sorbas Basin, SE Spain. Ph.D. Thesis, University of Plymouth, Plymouth, UK, 2013.
39. García-Meléndez, E.; Goy, J.L.; Zazo, C. Neotectonics and Plio-Quaternary landscape development within the eastern Huércal-Overa Basin (Betic Cordilleras, Southeast Spain). *Geomorphology* **2003**, *50*, 111–133. [CrossRef]
40. Stokes, M. Plio-Pleistocene drainage development in an inverted sedimentary basin: Vera basin, Betic Cordillera, SE Spain. *Geomorphology* **2008**, *100*, 193–211. [CrossRef]
41. Farines, B.; Calvet, M.; Gunnell, Y. The summit erosion surfaces of the inner Betic Cordillera: Their value as tools for reconstructing the chronology of topographic growth in southern Spain. *Geomorphology* **2015**, *233*, 92–111. [CrossRef]
42. Centro Nacional de Información Geográfica. Available online: http://centrodedescargas.cnig.es (accessed on 14 June 2018).

43. Boulton, S.J.; Stokes, M. Which DEM is best for analyzing fluvial landscape development in mountainous terrains? *Geomorphology* **2018**, *310*, 168–187. [CrossRef]

44. ESRI. How To: Identify Ridgelines from a DEM. Available online: https://support.esri.com/en/technical-article/000011289 (accessed on 14 June 2018).

45. Alexander, R.W.; Calvo-Cases, A.; Arnau-Rosalén, E.; Mather, A.E.; Lázaro-Suau, R. Erosion and stabilisation sequences in relation to base level changes in the El Cautivo badlands, SE Spain. *Geomorphology* **2008**, *100*, 83–90. [CrossRef]

46. Della Seta, M.; Del Monte, M.; Fredi, P.; Miccadei, E.; Nesci, O.; Pambianchi, G.; Piacentini, T.; Troiani, F. Morphotectonic evolution of the Adriatic piedmont of the Apennines: An advancement in the knowledge of the Marche-Abruzzo border area. *Geomorphology* **2008**, *102*, 119–129. [CrossRef]

47. Pérez-Peña, J.V.; Azañón, J.M.; Azor, A.; Tuccimei, P.; Della Seta, M.; Soligo, M. Quaternary landscape evolution and erosion rates for an intramontane Neogene basin (Guadix–Baza basin, SE Spain). *Geomorphology* **2009**, *106*, 206–218. [CrossRef]

48. Antón, L.; Muñoz-Martín, A.; De Vicente, G. Quantifying the erosional impact of a continental-scale drainage capture in the Duero Basin, northwest Iberia. *Quat. Res.* **2018**, 1–15. [CrossRef]

49. Geach, M.R.; Stokes, M.; Telfer, M.W.; Mather, A.E.; Fyfe, R.M.; Lewin, S. The application of geospatial interpolation methods in the reconstruction of Quaternary landform records. *Geomorphology* **2014**, *216*, 234–246. [CrossRef]

50. Rodés, Á.; Pallàs, R.; Ortuño, M.; García-Meléndez, E.; Masana, E. Combining surface exposure dating and burial dating from paired cosmogenic depth profiles. Example of El Límite alluvial fan in Huércal-Overa basin (SE Iberia). *Quat. Geochronol.* **2014**, *19*, 127–134. [CrossRef]

51. Hancock, G.S.; Anderson, R.S.; Chadwick, O.A.; Finkel, R.C. Dating fluvial terraces with 10Be and 26Al profiles: Application to the Wind River, Wyoming. *Geomorphology* **1999**, *27*, 41–60. [CrossRef]

52. Braucher, R.; Merchel, S.; Borgomano, J.; Bourlès, D. Production of cosmogenic radionuclides at great depth: A multi element approach. *Earth Planet. Sci. Lett.* **2011**, *309*, 1–9. [CrossRef]

53. Granger, D.E.; Smith, A.L. Dating buried sediments using radioactive decay and muogenic production of ^{26}Al and ^{10}Be. *Nucl. Instrum. Methods Phys. Res. Sect. B* **2000**, *172*, 822–826. [CrossRef]

54. Granger, D.E.; Muzikar, P.F. Dating sediment burial with in situ-produced cosmogenic nuclides: Theory, techniques, and limitations. *Earth Planet. Sci. Lett.* **2001**, *188*, 269–281. [CrossRef]

55. Balco, G.; Rovey, C.W. An isochron method for cosmogenic-nuclide dating of buried soils and sediments. *Am. J. Sci.* **2008**, *308*, 1083–1114. [CrossRef]

56. Balco, G.; Stone, J.O.H.; Lifton, N.; Dunai, T. A complete and easily accessible means of calculating surface exposure ages or erosion rates from Be and Al measurements. *Quat. Geochronol.* **2008**, *3*, 174–195. [CrossRef]

57. Lal, D. Cosmic ray labelling of erosion surfaces: In situ nuclide production rates and erosion models. *Earth Planet. Sci. Lett.* **1991**, *104*, 424–439. [CrossRef]

58. CRONUS Calculator 2.3. Available online: https://hess.ess.washington.edu/ (accessed on 4 June 2018).

59. Mather, A.E.; Harvey, A.M.; Stokes, M. Quantifying long-term catchment changes of alluvial fan systems. *Geol. Soc. Am. Bull.* **2000**, *112*, 1825–1833. [CrossRef]

60. Mather, A.E.; Westhead, K. Plio/Quaternary strain of the Sorbas Basin, SE Spain: Evidence from soft sediment deformation structures. *Quat. Proc.* **1993**, *3*, 57–65.

61. Nash, D.J.; Smith, R.F. Multiple calcrete profiles in the Tabernas Basin, southeast Spain: Their origins and geomorphic implications. *Earth Surf. Process. Landf.* **1998**, *23*, 1009–1029. [CrossRef]

62. King, L. The pediment landform: Some current problems. *Geol. Mag.* **1949**, *86*, 245–250. [CrossRef]

63. Dumas, B. Glacis et croutes calcaires dans le Levant espagnol. *Bull. Assoc. Géogr. Fr.* **1969**, *46*, 553–561. [CrossRef]

64. Strudley, M.W.; Murray, A.B. Sensitivity analysis of pediment development through numerical simulation and selected geospatial query. *Geomorphology* **2007**, *88*, 329–351. [CrossRef]

65. Hodge, E.J.; Richards, D.A.; Smart, P.L.; Andreo, B.; Hoffmann, D.L.; Mattey, D.P.; González-Ramón, A. Effective precipitation in southern Spain (~266 to 46 ka) based on a speleothem stable carbon isotope record. *Quat. Res.* **2008**, *69*, 447–457. [CrossRef]

66. Carrión, J.S.; Fernández, S.; Jiménez-Moreno, G.; Fauquette, S.; Gil-Romera, G.; González-Sampériz, P.; Finlayson, C. The historical origins of aridity and vegetation degradation in southeastern Spain. *J. Arid Environ.* **2010**, *74*, 731–736. [CrossRef]

67. Martrat, B.; Jimenez-Amat, P.; Zahn, R.; Grimalt, J.O. Similarities and dissimilarities between the last two deglaciations and interglaciations in the North Atlantic region. *Quat. Sci. Rev.* **2014**, *99*, 122–134. [CrossRef]
68. Pla-Pueyo, S.; Viseras, C.; Soria, J.M.; Tent-Manclús, J.E.; Arribas, A. A stratigraphic framework for the Pliocene–Pleistocene continental sediments of the Guadix Basin (Betic Cordillera, S. Spain). *Quat. Int.* **2011**, *243*, 16–32. [CrossRef]
69. Braga, J.C.; Martín, J.M.; Quesada, C. Patterns and average rates of late Neogene-Recent uplift of the Betic Cordillera, SE Spain. *Geomorphology* **2003**, *50*, 3–26. [CrossRef]
70. Stokes, M.; Mather, A.E.; Belfoul, A.; Farik, F. Active and passive tectonic controls for transverse drainage and river gorge development in a collisional mountain belt (Dades Gorges, High Atlas Mountains, Morocco). *Geomorphology* **2008**, *102*, 2–20. [CrossRef]
71. Harvey, A.M.; Silva, P.; Mather, A.E.; Goy, J.; Stokes, M.; Zazo, C. The impact of Quaternary sea level and climatic change on coastal alluvial fans in the Cabo de Gata ranges, southeast Spain. *Geomorphology* **1999**, *28*, 1–22. [CrossRef]
72. Pastor, A.; Babault, J.; Teixell, A.; Arboleya, M.L. Intrinsic stream-capture control of stepped fan pediments in the High Atlas piedmont of Ouarzazate (Morocco). *Geomorphology* **2012**, *173*, 88–103. [CrossRef]
73. Völk, H.R. *Quartare Reliefentwicklung in Sudost-Spanien*; Heidelberger Geographische Arbeiten: Heidelberger, Germany, 1979; 143p.
74. Stokes, M. Plio-Pleistocene Drainage Evolution of the Vera Basin, SE Spain. Ph.D. Thesis, University of Plymouth, Plymouth, UK, 1997.
75. Stokes, M.; Mather, A.E. Response of Plio-Pleistocene alluvial systems to tectonically induced base-level changes, Vera Basin, SE Spain. *J. Geol. Soc.* **2000**, *157*, 303–316. [CrossRef]
76. Wenzens, G. Mittelquartäre klimaverhältnisse und reliefentwicklung im semiariden becken von Vera (Südostspanien). *Eiszeitalt. Ggw.* **1992**, *42*, 121–133.
77. Wenzens, G. The influence of tectonics and climate on the Villafranchian morphogenesis in semiarid Southeastern Spain. *Z. Geomorphol.* **1992**, *84*, 173–184.
78. Wenzens, G. Die Quartäre küstenentwicklung im mündungsbereich der flüsse Aguas, Antas und Almanzora in Südostspanien. *Erdundliches Wissen* **1991**, *105*, 131–150.
79. Meikle, C.; Stokes, M.; Maddy, D. Field mapping and GIS visualisation of Quaternary river terrace landforms: An example from the Rio Almanzora, SE Spain. *J. Maps* **2010**, *6*, 531–542. [CrossRef]
80. Silva, P.G.; Roquero, E.; López-Recio, M.; Huerta, P.; Martínez-Graña, A.M. Chronology of fluvial terrace sequences for large Atlantic rivers in the Iberian Peninsula (Upper Tagus and Duero drainage basins, Central Spain). *Quat. Sci. Rev.* **2017**, *166*, 188–203. [CrossRef]
81. Santisteban, J.I.; Schulte, L. Fluvial networks of the Iberian Peninsula: A chronological framework. *Quat. Sci. Rev.* **2007**, *26*, 2738–2757. [CrossRef]
82. Antón, L.; De Vicente, G.; Muñoz-Martín, A.; Stokes, M. Using river long profiles and geomorphic indices to evaluate the geomorphological signature of continental scale drainage capture, Duero basin (NW Iberia). *Geomorphology* **2014**, *206*, 250–261. [CrossRef]
83. Gibbard, P.L.; Lewin, J. River incision and terrace formation in the Late Cenozoic of Europe. *Tectonophysics* **2009**, *474*, 41–55. [CrossRef]
84. Bellin, N.; Vanacker, V.; Kubik, P.W. Denudation rates and tectonic geomorphology of the Spanish Betic Cordillera. *Earth Planet. Sci. Lett.* **2014**, *390*, 19–30. [CrossRef]

quaternary

MDPI

Article

Episodic Sedimentary Evolution of an Alluvial Fan (Huangshui Catchment, NE Tibetan Plateau)

Linman Gao [1], Xianyan Wang [1,*], Shuangwen Yi [1], Jef Vandenberghe [1,2], Martin R. Gibling [3] and Huayu Lu [1]

[1] School of Geography and Ocean Science, Nanjing University, Nanjing 210023, China; mg1627003@smail.nju.edu.cn (L.G.); ysw7563@nju.edu.cn (S.Y.); jef.vandenberghe@vu.nl (J.V.); huayulu@nju.edu.cn (H.L.)
[2] Institute of Earth Sciences, VU University Amsterdam, 1081HV Amsterdam, The Netherlands
[3] Department of Earth Sciences, Dalhousie University, Halifax, NS B3H4R2, Canada; Martin.Gibling@Dal.Ca
* Correspondence: xianyanwang@nju.edu.cn

Academic Editors: David Bridgland and Valentí Rull
Received: 5 July 2018; Accepted: 23 August 2018; Published: 3 September 2018

Abstract: Alluvial-fan successions record changes in hydrological processes and environments that may reflect tectonic activity, climate conditions and changes, intrinsic geomorphic changes, or combinations of these factors. Here, we focus on the evolution of a stream-dominated fan in a tectonic depression of the Xining basin of China, laid down under a semi-arid climate in the northeastern Tibetan Plateau (NETP). The fan succession is composed of three facies associations, from bottom to top: (1) matrix to clast-supported, poorly sorted, planar cross-stratified to crudely stratified sheets of coarse-grained sediments; (2) horizontal laminated sand, laminated layers of reddish fine silt and yellow coarse silt with stacked mounds of sand; and (3) clay-rich deposits with incipient paleosols. The succession shows rapid sediment aggradation from high-energy to low-energy alluvial fans and finally to a floodplain. The dating results using optically stimulated luminescence (OSL) method show that a gravelly, high-energy fan was deposited during MIS 6, after which a low-energy fan, mainly composed of sand and silt, was deposited and finally covered by flood loam during the MIS 6–5 transition and the warmer last interglacial. Stacked sand mounds are interpreted from their sediment structure and grain-size distribution as shrub-coppice dunes in low-energy fan deposits. They may be considered as a response to the interaction of alluvial and aeolian processes in a semi-arid environment.

Keywords: alluvial fan; fluvial facies; grain-size analysis; optical stimulated luminescence (OSL) dating; vegetation-induced sedimentary structures

1. Introduction

Quaternary terrestrial sediments record information on climatic conditions and changes, environmental evolution, and tectonic events. Sediment composition and structures reflect changes in the sedimentary environment, which may be indicative of particular climatic conditions [1]. Paleoclimate changes in semi-arid regions in China have been studied extensively from aeolian loess [2–8], lacustrine [9–13], and fluvial (alluvial) sediments [14–16], but research based on alluvial fans is relatively scarce. Alluvial sediments are sensitive to changes in hydrological conditions, and interpretation of the sediments may provide valuable information on past climate [17]. The intricacy of fluvial response, including aggradation/incision rhythms and terrace formation, to late Quaternary climate change, has been studied for several decades [16,18–21]. In addition, the interactions between fluvial and aeolian processes and their response to climate change in semi-arid areas have also been reported elsewhere in China and around the world [22–26].

Alluvial sequences allow process inference, which then can be used to infer environments, at the depositional site, as well as environmental change and related controls in the catchment area [17]. In the past few decades, understanding of how tectonic activity and climate change influence alluvial deposits has progressed considerably [27–29]. Tectonically active zones may preserve sediment sequences by providing accumulation space, whereas climate influences sediment supply by affecting water and sediment discharge and vegetation cover [30]. Climatic factors and autogenic effects appear to exert an overwhelming control on Quaternary fan sequences and sedimentary styles [31–34], especially under arid or semiarid conditions [35].

As an important alluvial sediment record, a fan can be formed as a stream-dominated alluvial fan, as part of the main trunk valley, with braided channels created where the valley widens downstream, and as a gravity flow (debris) dominated fan at a tributary junction [1,36–38]. The majority of flow-dominated alluvial fans documented in modern and Quaternary settings are from perennial or semi-permanent channelized rivers, and most are located in humid areas, such as the Kosi fan in Nepal and India [37]. Such fluvially dominated mega-fans are special cases of alluvial fans with a lower slope, the presence of a floodplain, and the absence of sediment-gravity flows [39,40]. However, there has only been limited work on stream-dominated alluvial fans in semi-arid environments. In modern arid and semi-arid environments, fans are commonly characterized by debris flows. However, Jolley et al. [36] proposed that ephemeral river deposits are dominant in modern syn-tectonic fans, and they concluded that such fans were formed by highly concentrated, turbulent tractional flows.

The Xining basin, in the northeastern Tibetan Plateau (NETP), has been most intensively studied at a large scale [41–44], whereas the small depressions within the basin have been mainly studied at a reconnaissance scale [45,46]. Most studies have paid attention to the river terraces, for instance, along the Huangshui River, and the sedimentary process response to climate change and tectonic activity. However, the number of alluvial fans documented in this area is small, which may reflect, in part, their low preservation potential as geomorphologic features. The presence of a large sand pit with extensive outcrops in this study provided an excellent opportunity for investigating an alluvial-fan succession.

Local debris fan may dam the main channel of a river, resulting in dam-collapse and huge related floods; these have been widely studied in this area [47]. However, a stream-dominated fan in a large catchment has not been reported from this area. Thus, the first main goal of this study was to document and interpret the sedimentary lithofacies and investigate the stratigraphic evolution of a stream-dominated alluvial fan, with an estimated width of 1.7 km and length of 3.5 km, at the margin of a tectonic depression, the Ping'an depression, in a semi-arid environment. Secondly, we interpreted the relationships between sedimentary environments and climate and also the influence of vegetation and internal dynamics, drawing on detailed facies analysis, sedimentary structures, and sedimentological evidence for the former presence of vegetation. To further this goal, we analyzed the grain-size distributions of sediments to assess the relative contribution of fluvial and aeolian processes in a fan environment. Finally, we dated the sediments using optically stimulated luminescence (OSL) method, to establish the timing of deposition in relation to long-term climate evolution in the region, and tectonic events.

2. Geographical Setting

The study site is located in the Xining basin on the NETP, which has undergone coeval crustal shortening and left-lateral strike-slip faulting [48] and has experienced dramatic and pronounced cyclic climate changes. The Xining basin was bounded by two main thrusts, the North Middle Qilian Shan Fault (NMQF) in the north and the Laji Shan Fault (LSF) in the south [45,49] (Figure 1a). Tectonic fragmentation of the Xining basin has resulted in the formation of several gorges between subsiding sub-basins or depressions (e.g., the Ping'an depression) and up to 16 river terraces, recording deep incision into ~600 m of basin filling [45,49]. At the outlet of one gorge (Xiao gorge), a stream-dominated fan, correlated with a terrace, was formed (Figure 1c) (see Section 5. Discussion below).

The climate of the NETP is at the boundary of influence from the Asian summer monsoon (ASM) and the westerlies [50–52]. The integrated influence of these climate systems, combined with the high elevation, makes the area sensitive to climate changes. The study area, a depression in the Xining basin, is currently located at an elevation between ~2000 m and ~3000 m (Figure 1) and is characterized by a cold semi-arid climate with a strongly variable temperature and precipitation. The regional average annual temperature and precipitation are 6.4 °C and 319–532 mm, respectively, and the average annual evaporation can reach up to ~1800 mm. At present, precipitation occurs mainly in summer, probably caused by invasions of the Asian summer monsoon. In winter, the climate of this region is controlled by the winter monsoon, with strong winds and cold and dry conditions. This framework is also typical for the Quaternary climate situation in this area.

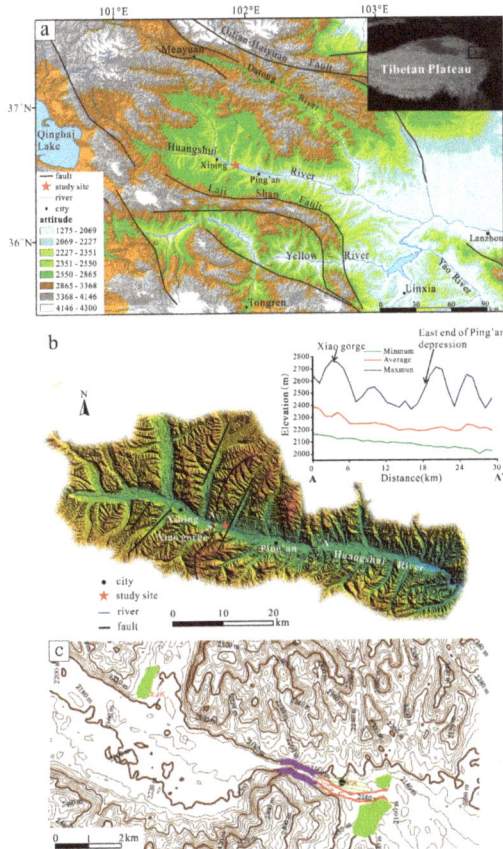

Figure 1. (**a**) Overview of the study area derived from ASTER Global Digital Elevation Model (ASTER GDEM) data. The black lines show the faults (modified after the 1:500,000 geological map). (**b**) Detailed morphological features (gorges and depressions) of the study area and the location of the study section (red star). The Huangshui River flows through a series of gorges and depressions. The swath profile (AA' in yellow rectangle) shows detailed elevation change of the study area and the Xiao gorge and Ping'an depression. The black lines show the faults (modified from [53]). (**c**) Geomorphological sketch map of the studied area, showing the terrace and fan (plain green: terrace; green hatched lines: fan; red: erosion of the terrace and the fan after deposition; purple: V-shaped incision of the Huangshui in the Xiao gorge).

The studied sediment sequence is present in a sand pit, excavated in the fan at the western side of the Ping'an depression (Figure 1c), with a section oriented northwest to southeast, around 22 m high and with a lateral extent of 500 m. The top of the basal unit at the west end of the sand-pit is situated 26 m above the modern floodplain of the Huangshui River, which is a first-order tributary of the Yellow River (Figure 1c). The river flows roughly W-E through the Xining basin and has deeply incised the Precambrian to Mesozoic bedrock in the gorges and Cenozoic clastic sediments in the basin [45,46]. The sediment sequence is located directly at the east of the Xiao gorge and the west end of the Ping'an depression (36°33′27.7″ N, 101°55′53.38″ E) (Figure 1b).

3. Methods

3.1. Field Work

In order to study the vertical and lateral variations of the sediments, and distinguish the sedimentary environment, eight sediment sections were logged in detail (Figure 2) and described. The facies codes of Miall [54] were used (Table 1). To allow an interpretation of the process and environment, some particular sedimentary structures/architectures were quantified by measuring individual maximum thickness and width, in the field and from photographs. Photomosaics were used to correlate units across the outcrop face. In addition, 50 samples were taken from different sediment units for grain-size analysis and 13 samples for OSL dating (Table 2) to supplement the interpretation of the sediment processes and transport dynamics and changes of the sedimentary environment.

3.2. Grain-Size Measurements

The grain-size samples were prepared largely according to the methods described by Konert and Vandenberghe [55]. A few grams of sediment were pre-treated with 10% H_2O_2 and 10% HCl to remove organic matter and carbonate, respectively. With $(NaPO_3)_6$ as a dispersing agent, the pure siliciclastic sediments were then measured using a Malvern master-size laser particle analyzer (Malvern Panalytical, Almelo, The Netherlands). The grain-size distribution shows 100 size classes, ranging from 0.02 to 2000 µm. In addition to modal values, distribution curves and additional grain-size parameters, including the proportion of specific grain-size intervals (e.g., clay and sand content), were calculated.

3.3. Optically Stimulated Luminescence Dating

OSL samples from different sediment units (Table 2) were taken by hammering a steel tube 25 cm long with a diameter of 5 cm into the cleaned profile. Pure quartz (grain size 63~90 µm, except samples XX-A90 and XX-A220, which extracted from the 40~63 µm fraction) was extracted from the intermediate material of the tubes using standard methods (30% H_2O_2, 10% HCl, wet sieving, 40% HF), and was used for an equivalent dose (D_e) measurement. For extracting K-rich feldspar, pre-treated 63–90 µm grains (30% H_2O_2, 10% HCl, wet sieving) were first cleaned using 10% HF to remove coatings and the outer alpha irradiated skin and then washed with 10% HCl to remove fluoride. K-rich feldspars were floated off using heavy liquid of a 2.58 g/cm^3 density.

All luminescence analyses used an automated Risø reader (DTU Nutech, Roskilde, Denmark) equipped with blue (470 nm; ~80 mW cm^2) LEDs and an IR laser diode (870 nm, ~135 mW cm^2). Quartz OSL signals were collected through a 7.5 mm Schott U-340 (UV) glass filter (emission 330 + 35 nm). K-feldspar pIRIR signals were collected through a combination of Corning 7–59 and Schott BG-39 glass filters (blue-violet part of the spectrum). Quartz equivalent doses (D_e) were measured using a standard single-aliquot regenerative-dose (SAR) protocol [56,57]. Preheating of natural and regenerative doses was for 10 s at 260 °C, and the response to the test dose was measured after a cut-heat to 220 °C. K-feldspar aliquots were measured using the post-IR IRSL (pIRIR$_{290}$) protocol [58–62]. Samples were preheated at 320 °C for 60 s followed by IR diode stimulation at 200 °C for 200 s and the IRSL signal was then measured at 290 °C for 200 s. The response to the test dose was measured in the same manner and was followed by IR illumination at 325 °C for 200 s at the end of each SAR cycle, to reduce recuperation.

Sample material from both outer ends of the tube was used for water-content and radioactive element analysis. The material (about 20 g) was first dried and then ground to powder to determine concentrations of U, Th, and K using neutron activation analysis (NAA). The water content varied during the long-term burial period, so we assumed 50% of the saturated water content as the average value during historical time according to previous work [16,26], with an absolute uncertainty of 7% of this value to allow for possible fluctuations on the basis of the local climatic characteristics. Based on applying conversion factors from Guérin [63] and beta attenuation factors from Mejdahl [64], the external beta and gamma dose rate was calculated using the radionuclide concentration. For K-feldspar dose rates, a K concentration of $12.5 \pm 0.5\%$ and Rb concentration of 400 ± 100 ppm was assumed [65]. A small internal dose rate contribution from U and Th of 0.03 ± 0.015 Gyka^{-1} and 0.06 ± 0.03 Gyka^{-1} was also included for quartz and K-feldspar, respectively [66–68].

4. Results

4.1. Facies Analysis and Grain-Size Analysis

The Ping'an site is characterized by a complex spatial facies distribution (Figure 2). The strata are subdivided into six units (Figure 2) and grouped into three facies associations from bottom to top: a high-energy alluvial fan, low-energy alluvial fan, and floodplain. The unit boundaries are typically abrupt but conformable, and the upward succession of facies associations reflects the evolution of the depositional environment, without any marked breaks in deposition.

4.1.1. High-Energy Alluvial Fan (Facies Association 1)

Description

The facies association forms unit 1 (U1) in the lower part of the study site. The basal contact of U1 is not exposed, and the unit is presumably below the exposed base of sections I and IV. The exposed unit varies between 1 and 1.6 m in thickness, with a high thickness of 4 to 5.5 m in section VII. U1 changes in dip (the dip of the boundary of U1 and U2) eastward from 9° in section VII to 7° and 5° in sections II and III, respectively, and the unit thins eastward.

The unit is composed of poorly sorted gravel, deposited as horizontally stratified sheets (Gh), planar cross-beds (Gp) which show an ESE flow direction, and disorganized, clast-supported and matrix-supported beds of facies Gcm and Gmm (Table 1) (Figure 3). Clasts in the Gp facies are locally imbricated. The gravel clasts are mainly sub-rounded and rounded, but also sub-angular, set in a matrix of fine-to-coarse grained sand and silt. Locally, lenticular sandy bodies with trough and planar cross-beds (St, Sp) are preserved (Figure 3a). Overall, the particle size of the gravels varies greatly, with small pebbles between 0.5 and 1 cm and cobbles ranging from 10 to 20 cm. A few large blocks (up to ~70 cm diameter) occur at the base. This unit fines upward in section VII, with particle size ranging from 60–70 cm at the bottom to 10–20 cm at the top (Figure 3b). The gravel is mainly composed of vein quartz, sandstone, granodiorite, and quartzite, as well as lesser amounts of granite and platy green schist. Sample II-1 (Table 2) is mainly composed of fine-to-medium sand (modal value of 225 μm) with a small amount of fine silt (Figure 3d).

Figure 2. Stratigraphical sketch of the main sediment units observed at the Ping'an study site. Each column represents a sediment log through the sediment sequences. Section VII is to the west of the photo, 145 m from section I. Section PA is 30 m from section VI in the west. The numbered horizontal line segments below the photo represent the distance between the two sections.

Table 1. Description and interpretation of sedimentary facies of the Ping'an site (after Miall [54] and Einsele [69]).

Facies Codes	Description	Interpretation
Gmm	Granules and pebbles, matrix-supported, granule-sand matrix, massive, poorly sorted, tabular-or lenticular-shaped, sub-rounded to rounded	Mass-flow deposition of hyper-concentrated or turbulent flow
Gcm	Granules and pebbles, clast-supported, massive, poorly sorted, sub-rounded to rounded	Rapid deposition by streamflow with concentrated clasts
Gh	Granules and pebbles, clast-supported, crudely horizontal beds, sub-rounded to rounded	Deposition of hyperconcentrated flow in unconfined sheet flood
Gp	Granules and pebbles, planar cross-beds, weak imbrication, tabular-shaped, sub-rounded to rounded	Transverse 2-D gravel bedforms
Gt	Granules and pebbles, trough cross-stratified, sub-angular to rounded, lenticular units	Minor channel fills with deposition from 3-D gravel dunes
Sh	Very fine to-coarse-grained sand, horizontal lamination	Deposition of planar bed flows in channels, or sheet flood, upper flow regime
Sp	Very fine-to-coarse-grained sand, planar cross-lamination	Migration of two-dimensional (2-D) dunes in channel
Sl	Very fine-to-very coarse-grained sand, low-angle (<15°) cross-lamination	Deposition of antidunes in upper flow regime
St	Very fine-to-coarse-grained sand, solitary or grouped trough cross-beds	Migration of sandy ripples/dunes in channels; or filling of scour hollows or channel pools
Sr	Very fine-to-coarse-grained sand with ripple cross-lamination, thinly bedded	Migration of ripples in shallow channels in lower flow regime; overland sheetflows
Sm	Very fine-to-coarse-grained sand, structureless	Rapid deposition of sand
Fl	Silt, fine lamination, very small ripple sets. Occasional lenses of granules and pebbles, calcareous cement and iron-manganese coating, mottled	Waning flood deposits or overbank, abandoned channel deposits
Fm	Silt, structureless, desiccation cracks. Occasionally calcareous cement and iron-manganese coating, bioturbation, mottled	Suspension deposits of waning flows in overbank or abandoned channel, weak pedogenesis
Fr	Silt, massive, roots, bioturbation	Root bed, weak pedogenesis

Table 2. Sediment types sampled for grain-size analysis and OSL dating (in bold and italics, respectively).

Sample	Sediment Type
Unit 1	
II-1	Lenticular sand body
Unit 2	
XX-II-1 **II-2**	Trough cross-stratified sand
II-3	Silt layer
XX-C490, XX-C600, XX-VI-1 **II-4; IV-1; VII-1**	Horizontally laminated sand
Unit 3	
XX-I-1, XX-II-2, XX-V-1, XX-V-2 **I-6, 7, 8; II-7, 8; III-5, 6, 7, 8; IV-4; V-4, 5, 6, 7**	Sand mounds
I-2, 4, 9; II-5; III-1, 3; IV-2; V-2; VII-2	Silt layers
XX-C200 **I-1, 3, 5; II-6; III-2, 4; IV-3; VII-3, 4**	Sand to silty sand layers
Unit 4	
XX-II-3	Planar cross-bedded sand
I-10; II-9, 10; V-3	Sand layers
Unit 5	
XX-B100 **I-11; II-12**	Sandy silt layers
II-11; VII-5	Silt layers
Unit 6	
XX-A90, XX-A220	Structureless silt
I-11; II-13, 14; VII-6	Silt and sandy silt layers

Interpretation

The disorganized and relatively rounded gravels, the sharp lateral changes of thickness (Figure 2), and the slope of unit 1 (5–9°) reflect deposition on a streamflow-dominated alluvial fan. The slope of U1 is relatively steep over a short distance but decreases eastward, which points to alluvial fan deposition with an eastward flow direction from the gorge to the depression. The flow direction is confirmed by the imbrication of gravels (e.g., Figure 3a) and also by the gravel composition, which reflects source areas in the upstream Huangshui catchment (including the nearby gorge), rather than local sources of Mesozoic to Cenozoic reddish sandstone that fill the Ping'an depression.

The sediments are interpreted as high-magnitude streamflow deposits on the basis of weakly developed imbrication, crude stratification, sub-rounded-rounded gravel shape, and the lack of basal inverse grading. The laterally extensive, sheet-like geometry of Gh beds within U1 is indicative of streamflow on low-relief longitudinal bars [70], and the planar cross-sets developed in Gp facies are interpreted as the foresets of low bars and dunes. The poorly sorted and disorganized gravels in facies Gmm and Gcm are indicative of rapid deposition from a highly concentrated sediment dispersion. High-energy streamflows can transport a mixture of gravel and sand in thin bedload sheets [35] and thick traction carpets [71]. Additionally, the interbedded sandy lenses could represent waning stages of the flows that laid down the gravel, or low-energy secondary overland flows across the gravel surface. The upward decrease in clast size indicates a declining energy towards the top of U1. The overall characteristics of this facies association are suggestive of a high-energy alluvial fan, which aggraded rapidly as a result of highly concentrated streamflows.

Figure 3. The basal stratigraphic unit (U1) consisting of high-energy alluvial environment facies: (**a**) clast-supported planar cross-bedded gravel (Gp) and a sand lens, and sampling point (II-1); the cross-bedded gravels are inclined to WNW, showing an ESE flow direction; (**b**) crudely stratified gravel sheet (Gh) of unit 1, and sampling points (III-1 to III-8); person (circled) for scale 155 cm; (**c**) clast-supported massive gravels (Gcm) and gravel sheet, forming crude fining-upward; person in white for scale 163 cm; (**d**) the grain-size distribution of sample II-1 taken from lenticular sand in section II.

4.1.2. Low-Energy Alluvial Fan (Facies Association 2)

Description

The middle part of the succession includes four units of mica-rich sand and silt (U2–U5, from bottom to top (Figure 2)). The overall thickness of facies association 2 ranges from 8.75 m to 13.6 m, and the top and basal parts contain a <5 cm thick layer of granules and pebbles of a ~1–3 cm size.

Unit 2 (U2) is dominated by pale-brown, silty sand and sand, locally interbedded with reddish-brown silt. The facies are mainly horizontal lamination (Sh) with minor trough cross-stratification (St), ripple cross-lamination (Sr), and planar cross-bedding (Sp) (Figure 4). U2 extends laterally for more than 150 m, and is 0.4 m thick in section II, absent from section III, and 2.8 m thick in section IV. Calcareous cement and a few nodules are present. Coarse sand is locally interbedded with lenses of 1–2 cm-sized pebbles (section VII). The beds show distinct boundaries and include local scours into the underlying gravel beds of U1.

Units 3 (U3), 3.2–5.4 m thick, and 5 (U5), 2.1–3.5 m thick, are characterized by laterally extensive alternate beds of reddish-brown silt and pale yellow-to-orange, micaceous sand and silty sand. The silt predominates (Figure 5), especially in U5. The thickness of individual reddish silt beds varies from 3 to 20 cm, and they typically have abrupt bases and tops, but fine upward locally, with stacked layers occupying slight hollows in places. The beds become thinner upwards in U3 in section III (~15–20 cm to ~3–10 cm). The reddish silt layers are structureless (Fm) or weakly laminated (Fl), and some show ripple structures (Fl) and weak soil formation (Fr). Mottling in some silt layers may be due to iron and manganese coatings or organic matter. The pale sand layers comprise horizontal lamination (Sh), minor low-angle cross-bedding (Sl), ripple cross-lamination (Sr), and structureless sand (Sm).

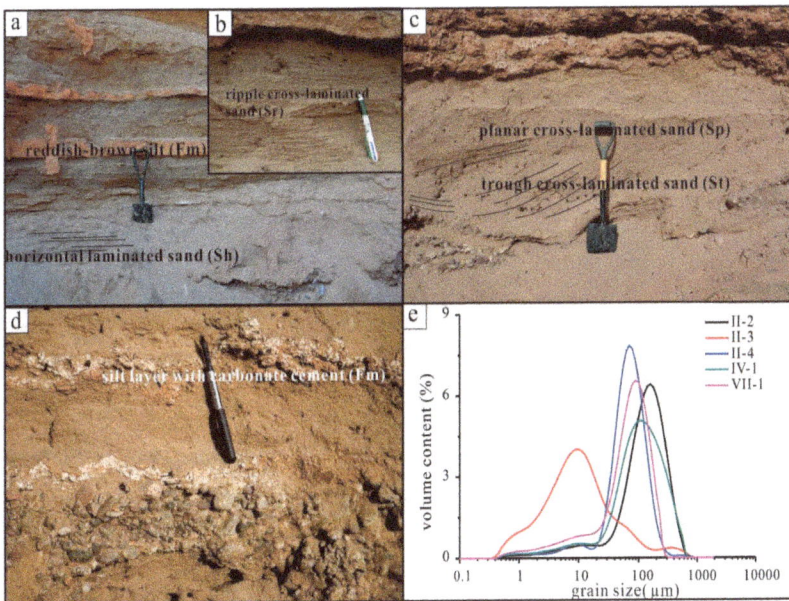

Figure 4. Sedimentary structures observed in unit 2 (U2): (**a**) horizontal laminated sand (Sh) interbedded with reddish-brown silt (Fm); length of the shovel is 58 cm; (**b**) minor ripple cross-laminated sand (Sr); pen length 14 cm; (**c**) planar cross-laminated sand (Sp) overlying trough cross-laminated sand (St); length of the shovel is 75 cm; (**d**) silt layer with carbonate cement; pen length 15 cm; (**e**) the grain-size distributions of samples (II-2 from the trough cross-stratified sand, II-4, IV-1, and VII-1 from the horizontal laminated sand, II-3 from a silt layer with carbonate cement) in U2 (see Figure 5, for the specific sampling points).

A distinctive feature of the sand layers in U3 (and locally in U4 and U5) is the presence of more than 50 connected and disconnected low mounds that were observed laterally over tens of metres in the outcrop and vertically over 7 m (Figures 5–7). The mounds have convex-up forms and contain form-concordant bedding that builds in height from a planar or gently inclined base; many contain ripple cross-lamination. The mounds are mainly symmetrical to slightly asymmetrical in cross-section, and although none were observed in three-dimensions, we infer from the similarity of cross-sectional form of many examples that the mounds are circular to slightly elongate in planform. In one case, a single mound divided into two and later recombined as the sand aggraded. The mounds are bordered by reddish clay layers that curve upward (upturned beds) and pinch out against the mounds. Successive clay layers extend progressively further across the mounds and eventually cover them. Most mounds are isolated (Figure 7a), but a few mounds at the same level are connected by thin sand layers (Figure 6). Root traces were observed in one mound (Figure 5).

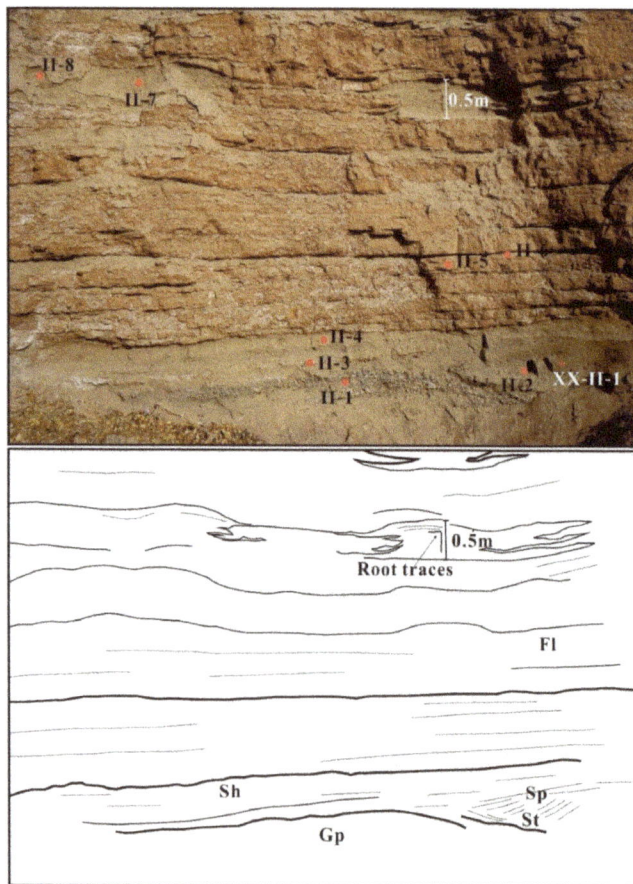

Figure 5. Line drawing of U1 to U3 in section II, showing alternate reddish-brown silt with sand layers that include mounds with local root traces. The red dots represent sampling points and the red triangle represents an OSL sample point. The thickness of the mound at the upper right is 0.5 m.

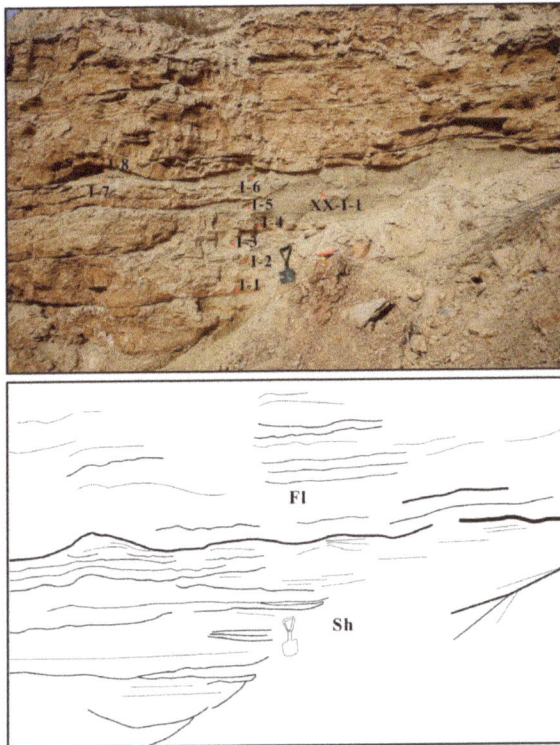

Figure 6. Line drawing of connected sand mounds with planar bases and silt layers that pinch out in the horizontal laminated sand (Sh). Overlying beds are fine laminated silt (Fl) in U3. The red dots represent sampling points. The length of the spade is 58 cm for scale.

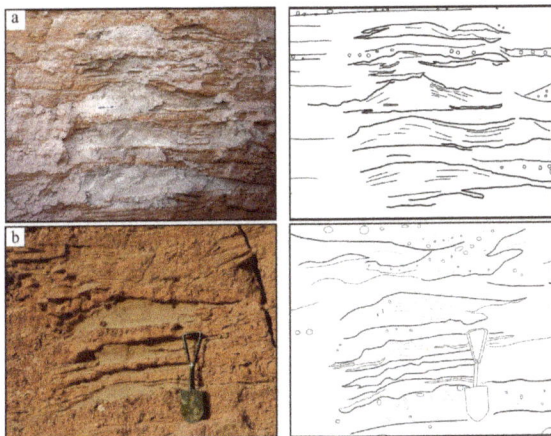

Figure 7. Line drawing of vertically stacked sand mounds interbedded with silt and gravel in U3. The sand bodies are flanked by upturned beds of clay and contain for-concordant stratification. The length of the pen and the spade are 14 cm and 58 cm, respectively.

For thirty-seven mounds that were low in the cliff and readily measured and photographed, the thickness ranges from 0.05 m to 1.14 m (mean 0.24 m), and the width ranges from 0.6 m to 5.14 m (mean 1.48 m), with a positive correlation of thickness and width (Figure 8). The aspect ratio (width/thickness) of the mounds averages 7.5, with most being between 5 and 10 (Figure 8).

At several levels, up to six mounds separated by thin clay layers are stacked near-vertically, to a maximum height of 2.3 m and width of 5.14 m, with a positive correlation between thickness and width (Figures 7 and 8). The width/thickness ratio of the stacked mounds ranges from 0.9 to 2.9. One stack of mounds shows a slight lateral drift in the crest position of individual mounds through time.

Gravel lenses, 10 cm thick and 3 m wide, are commonly present on the eastern side of vertically stacked mounds in section V, lapping in places against them. Some mounds are interbedded with 5–10 cm-thick silt layers that include 1–5 cm pebbles (Figure 7b). In places, U3 contains a higher proportion of dispersed, trough cross-stratified gravel lenses and sand mounds 1.5 m wide and 15 cm thick (Figure 9a,b). Both U3 and U5 show a small amount of white calcareous cement and nodules in silty layers and granules in coarse sand layers (Figure 9c), root traces (Figure 5), and bioturbation and desiccation cracks (Figure 9d).

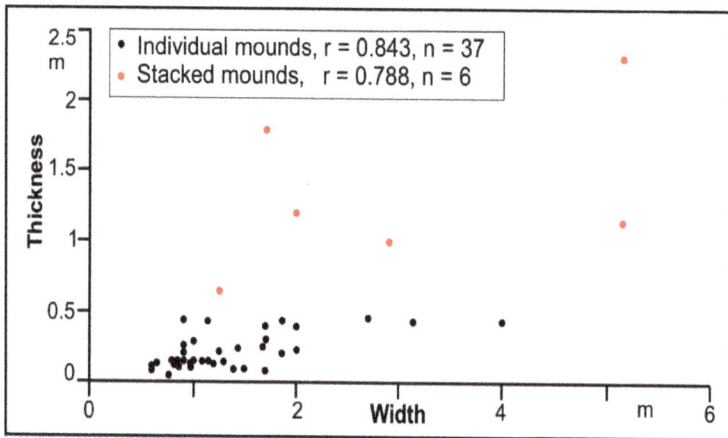

Figure 8. Dimensions of sand mounds, measured from two-dimensional outcrops in the Ping'an pit. The individual mounds include those that are part of stacked mounds.

Unit 4 (U4) is 2.4 to 3.6 m thick and is composed of silty sand and sand with horizontal lamination (Sh), planar cross-bedding (Sp), and low-angle cross lamination (Sl) (Figure 10a,b). Laterally, sand layers thin eastwards, where more and thicker reddish-brown silt layers appear; these layers are absent to the west in section VII. A sand mound and root traces with weak pedogenesis are present in the upper part of U4 in section II (Figure 10b). U4 in sections IV and PA and east of section V is characterized by large sand mounds interbedded with reddish-brown silt layers (Figure 2). In contrast with other sections further east, U4 in section VII contains gravel layers with trough cross-bedding (Gt) and channel forms 1–1.5 m wide and 0.6 m thick that contain clast-supported gravel (Gcm) with clast sizes of 10–15 cm (Figure 10d). Large angular blocks (up to ~1 m) are present locally. The small channel fills become thinner and their elevation in the area of section VII becomes lower when moving eastward.

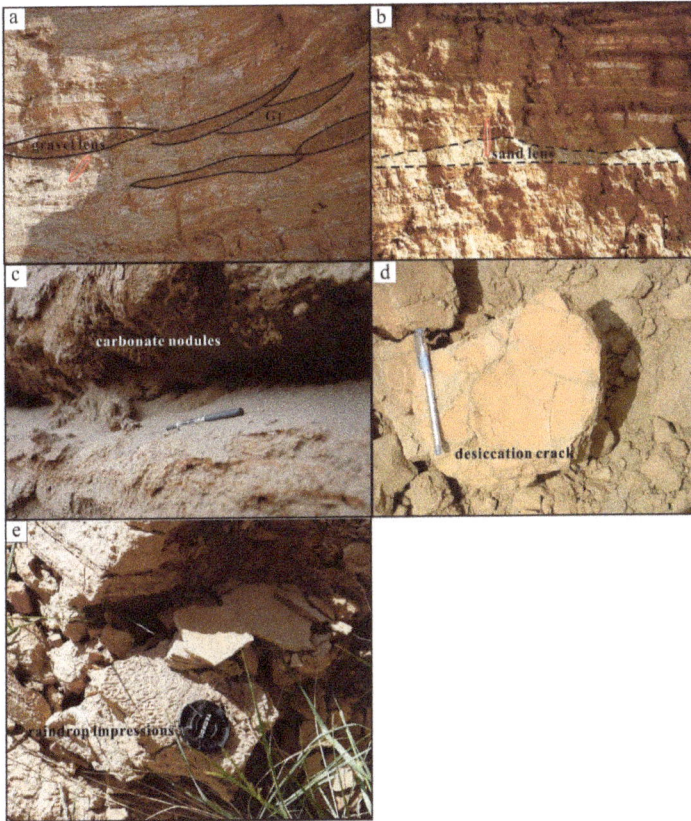

Figure 9. The characteristics of U3 deposits: (**a**) gravel lenses as trough fills; pencil length 17 cm for scale; (**b**) sand lens; pencil length 17 cm for scale; (**c**) carbonate nodules in coarse sand layer in section VII; pen length 14 cm for scale; (**d**) desiccation cracks; pen length 15 cm for scale; (**e**) raindrop impressions; diameter of the lens cap is 5.2 cm for scale.

OSL- and Grain-Size Analyses

In total, 11 OSL samples and 45 grain-size samples were taken from facies association 2 (Table 2), with the positions of some samples shown in Figures 4e and 11a–e.

U2: samples II-4 (from section II) and VII-1 (from section VII) are mainly coarse silt-to-fine sand with a modal value of 80 and 100 μm, respectively, and IV-1 is coarser with a modal size of 126 μm and is rather poorly sorted. Sample II-2 is very fine-to-medium sand (mode of 178 μm). Sample II-3 is poorly sorted silt with a modal value of 10 μm, mixed with a small amount of clay and sand. This silt is also present, although in very small amounts, in all other sediments of U2.

U3: all samples from sand mounds are closely similar (except IV-4), mainly consisting of well-sorted fine sand and coarse silt (major mode of 63–79 μm), minor fine silt (mode of 9–10 μm), and a very small amount of coarse sand (mode of 502 μm). Although sample IV-4 consists of very fine sand and coarse silt with a modal value of 71 μm, it also contains coarse to very coarse sand with a secondary modal value of 1262 μm, which may represent ferruginous or manganese concretions. The grain-size distribution of samples from silt layers is mostly bimodal, with main peaks of 32–45 μm and a secondary modal value of 7–8 μm. Sample III-1 is poorly sorted and only contains the very fine fraction with a modal value of 11μm. Compared with other samples in silt layers, samples IV-2 and VII-2 are relatively coarser grained, containing some sand. Samples I-2, I-3, I-5, and IV-3 from sand layers are very similar to samples taken from the sand mounds in this unit (very fine sand and coarse silt and minor amount of fine silt), but they are slightly finer grained with a modal value of 56–71 μm. Other samples from sections III and VII show grain-size distributions that are similar to equivalent ones in the other sections (Figure 11).

U4: samples I-10 and V-3 are mainly very fine sand and coarse silt with a modal value of 63 μm, mixed with a small amount of fine silt similar to many other samples. Samples II-9 and II-10 are fine-to-medium sand with a modal value of 200 μm, but II-9 shows a secondary modal value (1416 μm) in the very-coarse sand fraction.

U5: sample II-11 from silty laminated beds is almost identical to V-2 (in U3), presenting a bimodal distribution with a modal value of 45 and 8 μm. Sample VII-5, also from a silt bed, is similar to I-4 and VII-4 in U3, but shows a secondary modal value in the sand fraction (448 μm). Sample II-12 from a sand bed is mainly composed of very fine sand and coarse silt (major mode of 56 μm), but contains more fine silt and clay.

Figure 10. Main sedimentary structure observed in unit 4: (**a,b**) low-angle to horizontal laminated sand (Sl, Sh) with local roots; pen length 14 cm; (**c**) sketch of section VII, where the red dots represent sampling points; the dotted lines separating three facies associations; (**d**) minor channels which face ESE, containing pebble lenses in section VII; the person is of the height 155 cm for scale.

Interpretation

The horizontally laminated, trough and planar cross-bedded, and ripple cross-laminated facies (Sh, St, Sp, Sr) in U2 are interpreted as fluvial in origin. The abundance of horizontal lamination (Sh) indicates a common transition to an upper-flow regime. The local presence of cross-beds (St, Sp) may represent channel deposits of two- and three-dimensional dunes within low-energy channels, commonly as minor channels on gravel sheets at the top of unit 1 and within units 2–5 [72]. The ripple cross lamination (Sr), locally present at the bed tops, implies the gradual waning of flow, typically at the end of flood events [54,73].

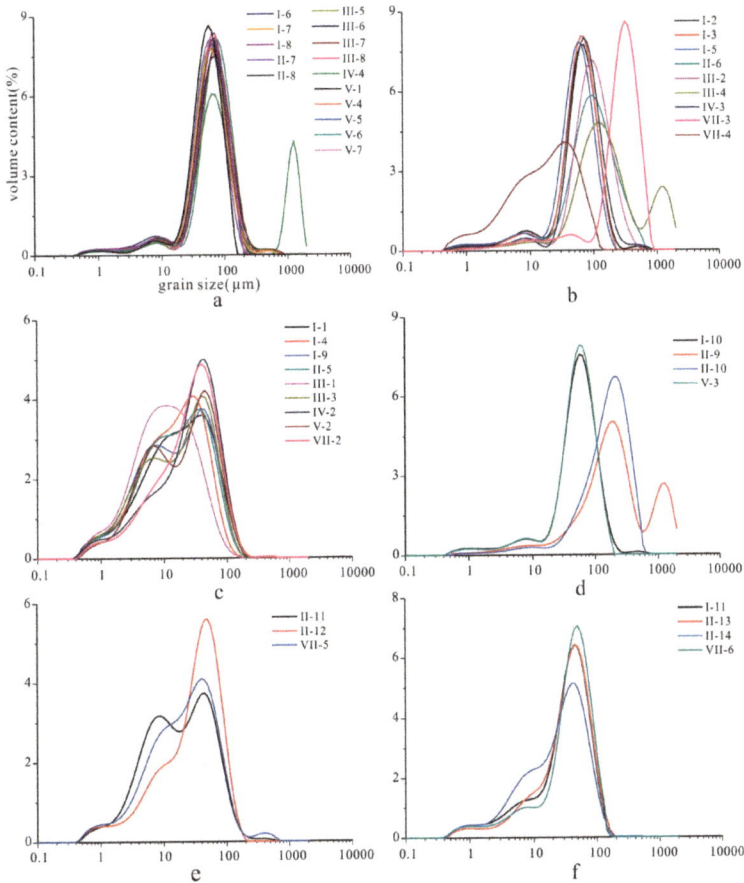

Figure 11. The grain-size distribution of samples taken from U3–U6: samples from (**a**) the sand mounds, (**b**) sand layers, and (**c**) silt layers in U3; from (**d**) sand sediments in U4; from (**e**) U5 (II-11 and VII-5 from silt layers and II-12 from a sand layer, and (**f**) massive silt sediments in U6.

The large differences in the grain-size of samples II-2, 3, and 4 (Figure 4d) in a 0.4 m thick sediment sequence in unit 2, show a variable transport energy. Sample II-2 (St) is coarser and contains more fine-to-medium sand than II-4, IV-1, and VII-1 (Sh), in line with the sedimentary structure, which represents a variable transport energy. Sample II-3 contains higher amounts of fine silt and clay, corresponding with the increase of carbonate cement in structureless silt that underwent weak pedogenesis. Planar to low-angle cross-bedded sand (Sl) points to sandy sheet flows. The presence of sand sheets a half meter to a few meters thick, showing little evidence of channelization and interbedded with silt with carbonate cement, is regarded as a clear indication of ephemeral, flash-flood sedimentation [54].

In U3 and U5, the alternate coarse- and fine-grained beds point to moderate- to low-energy fluvial transport. The interbedding of sand, thin silt layers, and gravel is considered the product of flood-wash over the surface of the alluvial fan or of fluctuations in flow strength during a single flow event [74]. The gravel stringers may represent diffuse channel lag sheets [75]. At some levels, the interspersed, incised gravel lenses in units U3, U4, and U5 are interpreted as shallow gravelly channels migrating on the surface of an alluvial-fan, and the eastward decrease in their scale and abundance in U4 indicates that wedges of coarser channelized sediment spread eastwards across the fan at this level, fining to silt away from the source. The presence of channel forms with bases scoured into sand sheets but coalescing above (Figure 10d), lenses of gravel, and some cross bedding is consistent with deposition from shallowly incised to poorly confined, braided-river channels on the fan surface, similar to those described by Nemec and Postma [75]. Sheets of gravel and sand lenses were the product of the lateral migration of the flow over the fan surface during a discharge event, or of the amalgamation of deposits laid down by several events. The bimodal grain-size distributions result from the mixture of clay with fine to coarse silt (I-9, II-11, III-3, V-2) and coarse silt to fine sand with very coarse sand (II-9, III-4, IV-4), resulting in poor sorting. These distributions reflect considerable hydrodynamic variation during deposition.

The beds of structureless and laminated silt (Fm, Fl) mark suspended-load deposition from low-velocity sheet flows. The presence of low-relief hollows filled with silt beds indicates that some flows covered a slightly irregular landscape, but there is little indication of strong scouring during these events. The abrupt silt beds, which locally fine upward, formed in single flow events, and their considerable thickness (typically 5–15 cm) suggests that large amounts of fine sediment were available during floods and that the successions aggraded rapidly. In places, reworking due to bioturbation and weak pedogenic processes contributed to the structureless nature and mottling of the Fm silt [74].

Sample VII-5, from fine laminated silt (U5), is similar to I-4, II-5, and VII-4 in U3, but the presence of coarse sand, analogous to samples I-2 and IV-3 in U3, indicates an occasional flood event. Samples II-6 (U3), III-2 (U2), and VII-1 (U2) are very-fine sand, which suggests streamflow or a low-energy sheet flow.

Considering the extent of horizontally laminated sand over tens of meters and the grain-size distributions of samples II-4, IV-1, and VII-1, we attribute the U2 sand sheets to moderate-energy, shallow sheet flows, capable of forming plane lamination in the upper flow regime. The predominant reddish mud layers represent lower energy flows from which fine suspended sediment settled out, and the occasional root traces and carbonate nodules indicate phases of stability and weak post-depositional pedogenic modification as temporary interruptions in deposition [76]. The observed poorly developed soil and desiccation cracks may also indicate the temporary sub-aerial exposure of shallow channels on the alluvial fan [77]. The small fine clay fraction (<1 μm) is probably due to soil formation in accordance with the presence of root traces and carbonate nodules. Sample II-3 from a silt bed with carbonate nodules contains a relatively high fraction of <1 μm clay, which supports this hypothesis.

The mounds have been previously attributed to several processes, including formation by termites [78]. However, the mounds in several units at Ping'an are interpreted as vegetation-induced sedimentary structures, formed as upturned beds with form-concordant stratification around in situ plants that decreased current velocity and promoted deposition [79]. Upright plants were not noted in association with the mounds, which may reflect decay of the vegetation and/or the presence of vegetation outside the plane of the cross-section. However, roots were observed in one instance (Figure 5). The positive correlation of mound width and thickness (Figure 8) indicates that the mounding process typically formed wider mounds as their height increased, as expected for the accumulation of soft sand. The mounds were fully developed as elevated bedforms prior to mud deposition, as indicated by the onlap of mud layers against the mounds. There is little indication of the erosion of sand where the mud drapes the mounds, suggesting that the clay settled from gently flowing or stagnant water. There is also little indication of deformation of the soft sand in the mounds, probably due to subsequent induration of the clay during sub-aerial exposure.

In general, the grain-size distributions of all samples from sand mounds (except IV-4) are almost identical, which implies a relatively similar transport energy. On the basis of the non-erosional bases of the stacked mounds and their closely similar grain-size distributions, we interpret the mounds as formed during aeolian deposition. The main modal value of 56~71 μm (see sediment type 1.a of Vandenberghe [80]) may point to saltation, which is analogous to the typical component of loess deposits overlying river terraces along the Huangshui River [51,81]. The secondary mode of 10 μm resembles that of an original fine-grained loess deposit (type 1c.2 of Vandenberghe [80]). The two fractions may occur in different relative percentages. Most sediments in the mounds (except IV-4) and some sediments in other sand layers (for example, I-2, 3, 5, 10, IV-3, V-3) show a small content of fine-silt components. Some samples (II-1, 9, III-4, IV-1, 4) contain a fraction of medium-coarse sand, interpreted as fluvial in origin.

Two other lines of evidence support an aeolian origin for the mounds. The planar bases of the mounds and the lack of scouring contrast with the abundance of scours around vegetation, which is typical of fluvial settings with high-velocity flow [79]. However, in other settings, scours are commonly associated with larger upright trunks, and smaller and less rigid plants would have been less likely to promote scour. The predominantly symmetric form in the cross-section implies mounding under gentle flow conditions, again in contrast to many fluvial examples elsewhere.

The presence in a few cases (Figure 6) of mounds of a similar height and quasi-regular spacing, connected by thin sand layers, might be interpreted as suites of mega-ripples. However, samples suggest a grain-size distribution compatible with an aeolian origin. Additionally, the examples lack unidirectional flow indicators such as planar cross-sets, and they are interpreted as mounds formed around closely spaced plants where enough sand was available to connect the mounds.

We interpret facies association 2 as the product of several processes. Higher energy flows laid down gravel and sand in shallowly incised and extensive braided-channel systems, especially in the western part of the outcrop belt. Associated sheet floods laid down thin sand and silt layers, which were locally affected by desiccation and incipient soil formation. The sand was redistributed by aeolian processes to form individual and stacked mounds around standing vegetation. In combination, these deposits represent an alluvial fan under moderate-energy, but predominantly low-energy, conditions [76].

4.1.3. Floodplain (Facies Association 3)

Description

The upper part of the succession in unit 6 (U6) is characterized by abundant layers of structureless sandy silt and silt (Fm) interbedded with thin reddish-brown fine laminated silt (Fl), forming a broadly fining-upward succession (Figure 12). Carbonate nodules occur locally within facies Fl. The thickness of each coarser-silt layer ranges from 1.7 m to 7 m and the thin finer-silt layers are up to 0.5 m thick. U6 is well exposed in section V, with a maximum thickness of 8.7 m.

Four samples (Table 2) are very fine sand and coarse silt with a modal value of 45–50 μm and small amounts of fine silt and clay (Figure 11f). II-14 is slightly finer grained and contains more fine silt and less very fine-grained sand than other samples. Two OSL samples were taken (Table 2).

Interpretation

Facies Fm reflects deposition from low-energy flows during waning-flow and channel abandonment on the alluvial fan [54,74]. The fine grain size and lack of distinct channel structures argue for floodplain deposits, which may have developed in an inactive part of the alluvial system that only received sediment during overbank floods. The structureless form and carbonate nodules suggest that the sediments were subject to weak pedogenic processes [72] on a floodplain that aggraded rapidly [82]. The grain-size distributions (Figure 11f) point to sediment that settled after the main flooding event or at the terminal calm period of the flooding event in pools on the floodplain. The fine silt may be interpreted as aeolian in origin but later reworked by the river and laid down in the standing water of pools on the floodplain [80,83]. The integration of all these characteristics points to floodplain processes in a low-energy fluvial system.

Figure 12. Sedimentary features of U6 in section IV and V, showing detailed sampling points. The dashed lines are the boundaries of the sediment units.

4.2. Dating Results

Table 3 shows all analytical data and quartz SAR-OSL ages and K-feldspar pIRIR$_{290}$ ages. The dates using each mineral show a relatively close grouping, without an apparent upward age trend (Table 3, Figure 13c).

For quartz SAR-OSL ages, the samples (XX-II-1, XX-C600, and XX-C490) in unit 2 were dated to around 78 ka, 86 ka, and 81 ka, respectively (Table 3). Samples XX-C200, XX-II-2, and XX-V-1 in unit 3, were dated to around 83 ka, 74 ka, and 83 ka, respectively (Table 3). Samples XX-II-3 from planar cross-bedded sand in unit 4 and XX-B100 from a sand layer in unit 5 were dated to around 89 ka and 87 ka, respectively (Table 3). Samples XX-A90 and XX-A220 in unit 6 were dated to around 72 ka and 92 ka, respectively (Table 3). The quartz SAR-OSL ages may be underestimated because the credible upper limit of an equivalent dose is ~200 Gy [84–86]. All samples calculated from quartz-D$_e$ values are more than 200 Gy (Table 3), and therefore, exercising prudence, these should be regarded as minimum ages. In summary, the quartz SAR-OSL ages demonstrate that these samples are older than 72 ka.

Table 3. Summary of U, Th, and K concentrations, estimates of saturated water content (W.C.) with the absolutely uncertainty of 7%, calculated OSL and pIRIR dose rates and equivalent doses (D_e). (N) denotes the number of aliquots used for the D_e data. Q: quartz, FK: K-feldspar.

Lab Code	Sample Code	Sediment Units	Depth (m)	U (ppm)	Th (ppm)	K (%)	W.C. (%)	Q-Dose Rate (Gy/ka)	FK-Dose Rate (Gy/ka)	Q-D_e (Gy)	FK-D_e (Gy)	Aliquots (N)	Q-Age (ka)	FK-Age (ka)
Nju-2579	XX-A90	U6	1.90	3.18 ± 0.04	11.70 ± 0.03	1.76 ± 0.03	16	3.15 ± 0.15	3.47 ± 0.15	226 ± 13	454 ± 9	6 [1]/6 [2]	72 ± 5	131 ± 7
Nju-2580	XX-A220	U6	3.20	3.02 ± 0.04	11.40 ± 0.03	1.68 ± 0.03	20	2.91 ± 0.13	3.23 ± 0.14	267 ± 9	435 ± 25	9 [1]/6 [2]	92 ± 6	135 ± 10
Nju-2582	XX-B100	U5	5.60	3.04 ± 0.04	11.30 ± 0.03	1.86 ± 0.03	20	2.97 ± 0.14		258 ± 11		8 [1]	87 ± 6	
Nju-2703	XX-II-3	U4	10.50	2.26 ± 0.04	9.06 ± 0.03	1.55 ± 0.03	17	2.46 ± 0.12	2.79 ± 0.12	218 ± 9	373 ± 11	14 [1]/7 [2]	89 ± 6	134 ± 8
Nju-2700	XX-I-1	U3	7.45	2.80 ± 0.04	10.30 ± 0.03	1.66 ± 0.03	19		3.03 ± 0.13		433 ± 8	10 [2]		143 ± 7
Nju-2583	XX-C200	U3	14.00	3.06 ± 0.04	11.70 ± 0.03	1.61 ± 0.03	17	2.78 ± 0.14		231 ± 8		16 [1]	83 ± 5	
Nju-2702	XX-II-2	U3	14.30	3.02 ± 0.04	11.00 ± 0.03	1.71 ± 0.03	21	2.72 ± 0.13	3.06 ± 0.13	203 ± 7	401 ± 8	16 [1]/4 [2]	74 ± 4	131 ± 7
Nju-2704	XX-V-1	U3	15.60	3.08 ± 0.04	11.40 ± 0.03	1.65 ± 0.03	20	2.72 ± 0.13	3.06 ± 0.13	225 ± 12	406 ± 9	14 [1]/6 [2]	83 ± 6	133 ± 7
Nju-2705	XX-V-2	U3	18.40	2.53 ± 0.04	10.20 ± 0.03	1.75 ± 0.03	20	2.97 ± 0.13	2.97 ± 0.13		416 ± 13	6 [2]		140 ± 8
Nju-2706	XX-VI-1	U2	14.00	2.41 ± 0.04	10.70 ± 0.03	1.50 ± 0.03	15	2.90 ± 0.13			458 ± 16	6 [2]		158 ± 10
Nju-2585	XX-C490	U2	16.90	2.77 ± 0.04	10.70 ± 0.03	1.76 ± 0.03	15	2.86 ± 0.14		232 ± 12		16 [1]	81 ± 6	
Nju-2586	XX-C600	U2	18.00	2.15 ± 0.04	9.95 ± 0.03	1.55 ± 0.03	14	2.53 ± 0.13		218 ± 9		8 [1]	86 ± 6	
Nju-2701	XX-II-1	U2	18.00	2.16 ± 0.04	9.50 ± 0.03	1.67 ± 0.04	15	2.64 ± 0.13	2.97 ± 0.14	204 ± 11	395 ± 6	13 [1]/6 [2]	78 ± 6	133 ± 7

Note: [1] Number of aliquots used in quartz SAR protocol; [2] Number of aliquots used in pIRIR$_{290}$ protocol.

For K-feldspar pIRIR$_{290}$ ages, the samples XX-II-1, XX-V-1, XX-II-2, XX-II-3, XX-A220, and XX-A90 were dated to 133 ± 7 ka, 133 ± 7 ka, 131 ± 7 ka, 134 ± 8 ka, 135 ± 10 ka, and 131 ± 7 ka, respectively (Table 3), which suggests rapid aggradation. Samples XX-I-1 and XX-V-2 from the sand mounds in unit 3 were dated to 143 ± 7 ka and 140 ± 8 ka, respectively, and sample XX-VI-1 from horizontal laminated sand in unit 2 was dated to 158 ± 10 ka (Table 3). It should be mentioned that, when a high-temperature post-IR IRSL (pIRIR) was carried out at 290 °C, less-fading or a non-fading signal may be supposed, and thus accurate fading corrections are not necessary, especially for older samples [58,62]. In summary, the K-feldspar pIRIR$_{290}$ ages appear in the range of ~168 to ~124 ka.

In assessing dates that were derived using protocols for quartz and feldspar, we consider the feldspar pIRIR dates to be reliable, and they form the basis for later interpretation. The quartz SAR dates with D$_e$ values in excess of 200 Gy provide minimum age for the samples, and they are helpful in constraining the age of the Ping'an sections.

5. Discussion

5.1. Origin of Stacked Sand Mounds, and the Interaction of Aeolian and Alluvial Process

Standing vegetation is able to slow sediment transport by hindering the flow in channels and overbank areas [87–91] and similarly decreasing the wind velocity. Downflow movement of sediment combined with in situ accumulation of organic material forms mounds around plants [79], and the local presence of root traces in the sections (Figures 5 and 10b) supports a link to vegetation. The nature of the nucleating vegetation is unclear, but the plants were probably thin shrubs or clumps of grass and herbaceous vegetation, based on the lack of several features (scour hollows with centroclinal cross strata, decay-related hollows with downturned beds, and stump casts [79,92]), which are commonly associated with thicker and stronger plants. This interpretation of the kind of vegetation matches palynological results at nearby sites on a terrace that might correlate with the studied sand-pit, around 10 km to the south, which shows herbaceous plants dominated by *Artemisia* and *Chenopodiaceae* growing in valleys [93].

Grain-size analysis provides important evidence for discriminating between fluvial and aeolian processes as agents responsible for the formation of the mounds. The grain-size distributions of all samples from different mounds and from sediments at different locations in a single mound are very similar, containing significant amounts of well-sorted very fine sand and coarse silt with modal values of 63–71 μm (Figure 11a). The low modal values and narrow grain-size range support accumulation by aeolian processes. Supporting sedimentological evidence for an aeolian origin includes the planar base of the mounds (Figure 7) and their mainly symmetric form.

The identification of vegetation-induced structures in deposits that largely lack preserved vegetation has important implications for the depositional setting, as does the interpretation of aeolian activity in their formation. The stacking of sandy mounds onlapped by finer sediment suggests that accumulation of water-laid deposits alternated with periods of aeolian reworking. Raindrop imprints and desiccation cracks (Figure 9d,e) also indicate temporary subaerial exposure on the alluvial fan. Sufficient time was available between inundation events for vegetation to become established, and the abundance of mounds suggests that the fan was well vegetated at times, although the plants were probably modest in scale. Stacked mounds with more than 2 m of sediment in aggregate suggest a rapid aggradation of sediment during the life of a single plant or group of plants on the fan.

During dry periods or seasons, the wind reworked the exposed alluvial material transported onto the fan. Under the effect of plants, the wind slowed and the material was redeposited, forming mounds as shrub-coppice dunes. The mounds were dated at a roughly similar age of ~143–131 ka. The mounds thus developed from aeolian processes over a short period, alternating with alluvial deposition, at the transition from marine isotope stage (MIS) 6 to MIS 5.

Most mound sediments are composed of two major components, fine silt (mode at 7~10 μm) and very fine sand or coarse silt (mode at 56~71 μm). These may reflect two typical aeolian components [80] that were reworked but preserve their original grain-size [83]. The finest silt component in all samples can be viewed as dust transported in high-level suspension clouds, as in westerlies over a long distance [94–96]. The coarse component can be regarded as derived from low-level suspension during spring-summer dust storms from proximal sources, for instance, dry interfluves or river terraces [5,26,83,97]. Combining the evidence from sedimentary structures and grain size, the deposits record the interaction of aeolian and alluvial processes on the fan, with wind action reworking local alluvial sediment and additions from distantly derived suspension clouds.

5.2. Alluvial-Fan Process and Links to Autogenic Models and Exterior Controls (Climate and Tectonics)

The upward-fining succession at Ping'an (Figure 13b) from a high-energy (gravel sheet of unit 1) to a low-energy facies association (sands and muds of units 2–5, with minor gravel), capped by a floodplain facies association (unit 6 muds), is attributed to deposition on a stream-dominated alluvial fan. Eastward fining in parts of the succession, along with paleoflow information from imbrication and the orientation of channel bodies, indicates sediment transport from the exit point of the Xiao gorge into the Ping'an depression. The complex intercalation of gravel, sand, and mud bodies; the rapid lateral fining; and the steep upper contact of unit 1 accord well with deposition on an alluvial fan, rather than deposition on the alluvial plain of the Huangshui river. In view of the transport direction, the fan succession is interpreted as a deposit of the mainstem Huangshui river, rather than a tributary valley. The units are conformable, and the succession is most simply interpreted as the product of a single alluvial-fan lobe (Figure 13a).

Many features of the Ping'an succession may be explained by an autogenic model for fan evolution in a situation of rapid sediment supply and buildup. Channel patterns in alluvial settings are mainly dependent on water discharge, sediment load, and basin slope [54], and these parameters change intrinsically as fans develop.

The succession is interpreted here in accord with this model (Figure 13a). As lobe 1 began to develop, the relatively steep slope of the alluvial fan led to high competency and deposition from the upper flow regime, laying down clast-supported, planar cross-stratified, and crudely horizontally stratified gravel of unit 1 (facies association 1). As accumulation proceeded, the slope gradually decreased. As the lobe built up vertically and extended downflow through time, the river exhibited a lower energy and transported finer sediment in repeated flood pulses, with occasional gravel from flash flooding (facies association 2). Streamflow was poorly confined on the fan, causing a decrease in capacity and velocity, and lenticular gravels were deposited in minor, shallow channels of a braided style. Rapid aggradation of the sediment sequence in units 2 to 5, as shown by the mound accumulations and thick fine-grained flood beds, was temporarily interrupted by weak pedogenesis of the deposits. As the elevation of the lobe increased, only the highest magnitude floods would have deposited sediment on the top of the succession. River incision and diversion would have transferred deposition to a successor lobe 2 (Figure 13a), leaving lobe 1 inactive.

This simplified model is subject to several caveats. The abrupt change from unit 1 gravel to finer sediment implies a sudden change in transport conditions. River incision and lobe abandonment may have commenced after the deposition of unit 5 (facies association 2), and the floodplain succession of facies association 3 may reflect flooding that topped lobe 1 as lobe 2 or lobe 3 built in elevation. Scouring along some surfaces (e.g., unit 1/2 contact), upward fining in several units, and the pronounced lateral facies change in unit 4 indicate that accumulation involved a series of events, and it is possible that the succession includes deposits of several superimposed or overlapping fans. These influenced the supply of water and sediment to the fan and were superimposed on the progressive growth of the fan lobe. The topmost part of unit 6 has been eroded, and it is not known whether a more mature paleosol or other evidence of a hiatus was originally present.

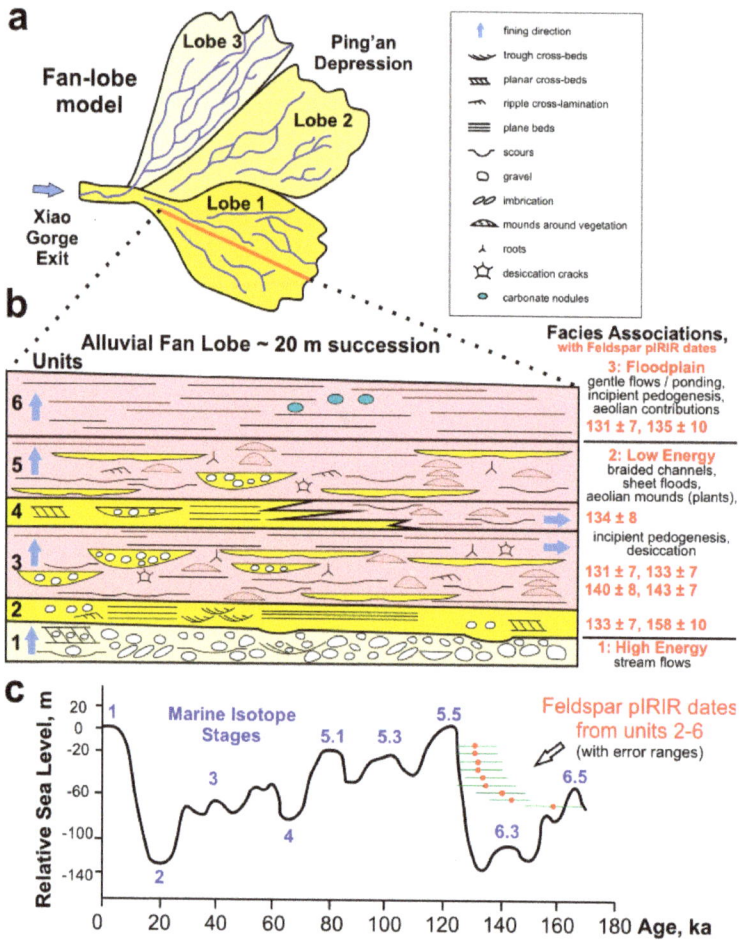

Figure 13. Model for alluvial fan development at Ping'an: (**a**) framework of sediment supply from the Xiao gorge exit, rapid accumulation, and intrinsic fan evolution as a result of lobe elevation and switching; (**b**) summary of the fan succession; and (**c**) correlation of OSL dates with marine isotope stages [98].

The studied sediment sequence started with high-energy alluvial gravels of unit 1, the top of which, in the western part of the pit, is 26 m above the flood plain of the Huangshui River. This unit may relate to terraces at 26 m above the floodplain upstream, at the west end of the Xiao gorge in the Xining basin [45,48], and downstream of the pit in the Ping'an depression [99]. Before the studied fan sequence was deposited, several higher terraces formed in the Ping'an depression [99]. The presence of these terraces indicates that the Xiao gorge had been cut and depressions upstream and downstream of the gorge (Xining and Ping'an respectively, Figure 1a) had been connected by Huangshui River before the fan began to form. These lines of evidence suggest that the start of alluvial fan deposition was a response to the uplift of Xiao gorge as a result of reactivation of the fault at the boundary of the gorge (Figure 1b), rather than a response to the cutting of the gorge. However, the tectonic evidence needs further investigation.

Climate change is one of the major factors controlling depositional processes and played a prominent role in the formation of Quaternary alluvial fans [31–34], affecting the sediment supply and (or) flow energy. The paleoclimate during Ping'an fan deposition was relatively arid, as shown by several features: episodic deposition with periods of exposure and desiccation, frequent aeolian activity in the reworking of fluvial sediment, and the presence of modest vegetation, which nucleated aeolian sediments around plants that were probably shrubs and herbs. Although a few carbonate nodules are present, they do not form prominent petrocalcic horzions that might indicate more humid conditions.

The OSL ages provide some constraint on the timing of deposition, in relation to climatic rhythms. Based on the principles laid out above, the quartz OSL dates are underestimated, and the feldspar pIRIR$_{290}$ dates are considered to provide a more reliable age. The dates are closely grouped without a clear upward stratigraphic trend (Figure 13c), and this may imply rapid deposition of the fan succession. As shown in Figure 13c, the feldspar dates with error ranges bracket Marine Isotope Stage (MIS) 6 and the transition to MIS 5. A reasonable interpretation of the dates is that the fan succession reflects rapid aggradation during the penultimate glacial (MIS 6) (units 1–5) and/or during the last interglacial (MIS 5), slowing down during stage 5 (unit 6). The fine-grained topmost floodplain beds may have been deposited at the beginning of the interglacial period (MIS 5). We speculate that during the next warm-cold transition (MIS 5 to 4), the temperature dropped and, with decreasing vegetation, lower evapotranspiration, and higher amounts of surface runoff [16,19], the river started to incise, and could no longer reach the fan surface. Thus, although some climatic response cannot be excluded, altogether, this succession is inferred to represent intrinsic fan evolution.

6. Conclusions

Based on sedimentary architecture, structures, and grain-size analysis, an alluvial-fan succession at Ping'an in the Xining basin shows the distinctive evolution of sediment processes and the environment. Three facies associations are recognized in upward succession: (1) a high-energy association with matrix- to clast-supported, disorganized, and organized gravel sheets, interpreted as highly concentrated streamflow deposits; (2) a low-energy association with laminated sand and interbedded laminated silt and sand, with vertically stacked sand mounds and dispersed gravel lenses, interpreted as braided-stream and sheet-flood deposits; and (3) a floodplain association composed of massive silt with incipient palaeosols, interpreted as the result of settling from floodwaters and weak pedogenesis. Vertically stacked mounds with form-concordant bedding are vegetation-induced sedimentary structures formed around in situ vegetation. Their grain-size distributions indicate that the mounds were the result of aeolian reworking of fluvial sediment.

Sediments were supplied from the west, from the Xiao gorge on the Huangshui River, forming a streamflow-dominated alluvial-fan system where the valley was locally enlarged. Based on sedimentary evidence and tightly grouped OSL ages, the Ping'an sediment sequence records rapid aggradation, with large volumes of fine detritus that were reworked by aeolian processes from fluvial sources. The succession yielded OSL dates of ~168–124 ka based on feldspar pIRIR$_{290}$ protocol, and the site was located in the semi-arid NETP during the transition from MIS 6 to 5, from the penultimate glacial to the last interglacial.

Author Contributions: Conceptualization, X.W. and J.V.; Data curation, S.Y. and M.R.G.; Formal analysis, J.V.; Funding acquisition, X.W.; Investigation, L.G., X.W., J.V., and M.R.G.; Methodology, S.Y.; Resources, H.L.; Supervision, S.Y., J.V., M.R.G., and H.L.; Validation, J.V., M.R.G., and H.L.; Visualization, L.G., X.W., and M.R.G.; Writing—original draft, L.G. and X.W.; Writing—review & editing, X.W., J.V., and M.R.G.

Funding: This research is supported by the National Natural Science Foundation of China (41522101), the National Key Research and Development Program (2016YFA0600500), the Joint KNAW-China Exchange Program (530-5CDP07), and the Discovery Grant of the Natural Sciences and Engineering Research Council of Canada (No. 03877).

Acknowledgments: Authors are grateful to FLAG members for discussion during the field for the FLAG conference in 2017. Yang Yu, Quanxu Hu, Zhengchen Li, and Bingling Wang are thanked for their assistance during the fieldwork. MRG is grateful for funding from a Discovery Grant of the Natural Sciences and Engineering Research Council of Canada. The authors thank three anonymous reviewers and the editor (David R. Bridgland) for their constructive suggestions.

Conflicts of Interest: The authors declare no conflicts of interest.

References

1. Mather, A.E.; Stokes, M.; Whitfield, E. River terraces and alluvial fans: The case for an integrated Quaternary fluvial archive. *Quat. Sci. Rev.* **2017**, *166*, 74–90. [CrossRef]
2. An, Z.; Kukla, G.; Porter, S.C.; Xiao, J. Late Quaternary dust flow on the Chinese loess plateau. *Catena* **1991**, *18*, 125–132. [CrossRef]
3. Ding, Z.; Liu, T.; Rutter, N.W.; Yu, Z.; Guo, Z.; Zhu, R. Ice-volume forcing of East Asian winter monsoon variations in the past 800,000 years. *Quat. Res.* **1995**, *44*, 149–159. [CrossRef]
4. Long, H.; Shen, J.; Chen, J.; Tsukamoto, S.; Yang, L.; Cheng, H.; Frechen, M. Holocene moisture variations over the semiarid-arid central Asia revealed by a comprehensive sand-dune record from the central Tian Shan, NW China. *Quat. Sci. Rev.* **2017**, *174*, 13–32. [CrossRef]
5. Prins, M.A.; Vriend, M.; Nugteren, G.; Vandenberghe, J.; Lu, H.; Zheng, H.; Jan Weltje, G. Late Quaternary aeolian dust input variability on the Chinese Loess Plateau: Inferences from unmixing of loess grain-size records. *Quat. Sci. Rev.* **2007**, *26*, 230–242. [CrossRef]
6. Sun, Y.; Chen, J.; Clemens, S.C.; Liu, Q.; Ji, J.; Tada, R. East Asian monsoon variability over the last seven glacial cycles recorded by a loess sequence from the northwestern Chinese Loess Plateau. *Geochem. Geophs. Geosyst.* **2006**, *7*. [CrossRef]
7. Wu, Y.; Qiu, S.; Fu, S.; Rao, Z.; Zhu, Z. Pleistocene climate change inferred from multi-proxy analyses of a loess-paleosol sequence in China. *J. Asian Earth Sci.* **2018**, *154*, 428–434. [CrossRef]
8. Zhang, W.; Lu, H.; Li, C.; Dodson, J.; Meng, X. Pollen preservation and its potential influence on paleoenvironmental reconstruction in Chinese loess deposits. *Rev. Palaeobot. Palynol.* **2017**, *240*, 1–10. [CrossRef]
9. Long, H.; Shen, J.; Wang, Y.; Gao, L.; Frechen, M. High resolution OSL dating of a late Quaternary sequence from Xingkai Lake (NE Asia): Chronological challenge of the "MIS3a Mega-paleolake" hypothesis in China. *Earth Planet. Sci. Lett.* **2015**, *428*, 281–292. [CrossRef]
10. Xu, H.; Ai, L.; Tan, L.; An, Z. Stable isotopes in bulk carbonates and organic matter in recent sediments of Lake Qinghai and their climatic implications. *Chem. Geol.* **2006**, *235*, 262–275. [CrossRef]
11. Xiao, J.; Chang, Z.; Si, B.; Qin, X.; Itoh, S.; Lomtatidze, Z. Partitioning of the grain-size components of Dali Lake core sediments: Evidence for lake-level changes during the Holocene. *J. Paleolimnol.* **2009**, *42*, 249–260. [CrossRef]
12. Wang, R.; Zhang, Y.; Wünnemann, B.; Biskaborn, B.K.; Yin, H.; Xia, F.; Zhou, L.; Diekmann, B. Linkages between Quaternary climate change and sedimentary processes in Hala Lake, northern Tibetan Plateau, China. *J. Asian Earth Sci.* **2015**, *107*, 140–150. [CrossRef]
13. Liu, X.; Vandenberghe, J.; An, Z.; Li, Y.; Jin, Z.; Dong, J.; Sun, Y. Grain size of Lake Qinghai sediments: Implications for riverine input and Holocene monsoon variability. *Palaeogeogr. Palaeoclimatol. Palaeoecol.* **2016**, *449*, 41–51. [CrossRef]
14. Wang, X.; Vandenberghe, D.; Yi, S.; Vandenberghe, J.; Lu, H.; Balen, R.V. Late Quaternary paleoclimatic and geomorphological evolution at the interface between the Menyuan basin and the Qilian Mountains, northeastern Tibetan Plateau. *Quat. Res.* **2013**, *80*, 534–544. [CrossRef]
15. Li, G.; Dong, G.; Wen, L.; Chen, F. Overbank flooding and human occupation of the Shalongka site in the Upper Yellow River Valley, northeast Tibet Plateau in relation to climate change since the last deglaciation. *Quat. Res.* **2014**, *82*, 354–365. [CrossRef]
16. Wang, X.; Vandenberghe, J.; Yi, S.; Van Balen, R.; Lu, H. Climate-dependent fluvial architecture and processes on a suborbital timescale in areas of rapid tectonic uplift: An example from the NE Tibetan Plateau. *Glob. Planet. Chang.* **2015**, *133*, 318–329. [CrossRef]
17. Harvey, A.M.; Mather, A.E.; Stokes, M. Alluvial fans: Geomorphology, sedimentology, dynamics—Introduction. A review of alluvial-fan research. In *Alluvial Fans: Geomorphology, Sedimentology, Dynamics*; Harvey, A.M., Mather, A.E., Stokes, M., Eds.; Geological Society: London, UK, 2005; pp. 1–7; ISBN 1-86239-189-0.
18. Vandenberghe, J. Timescales, climate and river development. *Quat. Sci. Rev.* **1995**, *14*, 631–638. [CrossRef]
19. Vandenberghe, J. Climate forcing of fluvial system development: An evolution of ideas. *Quat. Sci. Rev.* **2003**, *22*, 2053–2060. [CrossRef]
20. Bridgland, D.R. The record from British Quaternary river systems within the context of global fluvial archives. *J. Quat. Sci.* **2010**, *25*, 433–446. [CrossRef]

21. Vandenberghe, J. River terraces as a response to climatic forcing: Formation processes, sedimentary characteristics and sites for human occupation. *Quat. Int.* **2015**, *370*, 3–11. [CrossRef]

22. Hollands, C.B.; Nanson, G.C.; Jones, B.G.; Bristow, C.S.; Price, D.M.; Pietsch, T.J. Aeolian–fluvial interaction: Evidence for Late Quaternary channel change and wind-rift linear dune formation in the northwestern Simpson Desert, Australia. *Quat. Sci. Rev.* **2006**, *25*, 142–162. [CrossRef]

23. Belnap, J.; Munson, S.M.; Field, J.P. Aeolian and fluvial processes in dryland regions: The need for integrated studies. *Ecohydrology* **2011**, *4*, 615–622. [CrossRef]

24. Williams, M. Interactions between fluvial and eolian geomorphic systems and processes: Examples from the Sahara and Australia. *Catena* **2015**, *134*, 4–13. [CrossRef]

25. Han, G.; Zhang, G.; You, L.; Wang, Y.; Yang, L.; Yang, J.; Zhou, L.; Yuan, M.; Zou, X.; Cheng, H. Deflated rims along the Xiangshui River on the Xiliaohe Plain, Northeast China: A case of active fluvial–aeolian interactions. *Geomorphology* **2016**, *257*, 47–56. [CrossRef]

26. Wang, X.; Ma, J.; Yi, S.; Vandenberghe, J.; Dai, Y.; Lu, H. Interaction of fluvial and aeolian sedimentation processes and response to climate change since the last glacial in a semi-arid environment along the Yellow River. *Quat. Res.* **2018**. [CrossRef]

27. Armitage, J.J.; Duller, R.A.; Whittaker, A.C.; Allen, P.A. Transformation of tectonic and climatic signals from source to sedimentary archive. *Nat. Geosci.* **2011**, *4*, 231–235. [CrossRef]

28. Stokes, M.; Cunha, P.P.; Martins, A.A. Techniques for analysing Late Cenozoic river terrace sequences. *Geomorphology* **2012**, *165–166*, 1–6. [CrossRef]

29. Allen, P.A.; Armitage, J.J.; Carter, A.; Duller, R.A.; Michael, N.A.; Sinclair, H.D.; Whitchurch, A.L.; Whittaker, A.C. The Qs problem: Sediment volumetric balance of proximal foreland basin systems. *Sedimentology* **2013**, *60*, 102–130. [CrossRef]

30. Chen, L.; Steel, R.J.; Guo, F.; Olariu, C.; Gong, C. Alluvial fan facies of the Yongchong Basin: Implications for tectonic and paleoclimatic changes during Late Cretaceous in SE China. *J. Asian Earth Sci.* **2017**, *134*, 37–54. [CrossRef]

31. Ritter, J.B.; Miller, J.R.; Enzel, Y.; Wells, S.G. Reconciling the Roles of Tectonism and Climate in Quaternary Alluvial Fan Evolution. *Geology* **1995**, *23*, 245–248. [CrossRef]

32. Harvey, A.M.; Silva, P.G.; Mather, A.E.; Goy, J.L.; Stokes, M.; Zazo, C. The impact of Quaternary sea-level and climatic change on coastal alluvial fans in the Cabo de Gata ranges, southeast Spain. *Geomorphology* **1999**, *28*, 1–22. [CrossRef]

33. Hartley, A.J.; Mather, A.E.; Jolley, E.J.; Turner, P. Climatic controls on alluvial-fan activity, Coastal Cordillera, northern Chile. In *Alluvial Fans: Geomorphology, Sedimentology, Dynamics*; Harvey, A.M., Mather, A.E., Stokes, M., Eds.; Geological Society: London, UK, 2005; pp. 95–116; ISBN 1-86239-189-0.

34. Waters, J.V.; Jones, S.J.; Armstrong, H.A. Climatic controls on late Pleistocene alluvial fans, Cyprus. *Geomorphology* **2010**, *115*, 228–251. [CrossRef]

35. Frostick, L.E.; Reid, I.A.N. Climatic versus tectonic controls of fan sequences: Lessons from the Dead Sea, Israel. *J. Geol. Soc.* **1989**, *146*, 527–538. [CrossRef]

36. Jolley, E.J.; Turner, P.; Willams, G.D.; Hartley, A.J.; Flint, S. Sedimentological response of an alluvial system to Neogene thrust tectonics, Atacama Desert, northern Chile. *J. Geol. Soc.* **1990**, *147*, 769–784. [CrossRef]

37. Sinha, R.; Ahmad, J.; Gaurav, K.; Morin, G. Shallow subsurface stratigraphy and alluvial architecture of the Kosi and Gandak megafans in the Himalayan foreland basin, India. *Sediment. Geol.* **2014**, *301*, 133–149. [CrossRef]

38. Stokes, M.; Mather, A.E. Controls on modern tributary-junction alluvial fan occurrence and morphology: High Atlas Mountains, Morocco. *Geomorphology* **2015**, *248*, 344–362. [CrossRef]

39. Horton, B.K.; Decelles, P.G. Modern and ancient fluvial megafans in the foreland basin system of the central Andes, southern Bolivia: Implications for drainage network evolution in fold-thrust belts. *Basin Res.* **2001**, *13*, 43–63. [CrossRef]

40. Arzani, N. The fluvial megafan of Abarkoh Basin (Central Iran): An example of flash-flood sedimentation in arid lands. In *Alluvial Fans: Geomorphology, Sedimentology, Dynamics*; Harvey, A.M., Mather, A.E., Stokes, M., Eds.; Geological Society: London, UK, 2005; pp. 41–59; ISBN 1-86239-189-0.

41. Hoorn, C.; Straathof, J.; Abels, H.A.; Xu, Y.; Utescher, T.; Dupont-Nivet, G. A late Eocene palynological record of climate change and Tibetan Plateau uplift Xining basin, China. *Palaeogeogr. Palaeoclimatol. Palaeoecol.* **2012**, *344–345*, 16–38. [CrossRef]

42. Chi, Y.; Fang, X.; Song, C.; Miao, Y.; Teng, X.; Han, W.; Wu, F.; Yang, J. Cenozoic organic carbon isotope and pollen records from the Xining basin, NE Tibetan Plateau, and their palaeoenvironmental significance. *Palaeogeogr. Palaeoclimatol. Palaeoecol.* **2013**, *386*, 436–444. [CrossRef]

43. Zhang, C.; Guo, Z. Clay mineral changes across the Eocene–Oligocene transition in the sedimentary sequence at Xining occurred prior to global cooling. *Palaeogeogr. Palaeoclimatol. Palaeoecol.* **2014**, *411*, 18–29. [CrossRef]

44. Zan, J.; Fang, X.; Yan, M.; Zhang, W.; Lu, Y. Lithologic and rock magnetic evidence for the Mid-Miocene Climatic Optimum recorded in the sedimentary archive of the Xining basin, NE Tibetan Plateau. *Palaeogeogr. Palaeoclimatol. Palaeoecol.* **2015**, *431*, 6–14. [CrossRef]

45. Vandenberghe, J.; Wang, X.; Lu, H. Differential impact of small-scaled tectonic movements on fluvial morphology and sedimentology the Huang Shui catchment, NE Tibet Plateau. *Geomorphology* **2011**, *134*, 171–185. [CrossRef]

46. Wang, X.; Lu, H.; Vandenberghe, J.; Zheng, S.; Balen, R.V. Late Miocene uplift of the NE Tibetan Plateau inferred from basin filling, planation and fluvial terraces in the Huang Shui catchment. *Glob. Planet. Chang.* **2012**, *88–89*, 10–19. [CrossRef]

47. Wu, Q.; Zhao, Z.; Liu, L.; Granger, D.E.; Wang, H.; Cohen, D.J.; Wu, X.; Ye, M.; Bar-Yosef, O.; Zhang, J.; et al. Outburst flood at 1920 BCE supports historicity of China's Great Flood and the Xia dynasty. *Science* **2016**, *353*, 579–582. [CrossRef] [PubMed]

48. Wang, X.; Vandenberghe, J.; Lu, H.; Balen, R.V. Climatic and tectonic controls on the fluvial morphology of the Northeastern Tibetan Plateau China. *J. Geogr. Sci.* **2017**, *27*, 1325–1340. [CrossRef]

49. Lu, H.; Wang, X.; Sun, X.; Wang, X.; Yi, S.; Zhou, Y.; Liu, Q. Loess stratigraphy and palaeoclimate changes during Quaternary in northeastern Tibetan Plateau revealed by loess core. *Quat. Res.* **2007**, *27*, 230–241. (In Chinese with English Abstract)

50. An, Z.; Colman, S.M.; Zhou, W.; Li, X.; Brown, E.T.; Jull, A.J.T.; Cai, Y.; Huang, Y.; Lu, X.; Chang, H.; et al. Interplay between the Westerlies and Asian monsoon recorded in Lake Qinghai sediments since 32 ka. *Sci. Rep.UK* **2012**, *2*, 619. [CrossRef] [PubMed]

51. Vandenberghe, J.; Renssen, H.; Van Huissteden, K.; Nugteren, G.; Konert, M.; Lu, H.; Dodonov, A.; Buylaert, J. Penetration of Atlantic westerly winds into Central and East Asia. *Quat. Sci. Rev.* **2006**, *25*, 2380–2389. [CrossRef]

52. Henderson, A.C.G.; Holmes, J.A.; Leng, M.J. Late Holocene isotope hydrology of Lake Qinghai, NE Tibetan Plateau: Effective moisture variability and atmospheric circulation changes. *Quat. Sci. Rev.* **2010**, *29*, 2215–2223. [CrossRef]

53. Zhang, J.; Ma, Z.; Li, Z.; Li, W. Study on Deformational Characteristics of Northeastern Qinghai-Xizang (Tibetan) Plateau from Late Cenozoic Defomation in the Xining basin. *Geol. Rev.* **2009**, *55*, 457–472, (In Chinese with English Abstract).

54. Miall, A.D. *The Geology of Fluvial Deposits: Sedimentary Facies, Basin Analysis, and Petroleum Geology*; Springer: Berlin/Heidelberg, Germany, 1996; ISBN 3-540-59186-9.

55. Konert, M.; Vandenberghe, J. Comparison of laser grain size analysis with pipette and sieve analysis: A solution for the underestimation of the clay fraction. *Sedimentology* **1997**, *44*, 523–535. [CrossRef]

56. Murray, A.S.; Wintle, A.G. Luminescence dating of quartz using an improved single-aliquot regenerative-dose protocol. *Radiat. Meas.* **2000**, *32*, 57–73. [CrossRef]

57. Murray, A.S.; Wintle, A.G. The single aliquot regenerative dose protocol: Potential for improvements in reliability. *Radiat. Meas.* **2003**, *37*, 377–381. [CrossRef]

58. Thiel, C.; Buylaert, J.; Murray, A.; Terhorst, B.; Hofer, I.; Tsukamoto, S.; Frechen, M. Luminescence dating of the Stratzing loess profile (Austria)–Testing the potential of an elevated temperature post-IR IRSL protocol. *Quat. Int.* **2011**, *234*, 23–31. [CrossRef]

59. Wintle, A.G. Luminescence dating: Where it has been and where it is going. *Boreas* **2008**, *37*, 471–482. [CrossRef]

60. Buylaert, J.P.; Jain, M.; Murray, A.S.; Thomsen, K.J.; Thiel, C.; Sohbati, R. A robust feldspar luminescence dating method for Middle and Late Pleistocene sediments. *Boreas* **2012**, *41*, 435–451. [CrossRef]

61. Li, B.; Li, S. A reply to the comments by Thomsen et al. on "Luminescence dating of K-feldspar from sediments: A protocol without anomalous fading correction". *Quat. Geochronol.* **2012**, *8*, 49–51. [CrossRef]

62. Yi, S.; Buylaert, J.; Murray, A.S.; Lu, H.; Thiel, C.; Zeng, L. A detailed post-IR IRSL dating study of the Niuyangzigou loess site in northeastern China. *Boreas* **2016**, *45*, 644–657. [CrossRef]

63. Guérin, G.; Mercier, N.; Adamiec, G. Dose-rate conversion factors: Update. *Ancient TL* **2011**, *29*, 5–8.

64. Mejdahl, V. Thermoluminescence Dating: Beta-dose attenuation in quartz grains. *Archoeometry* **1979**, *21*, 61–79. [CrossRef]

65. Huntley, D.J.; Baril, M.R. The K content of the K-feldspars being measured in optical dating or in thermoluminescence dating. *Ancient TL* **1997**, *15*, 11–13.

66. Mejdahl, V. Internal radioactivity in quartz and feldspar grains. *Ancient TL* **1987**, *5*, 10–17.

67. Zhao, H.; Li, S. Internal dose rate to K-feldspar grains from radioactive elements other than potassium. *Radiat. Meas.* **2005**, *40*, 84–93. [CrossRef]

68. Vandenberghe, D.; De Corte, F.; Buylaert, J.P.; Kučera, J. On the internal radioactivity in quartz. *Radiat. Meas.* **2008**, *43*, 771–775. [CrossRef]

69. Einsele, G. *Sedimentary Basins: Evolution, Facies, and Sediment Budget*; Springer: Berlin/Heidelberg, Germany, 2000; ISBN 3-540-66193-X.

70. Brierley, G.J.; Liu, K.; Crook, K.A.W. Sedimentology of coarse-grained alluvial fans in the Markham Valley, Papua New Guinea. *Sediment. Geol.* **1993**, *86*, 297–324. [CrossRef]

71. Todd, S.P. Stream-driven, high-density gravelly traction carpets: Possible deposits in the Trabeg. *Sedimentology* **1989**, *36*, 513–530. [CrossRef]

72. Jo, H.R.; Rhee, C.W.; Chough, S.K. Distinctive characteristics of a streamflow-dominated alluvial fan deposit: Sanghori area, Kyongsang Basin Early Cretaceous, southeastern Korea. *Sediment. Geol.* **1997**, *110*, 51–79. [CrossRef]

73. Smith, N.D.; Cross, T.A.; Dufficy, J.P.; Clough, S.R. Anatomy of an avulsion. *Sedimentology* **1989**, *36*, 1–23. [CrossRef]

74. Uba, C.E.; Heubeck, C.; Hulka, C. Facies analysis and basin architecture of the Neogene Subandean synorogenic wedge, southern Bolivia. *Sediment. Geol.* **2005**, *180*, 91–123. [CrossRef]

75. Nemec, W.; Postma, G. Quaternary Alluvial Fans in Southwestern Crete Sedimentation Processes and Geomorphic Evolution. In *Alluvial Sedimentation*; International Association of Sedimentologists: Gent, Belgium, 1993; Volume 17, pp. 235–276.

76. Li, L.; Garzione, C.N.; Pullen, A.; Zhang, P.; Li, Y. Late Cretaceous–Cenozoic basin evolution and topographic growth of the Hoh Xil Basin, central Tibetan Plateau. *GSA Bull.* **2017**, *130*, 499–521. [CrossRef]

77. Mccarthy, P.J.; Martini, I.P.; Leckie, D.A. Anatomy and evolution of a Lower Cretaceous alluvial plain: Sedimentology and palaeosols in the upper Blairmore Group, south-western Alberta, Canada. *Sedimentology* **1997**, *44*, 197–220. [CrossRef]

78. Bennett, M.R.; Mather, A.E.; Glasser, N.F. Earth hummocks and boulder runs at Merrivale, Dartmoor. In *Devon and East Cornwall: Field Guide*; Charman, D.J., Newnham, R.W., Croot, D.W., Eds.; Quaternary Research Association: London, UK, 1996; pp. 81–96; ISBN 090778027X.

79. Rygel, M.C.; Gibling, M.R.; Calder, J.H. Vegetation-induced sedimentary structures from fossil forests in the Pennsylvanian Joggins Formation, Nova Scotia. *Sedimentology* **2004**, *51*, 531–552. [CrossRef]

80. Vandenberghe, J. Grain size of fine-grained windblown sediment: A powerful proxy for process identification. *Earth-Sci. Rev.* **2013**, *121*, 18–30. [CrossRef]

81. Vriend, M.; Prins, M.A. Calibration of modelled mixing patterns in loess grain-size distributions: An example from the north-eastern margin of the Tibetan Plateau, China. *Sedimentology* **2005**, *52*, 1361–1374. [CrossRef]

82. Bentham, P.A.; Talling, P.J.; Burbank, D.W. Braided stream and flood-plain deposition in a rapidly aggrading basin: The Escanilla formation, Spanish Pyrenees. *Geol. Soc. Lond. Spec. Publ.* **1993**, *75*, 177–194. [CrossRef]

83. Vandenberghe, J.; Sun, Y.; Wang, X.; Abels, H.A.; Liu, X. Grain-size characterization of reworked fine-grained aeolian deposits. *Earth-Sci. Rev.* **2018**, *177*, 43–52. [CrossRef]

84. Chapot, M.S.; Roberts, H.M.; Duller, G.A.T.; Lai, Z. A comparison of natural-and laboratory-generated dose response curves for quartz optically stimulated luminescence signals from Chinese Loess. *Radiat. Meas.* **2012**, *47*, 1045–1052. [CrossRef]

85. Buylaert, J.P.; Vandenberghe, D.; Murray, A.S.; Huot, S.; Corte, F.D.; Haute, P.V.D. Luminescence dating of old >70 ka. Chinese loess: A comparison of single-aliquot OSL and IRSL techniques. *Quat. Geochronol.* **2007**, *2*, 9–14. [CrossRef]

86. Buylaert, J.P.; Murray, A.S.; Vandenberghe, D.; Vriend, M.; Corte, F.D.; Haute, P.V.D. Optical dating of Chinese loess using sand-sized quartz: Establishing a time frame for Late Pleistocene climate changes in the western part of the Chinese Loess Plateau. *Quat. Geochronol.* **2008**, *3*, 99–113. [CrossRef]

87. Li, R.W.; Shen, H.W. Effect of tall vegetations on flow and sediment. *J. Hydraul. Eng.* **1973**, *99*, 793–814.
88. Petryk, S.; Bosmajian, G.I. Analysis of flow through vegetation. *J. Hydraul. Div.* **1975**, *101*, 871–884.
89. Graeme, D.; Dunkerley, D.L. Hydraulic resistance by the river red gum, *Eucalyptus camaldulensis*, in ephemeral desert streams. *Geogr. Res.* **1993**. [CrossRef]
90. Zierholz, C.; Prosser, I.P.; Fogarty, P.J.; Rustomji, P. In-stream wetlands and their significance for channel filling and the catchment sediment budget, Jugiong Creek, New South Wales. *Geomorphology* **2001**, *38*, 221–235. [CrossRef]
91. Jordanova, A.A.; James, C.S. Experimental study of bed load transport through emergent vegetation. *J. Hydraul. Eng.* **2003**, *129*, 474–478. [CrossRef]
92. Rygel, M.C.; Calder, J.H.; Gibling, M.R.; Gingras, M.K.; Melrose, C.S.A. Tournaisian forested wetlands in the Horton Group of Atlantic Canada. *Geol. Soc. Am.* **2006**, 103–126.
93. Miao, Y.; Fang, X.; Dai, S.; Liu, W.; Yang, M.; Chen, C. Origin of the second terraces in the Huangshui drainage area, Qinghai, China, studied based on sporopollen records. *Geol. Bull. China* **2007**, *12*, 1697–1702.
94. Pye, K.; Zhou, L. Late Pleistocene and Holocene aeolian dust deposition in north China and the northwest Pacific Ocean. *Palaeogeogr. Palaeoclimatol. Palaeoecol.* **1989**, *73*, 11–23. [CrossRef]
95. Sun, D.; Bloemendal, J.; Rea, D.K.; Vandenberghe, J.; Jiang, F.; An, Z.; Su, R. Grain-size distribution function of polymodal sediments in hydraulic and aeolian environments, and numerical partitioning of the sedimentary components. *Sediment. Geol.* **2002**, *152*, 263–277. [CrossRef]
96. Sun, D.; Su, R.; Bloemendal, J.; Lu, H. Grain-size and accumulation rate records from Late Cenozoic aeolian sequences in northern China: Implications for variations in the East Asian winter monsoon and westerly atmospheric circulation. *Palaeogeogr. Palaeoclimatol. Palaeoecol.* **2008**, *264*, 39–53. [CrossRef]
97. Vriend, M.; Prins, M.A.; Buylaert, J.; Vandenberghe, J.; Lu, H. Contrasting dust supply patterns across the north-western Chinese Loess Plateau during the last glacial-interglacial cycle. *Quat. Int.* **2011**, *240*, 167–180. [CrossRef]
98. Waelbroeck, C.; Labeyrie, L.; Michel, E.; Duplessy, J.C.; McManus, J.F.; Lambeck, K.; Balbon, E.; Labracherie, M. Sea-level and deep water temperature changes derived from benthic foraminifera isotopic records. *Quat. Sci. Rev.* **2002**, *21*, 295–305. [CrossRef]
99. Wang, X.; Lu, H.; Vandenberghe, J.; Chen, Z.; Li, L. Distribution and Forming Model of Fluvial Terrace in the Huangshui Catchment and its Tectonic Indication. *Acta Geol. Sin.-Engl.* **2010**, *84*, 415–423. [CrossRef]

![Q quaternary logo]

MDPI

Article

Specifying the External Impact on Fluvial Lowland Evolution: The Last Glacial Tisza (Tisa) Catchment in Hungary and Serbia

Jef Vandenberghe [1,*], Cornelis (Kees) Kasse [1], Dragan Popov [2], Slobodan B. Markovic [2], Dimitri Vandenberghe [3], Sjoerd Bohncke [1,†] and Gyula Gabris [4]

[1] Department of Earth Sciences, Vrije Universiteit, 1081HV Amsterdam, The Netherlands; c.kasse@vu.nl
[2] Department of Geography, Tourism and Hotel Management, Faculty of Science, University of Novi Sad, 21000 Novi Sad, Serbia; dpopov@live.be (D.P.); baca.markovic@gmail.com (S.B.M.)
[3] Laboratory of Mineralogy and Petrology, Department of Geology, Ghent University, B-9000 Gent, Belgium; davdenbe@gmail.com
[4] Department of Physical Geography, Eötvös University of Budapest, H-1117 Budapest, Hungary; ferencgyula@caesar.elte.hu
*   Correspondence: jef.vandenberghe@vu.nl
†   Deceased.

Academic Editors: David Bridgland and Valentí Rull
Received: 26 June 2018; Accepted: 9 August 2018; Published: 16 August 2018

Abstract: External impact on the development of fluvial systems is generally exerted by changes in sea level, climate and tectonic movements. In this study, it is shown that a regional to local differentiation of fluvial response may be caused by semi-direct effects of climate change and tectonic movement; for example, vegetation cover, frozen soil, snow cover and longitudinal gradient. Such semi-direct effects may be responsible for specific fluvial activity resulting in specific drainage patterns, sedimentation series and erosion–accumulation rates. These conclusions are exemplified by the study of the fluvial archives of the Tis(z)a catchment in the Pannonian Basin in Hungary and Serbia from the middle of the last glacial to the Pleistocene–Holocene transition. Previous investigations in that catchment are supplemented here by new geomorphological–sedimentological data and OSL-dating. Specific characteristics of this catchment in comparison with other regions are the preponderance of meandering systems during the last glacial and the presence of very large meanders in given time intervals.

Keywords: Tisza; Tisa; Pannonian Basin; fluvial evolution; terrace development; tectonic impact; local conditions; last glacial; OSL-dating

1. Introduction

The impact of climate on fluvial activity and subsequent drainage patterns has long been recognized [1–3] in addition to other external forcing factors, such as tectonic movements, base-level changes and anthropogenic activity. The traditional correspondence between glacial–interglacial cycles and fluvial evolution has been modified over the last few decades, focusing particularly on the important fluvial morphological changes occurring at climate transitions, originally in temperate to periglacial environments [4–6], and later slightly adapted by the same and other authors [7–9] and confirmed by many field cases (e.g., [10] and references therein). Meanwhile, the timing of considerable changes of fluvial evolution at climatic transitions has also been recognized in other climatic zones, such as the monsoonal environment, for instance in China [11,12] and Australia [13]. It has been called a model of non-linearity as the fluvial action is influenced by the delayed effect of vegetation development with regard to the climatic change driver.

However, progressive fluvial research has demonstrated that the general validity of the climatically driven, non-linear fluvial model may also need some adaptation in specific climate circumstances, e.g., in glacial [14–16] and arctic environments [17,18], Mediterranean environments [19,20] and tropical conditions [21]. Often, external variables, climate and tectonic movements, may play indirect roles. Such semi-direct climate influences are, for example, the development of specific vegetation cover, the degradation of thick snow cover and glacier ice or the presence of frozen ground. Similarly, climate may have a direct impact on the power supplied by the water flow characteristics that determine fluvial discharge, but also indirectly on the sediment delivery to the drainage system (by way of vegetation). Also, the morpho-structural framework and topography, as indirectly imposed by the tectonic setting, give a characteristic identity to the catchment [15,16,22,23]. In particular, the energy of the fluvial system is determined by the river gradient, which consequently represents a most influential factor in the eroding or accumulating character of a river. In addition, several studies consider the complexity of climate cycles with weakly expressed interstadials or stadials [24,25] or climatic episodes of short duration that hindered rivers crossing thresholds [26–28]. Finally, the ultimate morphological result of river action is determined by its preservation potential as an expression of the river's intrinsic evolution [9,26,29].

The present study concerns the Tisza catchment in the Pannonian (Carpathian) Basin of Hungary and Serbia ("Tisa" in Serbian, Romanian and Ukrainian; "Tisza" in Hungarian) (Figure 1), for which different phases of evolution are identified and dated. Some years ago, the first results were published for the middle Tisza in Hungary [30] and for the lower Tisa in Serbia [31]. We compare and extend that research in an adjacent area to the north in the middle Tisza and further downstream in the lower Tisa. This paper aims to illustrate the effects of a specific climatic and topographical setting on fluvial morphological development, namely that of temperate continental climate conditions, in contrast to previous studies, which generally dealt with more oceanic conditions. In addition, a low-relief and thus low-energy topography is considered. This means, with regard to vegetation cover, that a more steppic environment rather than tundra is involved and, with regard to topography, that a low bedload–suspension load ratio is characteristic of this catchment due to its low gradient.

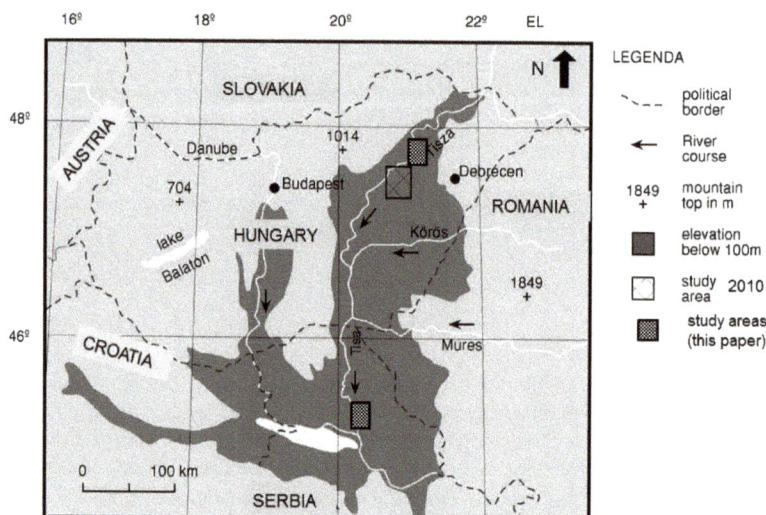

Figure 1. Location map of the Tis(z)a catchment in Hungary and Serbia.

2. Study Region and Previous Research

The Tis(z)a is a major tributary of the Danube, starting its course from the Eastern Carpathian Mountains in Romania, flowing through Ukraine and arriving in the Pannonian Basin in Hungary (middle Tisza) and finally in Serbia (lower Tisa), which is the lowest part of the Basin [32,33]. Both the Danube and Tis(z)a formed their valleys in thick Quaternary basin deposits. Their wide floodplains show an intriguing pattern of successive meandering systems and impressive fluvial deposits from the last glacial and Holocene, extensively described previously [30,34–36].

In comparison with the periglacial-temperate regions further north, the environmental setting in the loess-covered Tisza catchment differed in a few aspects during the late Weichselian. Firstly, the vegetation cover was less dramatically reduced; probably a steppic cover including some persistent tree refugia, in contrast to the bare tundra landscape in the northern coversand regions [37,38]. Secondly, the climate was more continental, with colder winters, warmer summers and less precipitation [39]. Thirdly, the study region coincides with the marginal zone of permafrost at the end of the last glacial, in contrast to NW Europe, which was invariably within the permafrost belt [40,41].

Structurally, the Tisza catchment is situated in the Pannonian basin, a zone of considerable tectonic subsidence [42–44]. On average, the longitudinal gradients in the drainage system of the tectonic basin are very low, from c. 1.86 cm/km in the lower Tisa to 5–10 cm/km in SE Hungary and to 15 cm/km in central–northern Hungary, leading almost entirely to suspension transport currently [30,31,35]. The evolution of the drainage system was obviously determined by the particular tectonic subsidence pattern during the Pleistocene. In addition, this tectonic impact was supplemented by evident climatic overprints [30,35,36,45–47].

3. Methods

Following the original research set-up applied by Kasse et al. [30], information on the sedimentary architecture was acquired by topographic (DEM) analysis and field investigations, mainly consisting of transects across the present floodplain using hand-drill data and a few geoelectric sections. Detailed coring enabled the reconstruction of the precise geometry of former, now abandoned channels. A semi-closed gouge and suction-tube corer were used in the water-saturated sediments. Drill coring was supplemented with exposure data from sand or clay exploitation pits. Apart from palynological analyses [30], sedimentological analysis involved grain-size measurements by laser diffraction. The preparation method of the grain-size analyses involved, for instance, the removal of calcium carbonate and organic material by adding HCl and H_2O_2, respectively [48]. It initiated, amongst others, a clay–silt boundary at 8 μm, which we apply here.

Our previous work in the Tisza basin demonstrated the complexity of interpreting chronologies derived from radiocarbon dating because of the possible and uncertain degree of reworking [30]. The present project therefore explores the possibilities of luminescence dating. Luminescence research in the Hungarian fluvial domain has been summarised previously [36]. The first results in the Serbian domain focused on the methodology (involving infrared stimulated luminescence—IRSL—signals from K-feldspar) that was used for dating a sequence of aeolian and fluvial deposits as exposed in the sandpit at Mužlja [49].

In this study, four additional optically stimulated luminescence (OSL) samples were collected from a sequence of floodplain deposits exposed at a section at Zrenjanin, of which three could be dated using OSL signals from fine-sand sized (63–90 μm) grains of quartz. These samples were prepared in the usual manner (HCl, H_2O_2, sieving, density separation, HF) and mounted on the inner 8 mm of 9.7 mm stainless steel discs using silicon oil as adhesive. Luminescence measurements were made using automated Risø readers (Risø National Laboratory, Roskilde, Denmark) equipped with blue (470 nm) and IR (870 or 875 nm) LEDs; all luminescence signals were detected through a 7.5 mm thick Hoya U-340 filter (details of the luminescence equipment can be found in Bøtter-Jensen et al. [50] and Lapp et al. [51]). The purity of the quartz grains was confirmed by an OSL IR depletion ratio consistent with 1.0 ± 0.1 [52]. Equivalent dose (D_e) determination was carried out using the single-aliquot regenerative-dose (SAR) protocol [53,54]. Optical stimulation with blue diodes was undertaken for 38 s at 125 °C. The initial 0.31 s of the decay curve, minus a background evaluated from the 0.31–1.08 s interval, was used in the calculations. The measurement of the test-dose signal was preceded by a reduction in heat (cut-heat) to 160 °C and followed by a high-temperature cleanout by stimulation with the blue diodes for 38 s at 280 °C. D_es were calculated from the flat 160–260 °C region in a plot of D_e versus preheat temperature ("preheat plateau"; the duration of the preheat was 10 s). For one of the samples, the dependence of the measured dose on preheat temperature was also investigated through a dose recovery test [54]. Natural aliquots were bleached using the blue LEDs at room temperature (two times 250 s, with a 10 ks pause between), given a dose close to the estimated D_e, and measured using the SAR protocol as outlined above. Three aliquots were measured at each of seven different preheat temperatures in the range of 160–280 °C. For all samples, a dose recovery test was also performed for a preheat of 220 °C only, but using a higher cut-heat to 200 °C. Low-level high-resolution gamma-ray spectrometry was used for the determination of the natural dose rate (see [55,56] for details). For this, the sediment collected around the OSL tubes was dried at 110 °C (until at a constant weight), homogenized and pulverized. A subsample of typically ~140 g was then cast in wax [57] and stored for one month before being measured. The radionuclide activities were converted to dose rates using tabulated conversion factors [58]. Based on Mejdahl [59] and Aitken [60], a factor of 0.9 (±5% relative uncertainty) was adopted to correct the external beta dose rates for the effects of attenuation and etching. An internal dose rate of 0.013 ± 0.003 Gy ka^{-1} was assumed [61]. Dose rates were corrected for the effect of moisture, assuming a time-averaged moisture content of 20 ± 5%. The contribution from cosmic rays was calculated [62]. Uncertainties on the luminescence ages were calculated following the error assessment system proposed by Aitken [63,64], using contributions from quantified systematic sources of uncertainty [56,65].

4. New Results

4.1. Middle Tisza (NE Hungary)

We complete the former description of the middle Tisza system [30] by adding the sedimentology and morphology of a large meander further upstream near Tiszacsege (Figures 2 and 3). That meander, belonging to system 3b (see below), together with a few neighbouring meanders of similar size (e.g., at Ujszentmargita; Figure 3), is eroded into the sediments of the previous meandering system 1 comprising, for instance, the Meggyes meander (Figures 2 and 3). A detailed drilling section along the axis of the Tiszacsege meander bend shows the development and appearance of a typical meander bend. It consists morphologically of a series of ridges and swales and an outer deep abandoned channel (Figure 3). The latter channel marks the final stage of that meander development before it was cut off. The entire area is covered with a veneer of clayey silt, often consisting of alternating laminae of sandy silt and clay, interpreted as flood sediment from the contemporaneous and/or younger river. The top of the ridges may be up to 3 m higher than the swales in between. They consist of vertical sets of weakly laminated, fining-upwards sediment sequences from coarse (gravelly) to fine silty sands (Figure 4), with each individual set reaching a maximum thickness (when not eroded) of 2–3 m. The depressions between them show a similar sediment sequence, except that the upper clayey–silty cover is generally thicker than on top of the ridges, which means that the topography predating the ultimate flooding of the areas was even more pronounced than it is at present. From the morphology, sedimentology and position within the meander bend, these forms and sediments are obviously interpreted as a series of (high) scroll bars (point bars) leaving a depression successively between each of them and its neighbours. The youngest channel of the system is filled, after abandonment, with up to 15 m of similar fine-grained sediment as the upper flood sediment that covers the point bars, but more humic and containing molluscs. The pollen diagram of Tiszacsege is located at the maximum curvature of this abandoned channel; it starts in the final phase of the Pleniglacial [30].

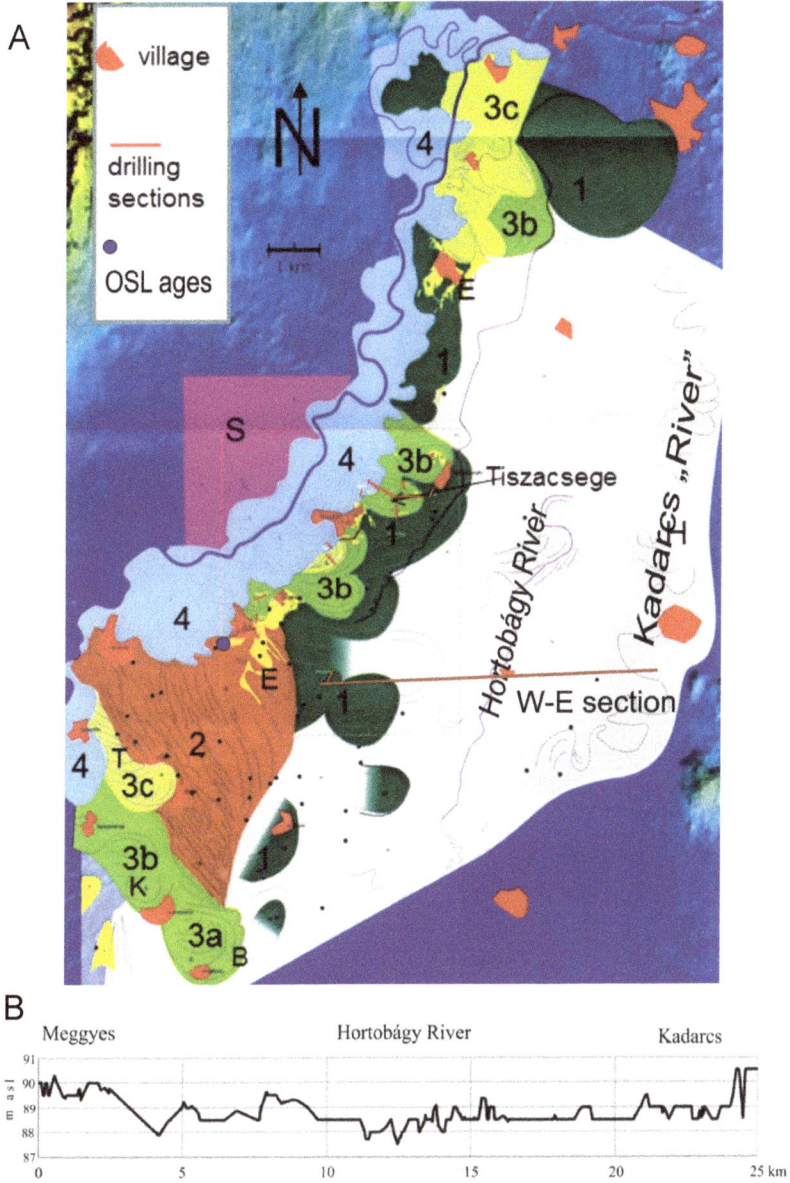

Figure 2. (**A**) Morphological map of the middle Tisza area. H: Hortobagy system; E: aeolian forms (yellow); 1–4: fluvial systems (see text); 3a-b-c: Berekfurdo (B), Kunmadaras (K), Tiszaörs (T) meanders in the lower left corner (from Mol et al. [10]); S: Sajo alluvial fan; W-E section, see B; (**B**) topographic W-E section through the Hortobagy plain (white color in A), including the old Kadarcs system, and the Meggyes meander (system 1 of the younger valley belt).

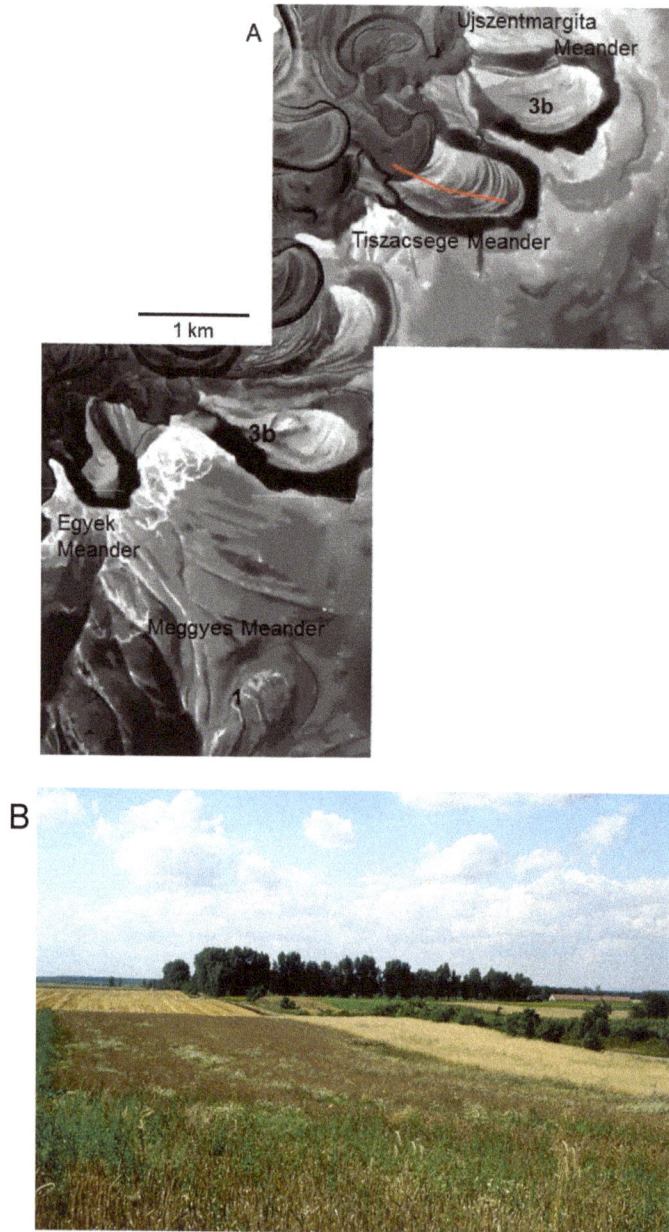

Figure 3. (**A**) Orthophoto of the study area in the middle Tisza (location see Figure 1). (**B**) Photo shows the swale and ridge topography in the meander bend of Tiszacsege.

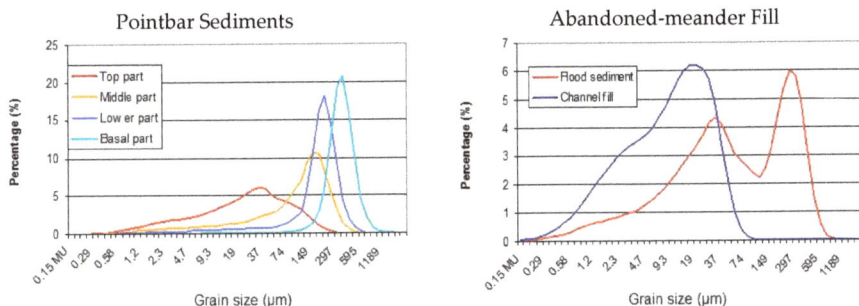

Figure 4. Typical grain-size distributions of point bar, flood and channel fill sediments of the Tiszacsege meander (middle Tisza). The left diagram shows a representative grain-size evolution in one individual point bar (at depths of 6.50, 5.20, 4.60 and 3.80 m). The right diagram shows the grain-size distribution in the centre of the last abandoned channel of max. 15.5 m depth (flood sediment from the upper 2 m, channel fill at c. 8 m depth).

Detailed grain-size information from the point-bar series and the channel fill is represented in Figure 4. An upward decreasing grain size typifies the sands of point bars [66]. The upper flood sediments generally contain a fraction with a mode around 40 μm (the red (top sediment) curve in the left diagram of Figure 4), which is the typical mode for loess deposits in this region [67,68], thus easily interpreted as reworked (alluvial) loess deposits [69]. In addition, the flood sediment often also contains a sand fraction (most obvious in the right panel of Figure 4) indicating some flow on top of the previous floodplain [69,70]. The channel fill shows a clear bimodality (Figure 4, right panel): one fraction is a fine silt of c. 19 μm, corresponding with the fine component in the regional loess cover [67,68] and preferentially settled in the standing water of the abandoned meander, in contrast to the coarser-grained silt (40 μm) which is dominant in the flood sediments; the second fraction is a clay (c. 2 μm) which can only be deposited by settling in standing water ('lacustro-aeolian loess' [69]).

4.2. Lower Tisa (Serbia)

The former morpho-sedimentary research [31] is extended here by additional field and laboratory investigations. Three sections approximately perpendicular to the present Tisa River were investigated mainly by hand drilling supplemented with some exposures [71]. A series of meander systems may be distinguished, sharply dissecting the Titel plateau in the west (at 110–130 m a.s.l.) and the Tamiš plateau in the east (at >82 m a.s.l.), both consisting of Late Pleistocene loess [68,72–74]. The Zrenjanin section is the most representative, crossing three terrace levels (location Figure 5). As in the case of the middle Tisza, the different systems were distinguished by elevation and are separated by distinct erosive meander scarps.

The oldest fluvial series, mainly consisting of large meanders, may be subdivided into systems (1), (3a) and (3b), all situated approximately at the same elevation of c. 79–81 m a.s.l.; a system (3c) slightly lower at 78–79 m a.s.l.; and a youngest one a few metres lower (system 3d at c. 75 m a.s.l.). Clearly separated from these series of large meanders is a series of small meanders (system 4), which are not further subdivided, at 72–75 m a.s.l. elevation (Figure 5).

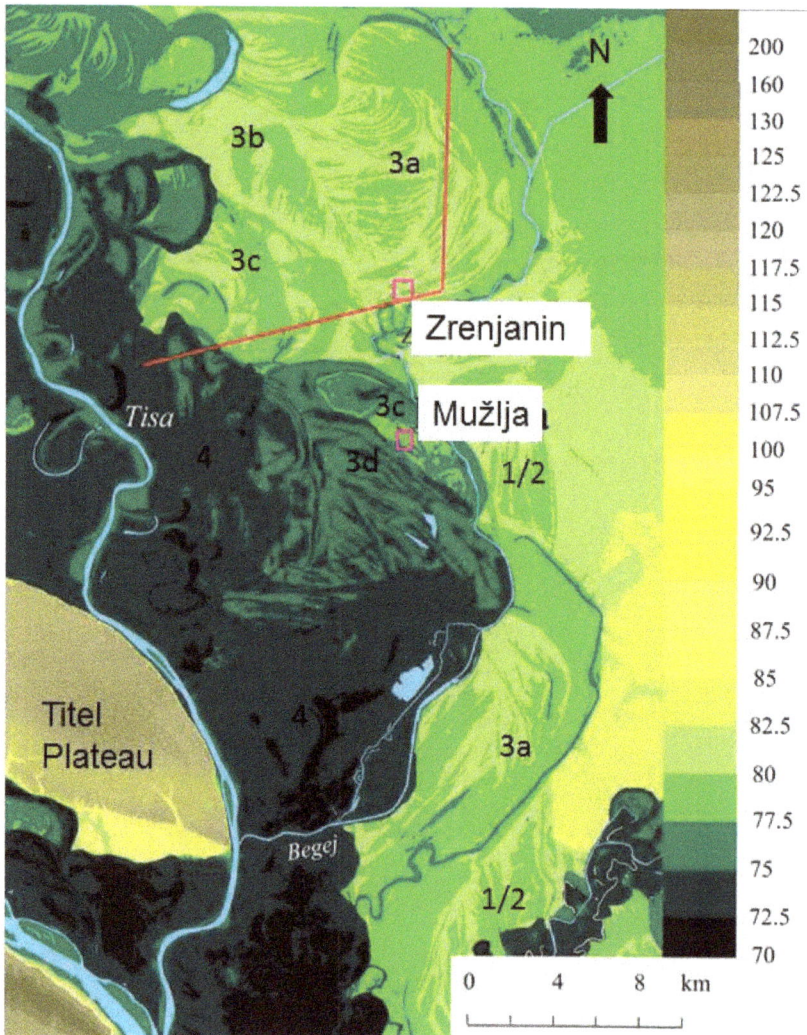

Figure 5. Morphological map of the study area in the lower Tisa comprising the terrace systems 1, 3 and 4 incised in the loess plateaus at both sides of the river. The system (2) braided pattern from Hungary (level B, unit 2) is absent here. The squares indicate the locations of the exposures at Zrenjanin and Mužlja. Height scale in m a.s.l.

In general, each terrace level shows the same morpho-sedimentary expression as in the Tiszacsege meander described above. Morphologically, there is a clear succession of ridges and depressions, interpreted as point bars and inter-bar swales, respectively. Both are covered by a sheet of clayey flood sediments. somewhat thicker than their equivalents in the middle Tisza (1.5 to 3.5 m). The underlying point-bar facies consist of individual sets of 1 to 3 m thick fining-upward sediment ending at their top in gyttja or clay (in the middle Tisza, these very fine capping sediments are missing, except in the abandoned channels) and occasionally containing shell debris and coarse sand grains at the bottom. They are interrupted by channels to a depth of 12 m, often at the outer edge of a terrace level. Following abandonment, the latter channels have been filled mainly with similar fine-grained sediments as the flood cover, i.e., (clayey) silt or fine sand at the base covered by (humic) silt-clay with sand laminae and ultimately clay–silt without sand. The lowest fill-sediments may occasionally contain poorly sorted sand beds.

An exposure at Zrenjanin (location Figure 5) enabled the detailed study and sampling of the sedimentary structures in the 2–6.5 m thick, fine-grained sediments that occur at the top of the series in meander system 3a (Figure 6). There is a gradual transition to the underlying sediments, which contain more sand and were cored by hand-drilling to a depth of 10.7 m below the surface. The sediment series is described according to sedimentary structures and grain-size composition:

- 0–2.30 m: clay-rich silt (c. 30% clay) with uniform grain-size composition, a modal size of c. 30 μm and almost completely free of sand (Figure 7A). A chernozem soil (1 m thick) is formed at the top while a brown-rusty soil with prismatic structure is present between 1.70 and 2.00 m (Figure 6);
- 2.30–2.80 m: a clay-silt sediment without any sand and a modal size of 10 μm, horizontal, and finely laminated (Figures 7A and 8A). The transition to the overlying sediment is gradual with bimodal laminae, and similarly the transition to the underlying layer is gradual;
- 2.80–5.20 m: alternating thin beds consisting of clay-rich fine-to-medium or coarse silt without sand (350 cm in Figure 7A) and obliquely bedded silty sand layers (up to 20% sand, modal values of 20 and 50 μm) (Figure 8A). The sand beds become thicker towards the base of this unit (ranging from a few centimetres to a few tens of centimetres thick), showing small but clear cross-laminated ripples pointing to a westward water flow (Figure 8B). The lower boundary of these sand beds is mostly sharp (erosive). Calcium carbonate nodules are frequently present, as well as mica grains on the bedding surfaces. A layer between 5.20 and 5.80 m shows a transitional grain size to the underlying layer;
- 5.80–6.50 m: horizontal, finely laminated, silty fine sands with modal values between 55 and 80 μm (Figure 7B) alternating with silt–clay beds (a few tens of centimetres thick) with sharp upper boundaries. Convoluted deformation at the scale of tens of centimetres resulting from liquefaction is seen in the silt–clay beds; a periglacial origin is not obvious. Gradually changing (6.50–7.00 m) to 7.00–8.70 m: sandy beds coarsening to >100 μm modal values with occasional coarse-sand fractions and decreasing silt and clay content, alternating with silt–clay beds (Figure 7B);
- 8.70–10.70 m: two clearly developed fining-up sand series (Figure 7C).

Figure 6. *Cont.*

Figure 6. Top of the section at Zrenjanin in the upper sediments of a large meander. Notice the chernozem soil at the surface and a soil at 1.7–2.00 m depth with prismatic structure. Sampling depth (base of exposed section) is at 6.90 m.

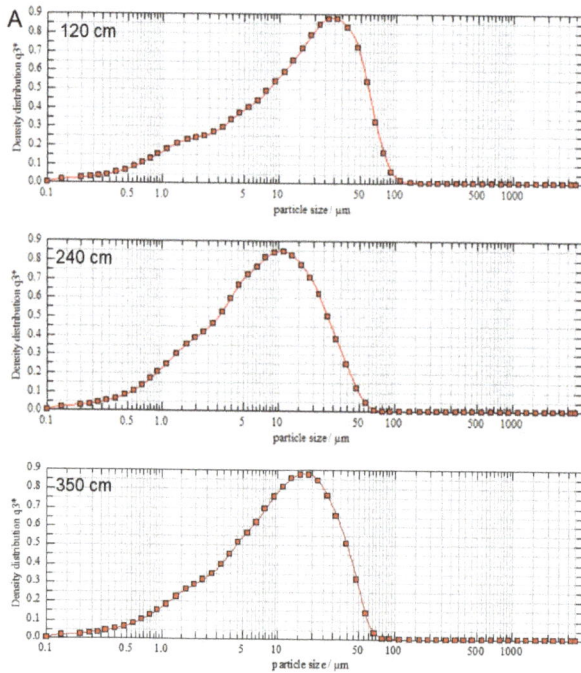

Figure 7. *Cont.*

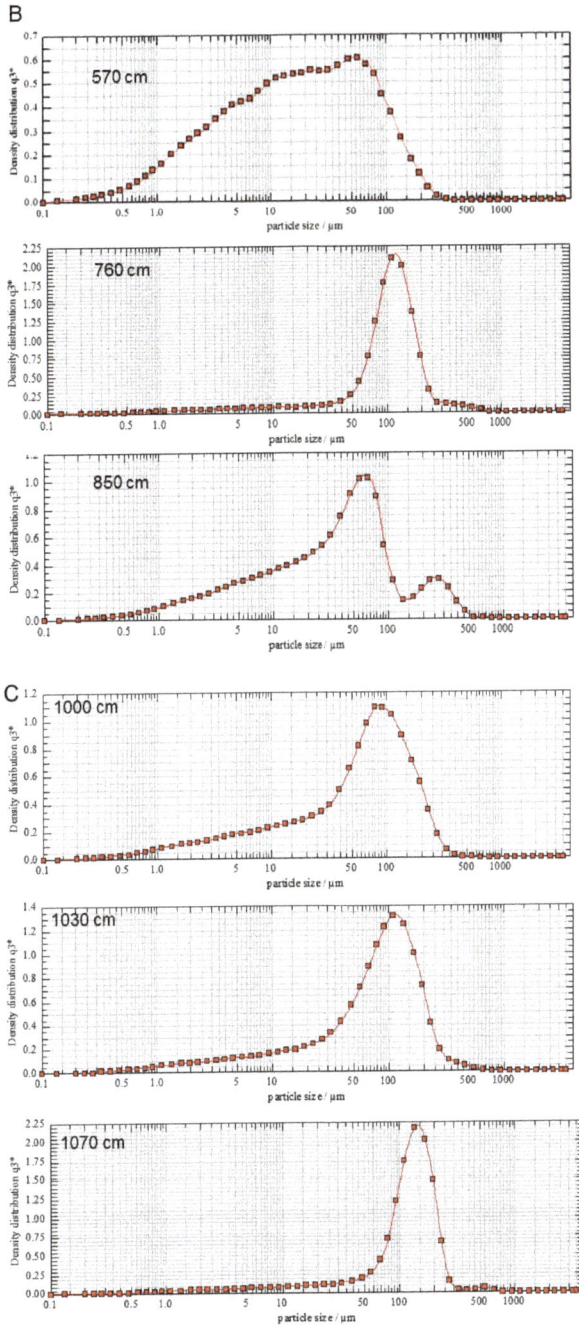

Figure 7. Grain-size distribution curves of the sediments in the meander system of Zrenjanin (location in Figure 5); depths in cm below surface. (**A**) different facies in the exposed upper silty floodplain sediment; (**B**) grain-size distributions from drilled laminated sandy–silty sediment (channel fill) at the same location; (**C**) grain-size distributions from a drilled fining-upward point-bar series. q3* is a volumetric density parameter equivalent with the distribution frequency.

(A)   (B)

Figure 8. Sedimentary structures in the section at Zrenjanin: (**A**) finely laminated clay–silt (2.30–2.80 m); (**B**) cross-laminated ripple bedding in fine silty sand at c. 4 m depth.

The lowest unit (8.70–10.70 m) is interpreted as a typical facies of channel and point-bar deposits within an actively meandering system. The facies of heterogeneous, poorly sorted sands and silts between 7.00 and 8.70 m shows characteristics of upper point-bar deposits. The upper 6.50 m fine-grained sediments are interpreted as channel fills or accumulation in the floodplain of a meandering system that formed in between the point bars of system 3a (see below and Figure 5).

Samples for quartz-based SAR-OSL dating were taken at 5.40 m in homogeneous silt–clay, 6.10 m and 6.60 m in cross-laminated sand, and at 6.90 m in a silt–clay bed. The uppermost sample yielded insufficient fine-sand quartz for OSL-analyses. The results for the other three samples are summarised in Table 1. An internally consistent set of OSL-ages was obtained (33–32 ka), showing no variation with depth. Given the depositional context of the samples and the multiple-grain mode of analysis (see above), the results are to be interpreted as maximum ages. It has been argued, however, that incomplete resetting is rarely a significant source of inaccuracy for the ages under consideration [75]. A dose-response and OSL-decay (inset) curve for an aliquot of sample GLL-150803 are shown in Figure S1. To describe the dose-response curves, we used either a single or the sum of two single saturating exponential functions. D_e values exhibit no systematic dependence on preheat temperatures in the range of 160–260 °C (see Figure S2); across this region, recycling ratios are generally consistent with 1.0 ± 0.1, and values for recuperation remain comfortably below 0.5% of the sensitivity-corrected natural signal. The dose-recovery test yielded an average measured to given dose ratio (± 1 standard error) of 1.13 ± 0.03 across the 160–260 °C preheat temperature range ($N = 18$; sample GLL-150803; Figure S3). This implies that a known laboratory dose, administered following bleaching but prior to heating, cannot be measured to within 10% and is systematically overestimated by about 10%. Increasing the cut-heat temperature to 20 °C below the preheat temperature did not improve our ability to recover a dose; a dose recovery test using a preheat temperature of 220 °C (chosen as the approximate middle of the preheat plateau) and a cut-heat to 200 °C yielded a value of 1.15 ± 0.05 ($n = 9$; three aliquots for each sample). This behaviour, and its relevance as to the accuracy of measurements of natural doses, remains to be understood. Combining these finds with our earlier results [49] indicates that the possibilities of OSL dating using quartz may be limited; future studies may therefore seek to investigate the potential of K-feldspar.

Table 1. Radionuclide concentrations used for dose rate evaluation, estimates of past water content, calculated dose rates, D$_e$s, and calculated ages (Zrenjanin). The dose rate includes the contributions from internal radioactivity and cosmic rays. The uncertainties mentioned with the D$_e$ and dosimetry data are random; the uncertainties on the ages are the overall uncertainties, which include the systematic errors. All uncertainties represent 1s.

Field Code	GLL Code	^{234}Th (Bq kg^{-1})	^{226}Ra (Bq kg^{-1})	^{210}Pb (Bq kg^{-1})	^{232}Th (Bq kg^{-1})	^{40}K (Bq kg^{-1})	Water Content (%)	Dose Rate (Gy ka^{-1})	D$_e$ (Gy)	Age (ka)
ZR 6.10	GLL-150802	33 ± 3	38.5 ± 0.6	35 ± 2	44.0 ± 0.3	605 ± 4	20 ± 5	2.82 ± 0.02	92 ± 5	32 ± 3
ZR 6.60	GLL-150803	29 ± 1	31.5 ± 0.5	34 ± 2	36.6 ± 0.4	445 ± 3	20 ± 5	2.27 ± 0.02	74 ± 3	33 ± 3
ZR 6.90	GLL-150804	31 ± 2	34.7 ± 0.5	33 ± 2	42 ± 0.4	624 ± 4	20 ± 5	2.80 ± 0.03	90 ± 6	32 ± 4

The lateral erosion scar of meandering system 3d enabled the study of an exposure in the sediments of system 3c at Mužlja (location Figure 5). This 4 m deep exposure is located only a few meters from the section that had previously been sampled for luminescence dating [49]. The sediments underlie a point bar of that system. In general, they are finely and horizontally laminated fine-grained sands, occasionally silty and cross-laminated.

Below a 1 m thick Holocene chernozem soil is a fine-grained sand, in general finely and horizontally laminated, with slightly more silty beds at 1.30–1.50 m and 1.70–1.80 m depth. Slightly coarser sand, mica-rich with low-angle cross-lamination and containing some pebbles of mm size, occurs at depths of 2.70–2.80 m and 3.10–3.25 m (Figure 9). The presence of a 2 cm thick, finely laminated, clayey silt bed is striking. The origin of this 4 m sediment series is not readily obvious: their spatial and stratigraphic position would suggest a fluvial point-bar origin, but no current flow structures were found, and a local dune blown up from point-bar deposits is not excluded. However, the nearby section used for the OSL-sampling (Figure 9) shows clearer channelling structures (concave-down discontinuous boundaries, cross-lamination of dm size) and ripple cross-lamination, thus favouring a (low-energy) fluvial interpretation of the deposits in the exposure.

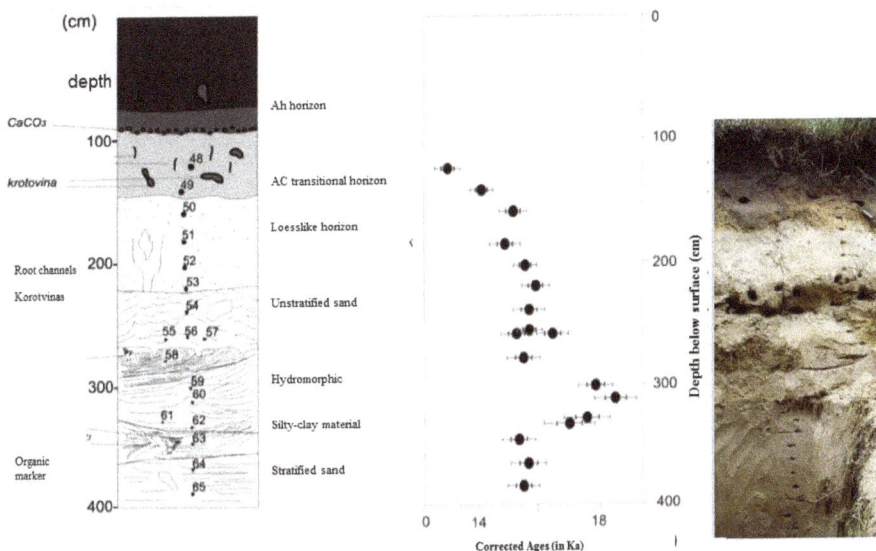

Figure 9. The Mužlja section showing sediments of the meander system 3c (location Figure 5); description of sedimentary structures and lithology, and IRSL dates from Popov et al. [49] with x-axis in ka.

IRSL-dating of K-feldspar provided an age of 15.3–15.6 ka for the deposition of the point bars of that specific meander system [49] (see Figure 9).

5. Identification and Age of the Different Phases of Fluvial Evolution

As a result of the dominant regional subsidence of the Pannonian Basin, the different phases of fluvial evolution are only weakly manifested by height differences, and no real terrace staircase landscape was formed. Some of these phases terminated their accumulation at a similar elevation and can only be distinguished from each other by erosional contacts or the dissection of the older fluvial morphological patterns. An additional difficulty in the sedimentary record is that, in general, all sediments of a specific evolution phase are covered by floodplain deposits of (a) younger phase(s) of fluvial activity as a consequence of the subdued morphology. Finally, as previously discussed extensively [30], both meander-fill and floodplain deposits may give radiocarbon ages that are too old due to the reworking of older organic deposits.

Originally, the Tisza occupied a position near to the eastern margin of the Pannonian basin during the Early–Middle Pleistocene [35]. Due to relatively more intense subsidence in the central part of the basin, the river shifted its position during the late Pleistocene in a westward direction to its current valley [33]. Remnants of the last drainage system before the Tisza reached its present-day valley occur in a subparallel belt east of the present valley (Kadarcs River), occupied in the Holocene by the Hortobagy River, and consist of a series of sinuous channels (grey shaded areas H in Figure 2).

In the Hungarian sector, the surface of the oldest part of the present valley (system 1; meander at Meggyes, Figure 3) is—similar to the precursor of the modern Tisza—morphologically weakly expressed due to the cover of younger vertical floodplain accretion (unit 1 and level A in Kasse et al. [30]). Remnants of sinuous channel scars and point bars with differing orientations suggest formation by a meandering river, as in the Kadarcs system. The humic clay with soil characteristics at the top of this floodplain deposit yielded uncalibrated radiocarbon ages of 24.7 to 22.2 ka BP for autochthonous organic matter (supported by dating of the alkali extract [30]). Since this alluvial clay–soil complex (unfortunately without pollen) probably represents the ultimate phase of floodplain deposition, we infer that the age of activity of this system was prior to that date: i.e., older than c. 28 ka cal BP. The oldest parts of the lower Tisa valley system in the Serbian sector are only represented by two very small areas at c. 80 m altitude (Figure 5: level 1/2). Morphologically, this oldest level of the present-day system shows alternating parallel ridges and swales with an indistinct or slightly curved pattern (Figure 5). No age information is available, preventing correlation with the middle Tisza.

The next and younger phase (2) in the middle Tisza region is represented by a braid plain consisting of straight to slightly curved ridges. The radiocarbon dates from these deposits range from 18,010 BP (bulk detritus) to 13,560 BP (snails) [30], which means a (calibrated) age of c. 22–17 ka (see also discussion below). This system cuts through the large meander system 1. More recently, aeolian sands occurring on top of these fluvial deposits were dated ([36], Figure 8; slightly modified by Novothny) at −27–21.5 ka (OSL/IRSL). System 2 has been previously interpreted as the southernmost extension of the Sajo alluvial fan and may thus be of local, rather than general, importance [33,76].

There then follows a characteristic series of meanders (3) in a belt of the present-day Tisza valley that is approximately at the same altitude as the oldest meanders (system 1; level C at 88–90 m a.s.l. in the middle Tisza [30] and at c. 80 m a.s.l. in Serbia) and the braided system 2. The morphology is characterized by well-developed point bars with ridge-and-swale topography and clear sinuous erosive scars, pointing to lateral migration which has often led to neck-cut-offs. The point bars consist of fine sandy deposits occurring in a series of fining-upward sequences. Abandoned channels are always filled with, occasionally humic, fine sandy to clayey silts, as in Figure 7A. Different generations (3a–d) may be distinguished morphologically in this belt by the successive lateral erosion of previous meander remnants and by progressively decreasing meander wave length. Lateral erosion is accompanied with distinct channel incision, and the deepest meander channels reach considerable depth (up to 15 m).

The correlation of remnants of meander generations is not easy, not least because of the difficulties associated with radiocarbon dating.

Based on the relative magnitude of the meanders and their morphologically relative age, we correlate the oldest and largest meanders of that phase (3a) in the lower and middle Tis(z)a with each other, i.e., the meander of Zrenjanin with the meander of Berekfurdo. The meanders that formed during that first phase have a remarkably large wavelength, although exact determination is difficult (in the Hungarian sector it is estimated at c. 6–10 km at Berekfurdo, and in the Serbian sector at c. 10 km at Zrenjanin); the palaeochannel is up to 600 m wide and locally more than 14 m deep at Berekfurdo in Hungary [30] and up to 2 km wide in Serbia. There seems to be a lateral morphological transition into a braided system (2) at Berekfurdo, which could mean that systems 2 and the terminal part of 3a were contemporaneous. The floodplain deposits at a depth of 6 to 7 m at Zrenjanin have been dated by OSL at 33–32 ka. The uncalibrated radiocarbon age of c. 29 ka BP for the meander fill in the Berekfurdo channel [30] fits with the OSL dates obtained at Zrenjanin, but conflicts with younger ages (25–22 ka) of adjacent dissected floodplain deposits (belonging to system 1 [30] and unpublished IRSL dates for the uppermost point-bar deposits at Berekfurdo (14.3 ± 1.1 ka to 9.4 ± 2.1 ka [77]). However, the latter point-bar deposits reflect only very weak flow, as evidenced by the dominance of clayey flood deposits overlying very thin beds of fine sand without any pebbles (except locally derived clay pebbles), the absence of channelling structures, but the frequent occurrence of small ripples draped by clay; thus, they may have been rejuvenated by younger (Lateglacial) fluvial action towards the final stage of activity of that system (see below), which would explain why their young age obviously deviates from the older radiocarbon ages. In addition, the older ages of the Berekfurdo meander system (3a) are also inconsistent with the younger radiocarbon age (22–17 ka cal BP; see above) of the older or contemporaneous braided system (2) [30], although older OSL-ages (27 to 20 ka) from braided system 2 were also obtained [36] (see above). At any rate, the different dates from Berekfurdo remain inconsistent with each other. Rather than trying to reconcile the different age determinations, we favour a more generalized pattern of the fluvial evolution of system 3a. It assumes continuous activity from at least 33–32 ka until 22–17 ka when river activity slowed down (top of Berekfurdo meander) (Table 2). At that same terminal phase of system 3a, the meandering system would seem to have interfingered with braided system 2, which probably represents the outermost zone of the Sajo alluvial fan (Figure 2). Continuous meandering alluviation over that long period from c. 33 to c. 17 ka may explain the occurrence of many beds dated in that interval (e.g., [36]), occurring at different positions in the system. They may even include dated material from 25–22 ka attributed to system 1 [30]. Moreover, transitions between a braided and meandering pattern exist at present at the Ukrainian–Hungarian border and occurred also during the Lateglacial in the Maros river (tributary of the Tisa near to the Hungarian–Serbian border) [78]. The advantage of this hypothesis of fluvial evolution is that we can assemble in a single framework all absolute dates (radiocarbon and OSL) from the two regions within their error bars.

Table 2. Summary of the fluvial systems 1 to 3d, and their periods of activity in the Tis(z)a River.

Fluvial Phase	Fluvial Systems with Type-Sites	Periods of Activity
3d	Meanders W of Mužlja (lower Tisa)	Late Glacial
3c	Meanders of Tiszaörs (middle Tisza)—Mužlja (lower Tisa)	15.6 ka to start of Younger Dryas
3b	Meanders of Kunmadaras (middle Tisza)	19–17 ka
3a	Meanders of Zrenjanin (lower Tisa)—Berekfurdo (middle Tisza)	From c. 33 to 17 ka
2	'Sajo alluvial fan'—braided	c. (27–)22–17 ka
1	Large meanders of Meggyes	>28 ka

A subsequent younger phase (3b) within this system of meanders clearly dissects the meanders of phase 3a. It has distinctly smaller wavelengths than the previous generation 3a, in the range of 4–5 km at Kunmadaras. In the lower Tisa, the oldest phase eroding the Zrenjanin terrace is represented by a terrace remnant of very limited extent W of Zrenjanin. The radiocarbon age of the Kunmadaras system is 19.26–16.25 ka BP, as derived from the meander fill, but 15.26–13.56 ka BP for adjacent sediments (c. 19–17 ka after calibration). It has been suggested that the ages of the meander fill are probably a few thousand years too old [30]. The Tiszacsege meander and a series of meanders nearby (of comparable size) may correspond with the Kunmadaras meander; the pollen diagram Tiszacsege (discussed by Kasse et al. [30]) shows the fill of a meander scar that started at the final part of the Pleniglacial). Fills of large meanders, radiocarbon-dated between 25 and 20 ka, were previously reported [34,36,79] in the neighbourhood of Polgar at c. 50 km north of our study area. They could be correlated with phase 3a or 3b.

The next, third phase (3c) has smaller meander wavelengths but still occurs at the same topographic level. It is IRSL-dated for the upper point-bar series at Mužlja, giving an age of 15.6–15.3 ka [49] (Figure 9), which is definitely younger than the Kunmadaras terrace. It may correspond with the Tiszaörs meander (wavelength 3–4 km) of the middle Tisza which is radiocarbon-dated between 17.9 and 13.8 ka BP using detritus from within point-bar deposits [30]. However, the pollen diagram of the meander fill at Tiszaörs shows that infill started only in the final part of the Lateglacial [30], which is certainly much younger than the Tiszacsege meander (system 3b). We assume that the meander may have been active from the end of the Pleniglacial until the Lateglacial after which the pollen registration started and thus—accepting the radiocarbon ages—we correlate the Tiszaörs meander with the Muzlja terrace. The period between 17 and 14 ka was a very dry interval that led to reduced vegetation and river activity [34,65,80].

In the lower Tisa, the next terrace system W of Mužlja (3d) occurs some 2 m lower (at c. 75 m elevation) than the previous one at Mužlja (3c). The meander size seems again to be somewhat reduced in comparison with the previous phase. It is not recognized in the study region of the middle Tisza. Taking into account the age of the Mužlja terrace, we infer a Lateglacial age for system 3d (Table 2).

A strikingly abrupt change towards the most recent phase, which dates from the Holocene, is found in the middle Tisza. In that area, this system of small meanders (4) is separated from the next older terrace by a sharp scarp of about 4 m. In both areas, there is a strong reduction in meander length, channel depth and width (which decreased to 500–400 m) and lateral and vertical erosion of the system. It persisted until pre-modern time, i.e., until human intervention.

Table 2 summarizes the different systems in the present valley belt with their derived ages. The most striking features are the clear phase of incision slightly prior to 33 ka, followed by a rather long period of alluviation in a system of very large meanders (system 3a). This system persisted until c. 17 ka, followed by relatively rapid (relative) decrease of meander size in several stages 3b–d ending in the Lateglacial. A sharp decrease of meander size occurred at the transition to the Holocene.

6. Discussion

6.1. General Characteristics of the Middle and Lower Tis(z)a River Systems

The general fluvial pattern is that of a dominantly meandering system in which accumulation and erosion seem to have alternated. The subsidence of the Pannonian Basin led to low river gradients and thus to generally low-energy conditions that favoured a meandering river pattern. The accumulative activity is interrupted once by the deep erosion of the meanders just before system 3a (Berekfurdo, Tiszaszege). Its age could not be established precisely but is estimated to be shortly before 33 ka. Similar to but possibly slightly earlier than in more northern regions [10], the onset of frozen-ground conditions and the delayed response of deteriorating vegetation cover in the Pannonian Basin may have induced considerable erosion before system 3 alluviation, leading to increased runoff with a still low sediment supply [8]. At the time of degradation of frozen soils (towards the end of the Weichselian and/or the transition to the Holocene) and subsequent increase of soil permeability, the runoff decreased, and meanders reduced their dimensions due to lower and more steady discharges [80].

The oldest recorded systems in the present-day valley belt (systems 1 and 3a) date from the end of the Weichselian Middle Pleniglacial and show meandering patterns that are relatively large compared to later patterns [30,33]. Because of a relatively thick cover of fine-grained flood-sediment, it is not possible to determine precise form parameters such as wavelength or curvature radius. A rough estimate of that system is between 6 and 10 km in wavelength. Postdating the meanders at Berekfurdo and Zrenjanin (3a), the size of the meanders decreased successively in the meander systems from 3b to 3c in the middle Tisza (Kunmadaras and Tiszaörs meanders). In the lower Tisa, this evolution from 3b to 3d is less distinct, as the preserved terrace remnants are too small to be measured (meanders W of Mužlja). In contrast, the Holocene meanders in both regions are considerably smaller than all pre-Holocene systems. This fact is very common in many river systems ([10,81,82] and references therein).

Comparable large meanders have been described from the Russian Plain and mid-Russian Uplands [83–85], as exemplified by the Late Valdaian (~Weichselian) Seim river with meander wavelengths of 5 to 6 km, an amplitude of 4 km and channel widths 15 times the recent ones [83]. In that river, the elevation difference between the highest levels and deepest point of the palaeo-channel amounts to 14 m. In that same river valley in the southern Russian Plain, younger meanders with Lateglacial valley fills show a reduction of the wavelength to 1.3 km, while the Holocene meander dimensions decreased considerably [84]. That Holocene size reduction is absent in the northeastern Russian Plain, which is still in the modern permafrost zone [83].

Explaining the large meander sizes requires consideration of the specific energy conditions in the fluvial systems during the Middle Pleniglacial (~MIS 3) and early Late Pleniglacial (~MIS 2). The available energy for meander development is especially determined by the discharge and the river gradient [26,86] as applied for the Tisza by Petrovszki et al. [87]. Due to the geographical position of the Tis(z)a within the Pannonian Basin, river slopes are very low (see below). With respect to the runoff in this region, it may further be assumed that in a continental environment, the soils were frozen and thus impervious, at least for a considerable part of the year, but especially after the relatively cold winters [2,28]. These conditions may have been present in the Middle Pleniglacial and even more pronounced during the Late Pleniglacial when this region was situated at the margins of sporadic to discontinuous permafrost [40,41]. In addition, peak discharges may have increased due to the thaw of glacier ice in the Carpathians, feeding the Tisza catchment. Also, the rapid and spatially extensive thaw of snow cover in the low-relief setting of the Hungarian Plain may have contributed to high discharges in spring, as they have been, e.g., up to eight times larger at that time than at present in the Russian Plain [83,84]. In the Russian systems, frozen-soil conditions were probably even harsher than in western regions, while the topography was subdued in an environment of thermokarst depressions [85].

Furthermore, it is striking that in the Serbian Tisa and the Russian systems, no braided-river patterns are present during the Late Pleniglacial. Also, in the middle Tisza, a braided system was only occasionally present (see below). This contrasts with most river systems in other West European cold-temperate regions [10,22,88] (more examples in Kasse et al. [30], p. 192). Although the impact of frozen soil may have been similar in both regions during that period, the river gradients in West Europe were generally higher, even in lowlands (e.g., 0.2–0.3 m.km^{-1} and c. 0.15 m.km^{-1} for the lower Maas river and the middle Warta river in Poland, respectively [81], versus 0.02–0.15 m.km^{-1} for the Tisza river). In addition, braided development requires, in general, a relatively high sedimentation bedload which is favoured by the absence of vegetation. As the tundra vegetation in West Europe was more sparsely developed, in contrast with the well-developed steppe or even forest-steppic vegetation in central and eastern Europe, it is logical to assume less bedload sediment supply to the rivers in the latter regions. A relatively denser vegetation cover, together with the very low river gradients, may thus explain the absent tendency for braiding pattern development in those regions during the Late Pleniglacial.

6.2. Intercomparison between the Middle and Lower Tis(z)a

The general characteristics of stream patterns towards the end of the Lateglacial and the transition to the Holocene are similar in both the middle and lower Tis(z)a in Hungary and Serbia. Despite the difficulties in absolute dating of individual systems and the discontinuous nature of fluvial deposition, the trends in the evolution of the Tis(z)a in Hungary and Serbia occur in a parallel manner.

There is one specific difference, however, which is the existence of a braided system (2) during the Late Pleniglacial in the middle Tisza in contrast to the lower Tisa. This braided system in the middle Tisza seems to be of local significance and restricted to a limited period of time. Apart from the theoretical possibility of the later removal of braided deposits and river patterns in the lower Tisa, the particular location of that braided system at the southern margin of the alluvial fan at the confluence of the Sajo and Tisza Rivers should be mentioned. In addition, different energy conditions exist in the two regions. Applying the same theories of drainage pattern development as described above to explain the dominant meandering systems during the last glacial period, topographical conditions may again provide a plausible explanation for the absence of a braided pattern in the lower Tisa. Especially striking is the strongly reduced longitudinal gradient in Serbia, which at present is 1.86 cm/km [31], when compared to the gradient in the middle Tisza of 15 cm/km in central to north Hungary [30], gradually diminishing to 5–10 cm/km in a zone of intense subsidence in south-eastern Hungary [35]. In contrast, it is not feasible to invoke a substantially different vegetation cover between the latter two regions. Thus, the strongly reduced stream power in a downstream direction is supposed to be the main reason for the persistent formation of a meandering pattern, uninterrupted by braiding, in the lower Tisa during the Late Pleniglacial.

As a general conclusion, and as a result of the further studies of the Tis(z)a system in Hungary and Serbia, the importance of timescale (cf. [5]) is confirmed here. In the long term, the tectonic framework (basin development) and resulting topography have determined the available energy in this fluvial system. At a shorter timescale, climatic conditions (including their effects on vegetation and soil frost) have left a distinct imprint. Finally, looking at more detailed spatial and temporal scales, local variations of topography and sediment availability may be responsible for further shaping the river morphology [22,81] next to intrinsic variables as thresholds and delay times.

7. Conclusions

Several characteristics of fluvial development found in the continental low-relief setting of the Tis(z)a river are identical to the cold-temperate conditions in the West-European lowlands. In this respect, the sharp decrease in meander size at the Weichselian to Holocene transition is striking. However, there are also a few differences, such as the near absence of braided river patterns and the existence of exceptionally large meanders during the Weichselian Middle and Late Pleniglacial.

Previous studies have frequently stressed the importance of climate and tectonic movement as external factors in fluvial morphological development. This study of the Tis(z)a valley illustrates that, for a reliable understanding of fluvial evolution, the impact of climate on fluvial processes and morphology cannot be limited to general estimates or averages but also needs to be effectively specified in detail, for instance including the continental conditions with cold winters and relatively warm summers in the Tis(z)a catchment and in the Russian plain. In addition, they have to be supplemented with indirect climate variables (as vegetation, frozen soil and snow cover), while the impact of tectonic subsidence, and its variability within one catchment as in the Tis(z)a, has to be specified in semi-direct variables such as the longitudinal river gradient and relief intensity. This study confirms the impact of varying conditions of climate and topography at a local and regional level on fluvial development.

Supplementary Materials: The following are available online at http://www.mdpi.com/2571-550X/1/2/14/s1. Figure S1. Illustrative dose-response and luminescence decay curve (inset) for an aliquot of quartz grains extracted from sample GLL-150803; Figure S2. Dependence of equivalent dose on preheat temperature; Figure S3. Ratios of measured to given dose (dose recovery data) as a function of preheat temperature for sample GLL-150803.

Author Contributions: J.V.: conceptualization, field data collection, interpretation, writing original draft; K.K. and G.G.: conceptualization, field data collection, interpretation; S.B.: field data collection, paleobotanical interpretation; S.M.: field data collection, interpretation; D.V. and D.P.: OSL-dating.

Funding: D.V. gratefully acknowledges financial support from the Research Foundation Flanders (FWO-Vlaanderen). This research received no other external funding.

Acknowledgments: D.V. gratefully acknowledges financial support from the UGent Laboratory of Mineralogy and Petrology, as well as the technical assistance of Ann-Eline Debeer. Field work has been carried out with the help of students from the Vrije Universiteit Amsterdam (VU) in the framework of their BSc or Msc thesis. The Laboratory of Sediment Analysis of the VU took care for the grain-size analyses. We thank M. Frechen for providing luminescence measurements from the middle Tisza in 2003, and A. Novothny for communicating recent unpublished OSL-dates from the middle Tisza region. Finally, two anonymous reviewers are thanked for their useful suggestions and also the editor David Bridgland for his editorial and linguistic corrections and modifications.

Conflicts of Interest: The authors declare no conflict of interest.

References

1. Büdel, J. *Klima-Geomorphologie*; Gebrüder Bornträger: Berlin, Germany, 1977; p. 304.
2. Vandenberghe, J. The relation between climate and river processes, landforms and deposits during the Quaternary. *Quat. Int.* **2002**, *91*, 17–23. [CrossRef]
3. Starkel, L. Climatically controlled terraces in uplifting mountain areas. *Quat. Sci. Rev.* **2003**, *22*, 2189–2198. [CrossRef]
4. Vandenberghe, J. Changing fluvial processes under changing periglacial conditions. *Z. Geomorphol.* **1993**, *88*, 17–28.
5. Vandenberghe, J. Timescales, climate and river development. *Quat. Sci. Rev.* **1995**, *14*, 631–638. [CrossRef]
6. Bridgland, D.R.; Allen, P. A revised model for terrace formation and its significance for the early middle Pleistocene terrace aggradations of north-east Essex, England. In *The Early Middle Pleistocene in Europe*; Turner, C., Ed.; Balkema: Rotterdam, The Netherlands, 1996; pp. 121–134.
7. Maddy, D.; Bridgland, D.; Westaway, R. Uplift-driven valley incision and climate-controlled river terrace development in the Thames Valley, UK. *Quat. Int.* **2001**, *79*, 23–36. [CrossRef]
8. Vandenberghe, J. The fluvial cycle at cold-warm-cold transitions in lowland regions: A refinement of theory. *Geomorphology* **2008**, *98*, 275–284. [CrossRef]
9. Vandenberghe, J. River terraces as a response to climatic forcing: Formation processes, sedimentary characteristics and sites for human occupation. *Quat. Int.* **2015**, *370*, 3–11. [CrossRef]
10. Mol, J.; Vandenberghe, J.; Kasse, C. River response to variations of periglacial climate. *Geomorphology* **2000**, *33*, 131–148. [CrossRef]
11. Pan, B.; Su, H.; Hu, Z.; Hu, X.; Gao, H.; Li, J.; Kirby, E. Evaluating the role of climate and tectonics during non-steady incision of the Yellow River: Evidence from a 1.24 Ma terrace record near Lanzhou, China. *Quat. Sci. Rev.* **2009**, *28*, 3281–3290. [CrossRef]
12. Wang, X.; Van Balen, R.; Yi, S.; Vandenberghe, J.; Lu, H. Differential tectonic movements in the confluence area of the Huang Shui and Huang He rivers (Yellow River), NE Tibetan Plateau, as inferred from fluvial terrace positions. *Boreas* **2014**, *43*, 469–484. [CrossRef]
13. Kemp, J.; Pietsch, T.; Gontz, A.; Oley, J. Lacustrine-fluvial interactions in Australia's Riverine Plains. *Quat. Sci. Rev.* **2017**, *166*, 352–362. [CrossRef]
14. Marsh, P.; Woo, M.K. Snowmelt, glacier melt and High Arctic streamflow regimes. *Can. J. Earth Sci.* **1981**, *18*, 380–1384. [CrossRef]
15. Cordier, S.; Frechen, M.; Harmand, D. Dating fluvial erosion: Fluvial response to climate change in the Moselle catchment (France, Germany) since the Late Saalian. *Boreas* **2014**, *43*, 450–468. [CrossRef]
16. Cordier, S.; Adamson, K.; Delmas, M.; Calvet, M.; Harmand, D. Of ice and water: Quaternary fluvial response to glacial forcing. *Quat. Sci. Rev.* **2017**, *166*, 57–73. [CrossRef]
17. Ballantyne, C.K. A general model of paraglacial landscape response. *Holocene* **2002**, *12*, 371–376. [CrossRef]
18. Owczarek, P.; Nawrot, A.; Migala, K.; Malik, I.; Korabiewski, B. Floodplain responses to contemporary climate change in small High-Arctic basins (Svalbard, Norway). *Boreas* **2014**, *43*, 384–402. [CrossRef]
19. Dogan, U. Fluvial response to climate change during and after the Last Glacial Maximum in Central Anatolia, Turkey. *Quat. Int.* **2010**, *222*, 221–229. [CrossRef]
20. Avcin, N.; Vandenberghe, J.; van Balen, R.T.; Kıyak, N.; Öztürk, T. Tectonic and Climatic Controls on Quaternary Fluvial Processes and Terrace Formation in a Mediterranean Setting: The Göksu River, Southern Anatolia. *Quat. Res.* **2018**. under review.

21. Hughes, K.; Croke, J. How did rivers in the wet tropics (NE Queensland, Australia) respond to climate changes over the past 30000 years? *J. Quat. Sci.* **2017**, *32*, 744–759. [CrossRef]
22. Kasse, C. Depositional model for cold-climate tundra rivers. In *Palaeohydrology and Environmental Change*; Benito, G., Baker, V.R., Gregory, K.J., Eds.; Wiley and Sons: Chichester, UK, 1998; pp. 83–97.
23. Rose, J.; Meng, X. River activity in small catchments over the last 140 ka, North-east Mallorca, Spain. In *Fluvial Processes and Environmental Change*; Brown, A.G., Quine, T.A., Eds.; Wiley & Sons: Chichester, UK, 1999; pp. 91–102.
24. Olszak, J. Evolution of fluvial terraces in response to climate change and tectonic uplift during the Pleistoce: Evidence from Kamienica and Ochotnica River valleys (Polish Outer Carpathians). *Geomorphology* **2011**, *129*, 71–78. [CrossRef]
25. Stange, K.M.; van Balen, R.T.; Carcaillet, J.; Vandenberghe, J. Terrace staircase development in the Southern Pyrenees Foreland: Inferences from 10Be terrace exposure ages at the Segre River. *Glob. Planet. Change* **2013**, *101*, 97–112. [CrossRef]
26. Schumm, S. *The Fluvial System*; Wiley-Interscience: New York, NY, USA, 1977; p. 338.
27. Van Huissteden, J. *Tundra Rivers of the Last Glacial: Sedimentation and Geomorphological Processes during the Middle Pleniglacial in the Dinkel Valley (Eastern Netherlands)*; Rijks Geologische Dienst: Haarlem, The Netherlands, 1990; Volume 44, pp. 3–138.
28. Vandenberghe, J. A typology of Pleistocene cold-based rivers. *Quat. Int.* **2001**, *179*, 111–121. [CrossRef]
29. Bridgland, D.R.; Westaway, R. Preservation patterns of Late Cenozoic fluvial deposits and their implications: Results from IGCP 449. *Quat. Int.* **2008**, *189*, 5–38. [CrossRef]
30. Kasse, C.; Bohncke, S.J.P.; Vandenberghe, J.; Gabris, G. Fluvial style changes during the last glacial—Interglacial transition in the middle Tisza valley (Hungary). *Proc. Geol. Assoc.* **2010**, *121*, 180–194. [CrossRef]
31. Popov, D.; Markovic, S.; Strbac, D. Generations of meanders in Serbian part of Tisa valley. *J. Geogr. Inst. Jovan Cvijic SASA* **2008**, *58*, 29–41. [CrossRef]
32. Nádor, A.; Lantos, M.; Tóth-Makk, Á.; Thamó-Bozsó, E. Milankovitch-scale multi-proxy records from fluvial sediments of the last 2.6 Ma, Pannonian Basin, Hungary. *Quat. Sci. Rev.* **2003**, *22*, 2157–2175. [CrossRef]
33. Gabris, G.; Nador, A. Long-term fluvial archives in Hungary: Response of the Danube and Tisza rivers to tectonic movements and climatic changes during the Quaternary: A review and new synthesis. *Quat. Sci. Rev.* **2007**, *26*, 2758–2782. [CrossRef]
34. Timar, G.; Sümegi, P.; Horvath, F. Late Quaternary dynamics of the Tisza river: Evidence of climatic and tectonic controls. *Tectonophysics* **2005**, *410*, 97–110. [CrossRef]
35. Nádor, A.; Thamó-Bozsó, E.; Magyari, A.; Babinszki, E. Fluvial responses to tectonics and climate change during the Late Weichselian in the eastern part of the Pannonian Basin (Hungary). *Sediment. Geol.* **2007**, *201*, 174–192. [CrossRef]
36. Gabris, G.; Horvath, E.; Novothny, A.; Ruszkiczay-Rüdiger, Z. Fluvial and Aeolian landscape evolution in Hungary-results of the last 20 years research. *Neth. J. Geosci.* **2012**, *91*, 111–128. [CrossRef]
37. Willis, K.J.; Rudner, E.; Sümegi, P. The Full-Glacial Forests of Central and Southeastern Europe. *Quat. Res.* **2000**, *53*, 203–213. [CrossRef]
38. Sümegi, P.; Krolopp, E. Quatermalacological analyses for modelling of the Upper Weichselian paleoenvironmental changes in the Carpathian Basin. *Quat. Int.* **2002**, *91*, 53–63. [CrossRef]
39. Bacso, N. *Climate of Hungary*; Akademiai Kiado: Budapest, Hungary, 1959; p. 302. (In Hungarian)
40. Fábián, S.A.; Kovács, J.; Varga, G.; Sipos, G.; Horváth, Z.; Thamó-Bozsó, E.; Tóth, G. Distribution of relict permafrost features in the Pannonian Basin, Hungary. *Boreas* **2014**, *43*, 722–732. [CrossRef]
41. Vandenberghe, J.; French, H.M.; Gorbunov, A.; Marchenko, S.; Velichko, A.A.; Jin, H.; Cui, Z.; Zhang, T.; Wan, X. The Last Permafrost Maximum (LPM) map of the Northern Hemisphere: Permafrost extent and mean annual air temperatures, 25–17 ka BP. *Boreas* **2014**, *43*, 652–666. [CrossRef]
42. Royden, L.H.; Horvath, F. The Pannonian Basin. A case study in basin evolution. *AAPG Memoirs* **1988**, *45*, 394.
43. Horvath, F.; Cloetingh, S. Stress induced late-stage subsidence anomalies in the Pannonian Basin. *Tectonophysics* **1996**, *266*, 287–300. [CrossRef]
44. Bada, G.; Horvath, F. On the structure and tectonic evolution of the Pannonian Basin and surrounding orogens. *Acta Geol. Hung.* **2001**, *44*, 301–327.

45. Pecsi, M. *Entwicklung und Morphologie des Donautales in Ungarn*; Akademiai Kiado: Budapest, Hungary, 1959; p. 359. (In Hungarian)

46. Thamo-Bozso, E.; Murray, A.A.; Nador, A.; Magyari, A.; Babinszki, E. Investigation of river network evolution using luminescence dating and heavy-mineral analysis of Late Quaternary fluvial sands from the Great Hungarian Plain. *Quat. Geochronol.* **2007**, *2*, 168–173. [CrossRef]

47. Gabris, G. Pleistocene evolution of the Danube in the Carpathian Basin. *Terra Nova* **1994**, *6*, 495–501. [CrossRef]

48. Konert, M.; Vandenberghe, J. Comparison of laser grain size analysis with pipette and sieve analysis: A solution for the underestimation of the clay fraction. *Sedimentology* **1997**, *44*, 523–535. [CrossRef]

49. Popov, D.; Vandenberghe, D.A.G.; Marković, S.B. Luminescence dating of fluvial deposits in Vojvodina, N Serbia: First results. *Quat. Geochronol.* **2012**, *13*, 42–51. [CrossRef]

50. Bøtter-Jensen, L.; Andersen, C.E.; Duller, G.A.T.; Murray, A.S. Developments in radiation, stimulation and observation facilities in luminescence measurements. *Radiat. Meas.* **2003**, *37*, 535–541. [CrossRef]

51. Lapp, T.; Kook, M.; Murray, A.S.; Thomsen, K.J.; Buylaert, J.-P.; Jain, M. A new luminescence detection and stimulation head for the Risø TL/OSL reader. *Radiat. Meas.* **2015**, *81*, 178–184. [CrossRef]

52. Duller, G.A.T. Distinguishing quartz and feldspar in single grain luminescence measurements. *Radiat. Meas.* **2003**, *37*, 161–165. [CrossRef]

53. Murray, A.S.; Wintle, A.G. Luminescence dating of quartz using an improved single-aliquot regenerative-dose protocol. *Radiat. Meas.* **2000**, *32*, 57–73. [CrossRef]

54. Murray, A.S.; Wintle, A.G. The single aliquot regenerative dose protocol: Potential for improvements in reliability. *Radiat. Meas.* **2003**, *37*, 377–381. [CrossRef]

55. Hossain, S.M. A Critical Comparison and Evaluation of Methods for the Annual Radiation Dose Determination in the Luminescence Dating of Sediments. Ph.D. Thesis, Ghent University, Gent, Belgium, 2003.

56. Vandenberghe, D. Investigation of the Optically Stimulated Luminescence Dating Method for Application to Young Geological Samples. Ph.D. Thesis, Ghent University, Gent, Belgium, 2004.

57. De Corte, F.; Vandenberghe, D.; De Wispelaere, A.; Buylaert, J.-P.; Van den haute, P. Radon loss from encapsulated sediments in Ge gamma-ray spectrometry for the annual radiation dose determination in luminescence dating. *Czechoslov. J. Phys.* **2006**, *56*, D183–D194. [CrossRef]

58. Adamiec, G.; Aitken, M. Dose-rate conversion factors: Update. *Ancient TL* **1998**, *16*, 37–50.

59. Mejdahl, V. Thermoluminescence dating: Beta-dose attenuation in quartz grains. *Archaeometry* **1979**, *21*, 61–72. [CrossRef]

60. Aitken, M.J. *Thermoluminescence Dating*; Academic Press Inc.: London, UK, 1985; p. 359.

61. Vandenberghe, D.; De Corte, F.; Buylaert, J.-P.; Kucera, J.; Van den haute, P. On the internal radioactivity in quartz. *Radiat. Meas.* **2008**, *43*, 771–775. [CrossRef]

62. Prescott, J.R.; Hutton, J.T. Cosmic ray contributions to dose rates for luminescence and ESR dating: Large depths and long-term time variations. *Radiat. Meas.* **1994**, *23*, 497–500. [CrossRef]

63. Aitken, M.J. Thermoluminescence age evaluation and assessment of error limits: Revised system. *Archaeometry* **1976**, *18*, 233–238. [CrossRef]

64. Aitken, M.J.; Alldred, J.C. The assessment of error limits in thermoluminescence dating. *Archaeometry* **1972**, *14*, 257–267. [CrossRef]

65. Vandenberghe, D.; Kasse, C.; Hossain, S.M.; De Corte, F.; Van den haute, P.; Fuchs, M.; Murray, A.S. Exploring the method of optical dating and comparison of optical and 14C ages of Late Weichselian coversands in the southern Netherlands. *J. Quat. Sci.* **2004**, *19*, 73–86. [CrossRef]

66. Vandenberghe, J.; Bohncke, S.; Lammers, W.; Zilverberg, L. Geomorphology and palaeoecology of the Mark valley (southern Netherlands). I Geomorphological valley development during the Weichselian and Holocene. *Boreas* **1987**, *16*, 55–67. [CrossRef]

67. Novotny, A.; Frechen, M.; Horvath, E.; Wacha, L.; Rolf, C. Investigating the penultimate and last glacial cycles of the Süttö loess section (Hungary) using luminescence dating, high-resolution grain size, and magnetic susceptibility data. *Quat. Int.* **2011**, *234*, 75–85. [CrossRef]

68. Bokhorst, M.; Vandenberghe, J.; Sümegi, P.; Lanczont, M.; Gerasimenko, N.P.; Matviishina, Z.N.; Markovic, S.B.; Frechen, M. Atmospheric circulation patterns in Central and Eastern Europe during the Weichselian Pleniglacial inferred from loess grain-size records. *Quat. Int.* **2011**, *234*, 62–74. [CrossRef]

69. Vandenberghe, J.; Sun, Y.; Wang, X.; Abels, H.A.; Liu, X. Grain-size characterization of reworked fine-grained aeolian deposits. *Earth Sci. Rev.* **2018**, *177*, 43–52. [CrossRef]

70. Wang, X.; Ma, J.; Yi, S.; Vandenberghe, J.; Dai, Y.; Lu, H. Interaction of fluvial and aeolian sedimentation processes and response to climate change since the last glacial in a semi-arid environment along the Yellow River. *Quat. Res.* **2018**, *1*. [CrossRef]

71. Verdonk, S. Fluvial development of lithology and geomorphology of Tisza river terraces since the last glacial maximum in Vojvodina, Serbia. BSc Thesis, Faculty of Earth and Life Sciences, Vrije Universiteit, Amsterdam, The Netherlands, 2003.

72. Marković, S.B.; Bokhorst, M.; Vandenberghe, J.; McCoy, W.D.; Oches, E.A.; Hambach, U.; Gaudenyi, T.; Jovanović, M.; Stevens, T.; Zöller, L.; et al. Late Pleistocene loess-palaeosol sequences in the Vojvodina region, north Serbia. *J. Quat. Sci.* **2008**, *23*, 73–84. [CrossRef]

73. Marković, S.B.; Stevens, T.; Kukla, G.J.; Hambach, U.; Fitzsimmons, K.E.; Gibbard, P.; Buggle, B.; Zech, M.; Guo, Z.T.; Hao, Q.Z.; et al. The Danube loess stratigraphy—New steps towards the development of a pan-European loess stratigraphic model. *Earth Sci. Rev.* **2015**, *148*, 228–258. [CrossRef]

74. Popov, D.; Marković, S.B.; Jovanović, M.; Mesaroš Arsenović, D.; Stankov, U.; Gubik, D. Geomorphological investigations and GIS approach of the Tamiš loess plateau, Banat region (northern Serbia). *Geogr. Pannonica* **2012**, *16*, 1–9. [CrossRef]

75. Jain, M.; Murray, A.S.; Bøtter-Jensen, L. Optically stimulated luminescence dating: How significant is incomplete light exposure in fluvial environments? *Quaternaire* **2004**, *15*, 143–157. [CrossRef]

76. Borsy, Z. Evolution of the alluvial fans of the Alföld. In *Alluvial Fans: A Field Approach*; Rachocki, A.H., Church, M., Eds.; Wiley and Sons Ltd.: Chichester, UK, 1990; pp. 229–246.

77. Frechen, M.; Leibniz Institute for Applied Geophysics, Hannover, Germany. Personal communication, 2003.

78. Berec, B.; Gábris, G. A Maros hordalékkúp bánsági szakasza (Alluvial fan of Maros River in Banat, Serbia–Romania). In *Kárpát-Medence: Természet, Társadalom, Gazdaság (Carpathian Basin: Nature, Society, Economy)*; Frisnyák, S., Gál, A., Eds.; Nyíregyháza–Szerencs: Nyíregyháza, Hungary, 2013; pp. 51–64. (In Hungarian)

79. Davis, B.A.S.; Passmore, D.G. Upper Tisza Project: Radiocarbon analyses of Holocene alluvial and lacustrine sediments. In *Interim Report on Current Analyses to the Excavation and Fieldwork Committee*; University of Newcastle: Callaghan, Australia, 1998; Volume 7.

80. Kasse, C. Cold-climate aeolian sand-sheet formation in north-western Europe (c. 14-12.4 ka): A response to permafrost degradation and increased aridity. *Permafr. Periglac. Process.* **1997**, *8*, 295–311. [CrossRef]

81. Vandenberghe, J.; Kasse, C.; Bohncke, S.; Kozarski, S. Climate-related river activity at the Weichselian-Holocene transition: A comparative study of the Warta and Maas rivers. *Terra Nova* **1994**, *6*, 476–485. [CrossRef]

82. Kasse, C.; Vandenberghe, J.; Bohncke, S. Climatic change and fluvial dynamics of the Maas during the Late Weichselian and Early Holocene. In *European River Activity and Climatic Change during the Lateglacial and Early Holocene*; Frenzel, B., Vandenberghe, J., Kasse, C., Bohncke, S., Gläser, B., Eds.; Gustav Fischer: Jena, Germany, 1995; Volume 14, pp. 123–150.

83. Sidorchuk, A.; Borisova, O.; Panin, A. Fluvial response to the Late Valdai/Holocene environmental change on the East European Plain. *Glob. Planet. Change* **2001**, *28*, 303–318. [CrossRef]

84. Sidorchuk, A.; Panin, A.; Borisova, O. Morphology of river channels and surface runoff in the Volga River basin (East European Plain) during the Late Glacial period. *Geomorphology* **2009**, *113*, 137–157. [CrossRef]

85. Borisova, O.; Sidorchuk, A.; Panin, A. Palaeohydrology of the Seim River basin, Mid-Russian Upland, based on palaeochannel morphology and palynological data. *Catena* **2006**, *66*, 53–73. [CrossRef]

86. Leopold, L.B.; Wolman, L.G.; Miller, J. *Fluvial Processes in Geomorphology*; W.H. Freeman and Co.: San Francisco, CA, USA, 1964; p. 522.

87. Petrovszki, J.; Timar, G.; Molnar, G. Is sinuosity a function of slope and bankfull discharge?—A case study of the meandering rivers in the Pannonian Basin. *Hydrol. Earth Syst. Sci. Discuss.* **2014**, *11*, 12271–12290. [CrossRef]
88. Van Huissteden, J.; Vandenberghe, J. Changing fluvial style of periglacial lowland rivers during the Weichselian Pleniglacial in the eastern Netherlands. *Z. Geomorphol.* **1988**, *71*, 131–146.

![Q quaternary]

MDPI

Article

Climatically-Controlled River Terraces in Eastern Australia

James S. Daley [1,2,*] and Tim J. Cohen [3,4]

[1] Griffith Centre for Coastal Management, Griffith University, Gold Coast, QLD 4215, Australia
[2] School of Earth and Environmental Science, University of Queensland, St. Lucia, QLD 4072, Australia
[3] ARC Centre of Excellence for Australian Biodiversity and Heritage–School of Earth and Environmental Sciences, University of Wollongong, Wollongong, NSW 2522, Australia; tcohen@uow.edu.au
[4] GeoQuEST Research Centre–School of Earth and Environmental Sciences, University of Wollongong, Wollongong, NSW 2522, Australia
*   Correspondence: j.daley@griffith.edu.au

Academic Editors: Jef Vandenberghe, David Bridgland and Xianyan Wang
Received: 23 August 2018; Accepted: 13 October 2018; Published: 16 October 2018

Abstract: In the tectonically stable rivers of eastern Australia, changes in response to sediment supply and flow regime are likely driven by both regional climatic (allogenic) factors and intrinsic (autogenic) geomorphic controls. Contentious debate has ensued as to which is the dominant factor in the evolution of valley floors and the formation of late Quaternary terraces preserved along many coastal streams. Preliminary chronostratigraphic data from river terraces along four streams in subtropical Southeast Queensland (SEQ), Australia, indicate regionally synchronous terrace abandonment between 7.5–10.8 ka. All optically stimulated luminescence ages are within 1σ error and yield a mean age of incision at 9.24 ± 0.93 ka. Limited samples of the upper parts of the inset floodplains from three of the four streams yield near-surface ages of 600–500 years. Terrace sediments consist of vertically accreted fine sandy silts to cohesive clays, while top stratum of the floodplains are comprised of clay loams to fine-medium sands. The inundation frequency of these alluvial surfaces depends on their specific valley setting. In narrow valley settings, where floodplains comprise <5% of the valley floor, terraces are inundated between the 20 and 50-year annual exceedance probability (AEP) flood, while in wide settings (floodplains >20%), the terraces are no longer inundated. Floodplain inundation frequencies also vary between these settings by an order of magnitude between 5- to 50-year AEP, respectively. The correlation of terrace abandonment within SEQ with fluvial and palaeoenvironmental records elsewhere in the subtropics, and more broadly across eastern Australia, are an indication that terrace abandonment has primarily been driven by climatic forcing. Contemporaneous channel incision in the early Holocene may have been driven by an increasingly warmer and wetter environment in SEQ, with a climate commensurate with the delivery of more extreme weather events. Following channel incision, many streams in SEQ have been largely confined to their entrenched "macrochannel" form that remains preserved within the valley floor.

Keywords: terrace; channel entrenchment; extrinsic controls; Holocene; climate; optically stimulated luminescence; OSL; eastern Australia

---

1. Introduction

River terraces typically form due to tectonic or climatic controls but can also develop due to intrinsic factors (within-system changes, such as slope threshold exceedance, channel avulsion, or stream capture) [1–3], as well as direct or indirect human disturbances [4,5]. In tectonically-controlled settings, persistent uplift can create terrace staircases several hundred of meters above the present channel, preserved over millennia [6,7]. However, even in such settings, climatic controls can remain critical factors that cause subsequent incision [8]. A global comparison of terrace sequences reveals that terrace formation is dominantly a climatically-induced process, commonly associated with transitions between glacial and interglacial cycles. Issues remain regarding the exact timing of terrace development within the climatic cycle, which may depend on other geoclimatic factors [9–13]. Generally, aggradation is initiated during glacial cycles and pronounced channel incision occurs during the subsequent deglacial (warming) phase. Channel incision and floodplain abandonment (e.g., Figure 1) may reflect changes in discharge or sediment regimes associated with rainfall, vegetation changes, or weathering/erosion production [13].

In Australia, low to negligible rates of tectonic uplift and denudation over the past ca. 30 Ma [14,15], along with high streamflow variability [16,17], ensure terraces are abandoned relatively slowly. Shallow alluvium in narrow valley settings often feature polycyclic alluvial units which are only partially removed under subsequent climatic regimes [18,19]. In southeastern Australia, debate has ensued as to whether terraces have formed in response to climatic controls or intrinsic properties [20,21]. In this region, chronological investigation of alluvial fills has identified a hiatus in sedimentary records from ca. 10–4.5 ka [21,22]. This gap was initially interpreted as an erosional void caused by episodic but laterally active river systems under a warmer and wetter climate, termed the Nambucca Phase and comparative in timing to the Holocene hypsithermal [22]. Contemporaneous terrace abandonment and erosion occurred in two basins on the subtropical eastern seaboard Nambucca and Bellinger basins [18,22] (Figure 2B) and a comparative alluvial gap has also been identified in the wet tropics [23–25].

Additional palaeoclimatic records (e.g., pollen, charcoal, and other biological indicators) suggest increased warming or precipitation between 10–6 ka in eastern Australia [26–29] accompanied by abrupt changes in vegetation cover, with a significant expansion of arboreal taxa recorded from 10 ka [30–32]. Increasing lake levels (relative to today) across the continent provide further indication of a broad increase in effective precipitation [33–35]. In the eastern highlands, wet conditions for peat formation occurred from ca. 9–6 ka [36] and a reduction in charcoal abundance on the subtropical North Stradbroke Island also indicates regionally wetter conditions at this time [31]. A general reduction in rainfall has been interpreted from ca. 6 ka [28,37].

In contrast, geomorphic responses to intrinsic drivers can produce a range of chronologically disjunct terraces along a channel network [18]. Indeed, upstream channel instability can result in downstream infilling and apparent asynchronicity in longitudinal and adjacent valley fills [38]. A focus on catastrophic floodplain formation and pseudo-cyclic river behavior has emerged from the Australian geomorphic literature [39–41]. The role of large floods has been implicit in the understanding of past and present river dynamics and form-process relationships in southeastern Australia [39–42]. Large-magnitude flood events often act as geomorphic perturbations that exceed thresholds of channel stability and adjust the channel boundaries [42,43]. As such, intrinsic factors such as within-channel sedimentation, channel avulsion, and stochastic events have been suggested as the dominant mechanism in the dynamic evolution of late Quaternary alluvial units in eastern Australia [20,41].

Research in Southeast Queensland (SEQ) has highlighted that the riverine response to extensive high-magnitude flooding in recent years was largely affected by the presence of entrenched river channels, termed macrochannels [44–49]. Macrochannels were first described in the South African Sabie River [50,51], but similar channel forms have been identified elsewhere [52–54]. Such streams feature large, compound river channels where multiple alluvial units are inset within a broader, entrenched channel (Figure 1). An active low-flow channel lies within this much larger "macrochannel", which can accommodate flood waters of extreme magnitude. Along Lockyer Creek in SEQ, synchronous incision of a Pleistocene terrace after 11.5–9.3 ka suggests the macrochannel formed rapidly as a basin-wide response [46]. However, given the extent of these findings are limited to the partly-confined, mid-reaches of Lockyer Creek, it remains unclear whether channel entrenchment and terrace abandonment was a single-basin phenomenon or reflects a regional scale response.

This is a critical question with regard to understanding regional stream patterns and the inherent drivers of macrochannel formation. Regional integration of fluvial processes, such as channel entrenchment, can provide substantial evidence of the factors driving channel change [11,55]. Did channel entrenchment in SEQ occur coevally or have intrinsic factors driven an asynchronous response? This paper investigates adjacent alluvial units at four sites throughout SEQ exhibiting macrochannel morphologies to determine the chronological characteristics of the channel boundary and whether Quaternary regional climate change has had an impact on contemporary river systems.

Figure 1. Field photo from the left bank of the incised Lockyer Creek macrochannel, exhibiting inset benches, steep banks, and a terrace surface beyond the macrochannel boundary. Bankfull capacity of the macrochannel exceeds the 50-year recurrence interval.

2. Materials and Methods

2.1. Site Selection and Fieldwork

SEQ is a biogeographical region ranging in latitude from 26° to 28° south, along the central eastern coast of Australia (Figure 2), consisting of four coastal basins with a total area of approximately 20,400 km^2 that rise along the eastern slopes of the Great Dividing Range. The region includes the major metropolitan areas of Brisbane and the Gold Coast (Figure 2). The geology of SEQ has primarily developed from a complex series of tectonic processes from the Late Palaeozoic onwards, primarily associated with the development of the New England Orogen. The dominant lithologies are broadly comprised of Palaeozoic to early Triassic metasediments, Mesozoic intracratonic sedimentary basins and intrusive granites, and early Miocene volcanics capping the highland regions along the basin boundaries (Table 1). Erosion-resistant Palaeozoic metamorphic and igneous lithologies dominate much of the highlands throughout the region while the lowland and coastal regions are mostly comprised of Mesozoic sedimentary basins and more recently deposited sediments.

The SEQ region has a subtropical climate with mean monthly temperatures between 21 to 29 °C and annual rainfall between 700–2200 mm. High seasonal, inter-annual and inter-decadal variability in rainfall is driven by a number of Pacific climate phenomena that modulate ocean temperatures, pressure gradients, and precipitation across the Indo-Pacific region, including the El Niño-Southern Oscillation (ENSO) and the Inter-Decadal Pacific Oscillation (IPO) [56–59]. Most rainfall occurs in the summer months between October and February followed by a relatively dry winter. The most substantial sources of precipitation for the region are summer-dominated, east-coast, low-pressure systems and the southward migration of post-tropical cyclone storms.

Extreme climatic events are frequent and include long-term droughts, very intense rainfall, flooding, and flash flooding. Significant localized variations in annual rainfall are also driven by orographic effects related to the presence of highland ranges along basin divides. SEQ is broadly characterized by a high index of flood variability, particularly in the major river systems. A review of flood variability along the east coast of Australia found from three metrics that peak discharge variability occurs in SEQ, declining to the south in the relatively smaller Gold Coast streams [17].

Four study sites within SEQ were selected for geomorphic and chronostratigraphic analysis, including the Brisbane, Logan, and Albert Rivers and Lockyer Creek (Figure 2). Study sites were paired with stream gauging stations (with minimum record lengths of 58 years) to provide an association of alluvial surfaces with contemporary inundation frequencies. Flood frequency analyses utilized the Generalised Extreme Value of the annual maximum series for water years (September–August) and is further detailed in [60]. All sites are located in the mid-reaches of their valleys and feature macrochannel morphologies with varying degrees of inset alluvial units, but all are characterized as having Quaternary alluvium on both channel margins.

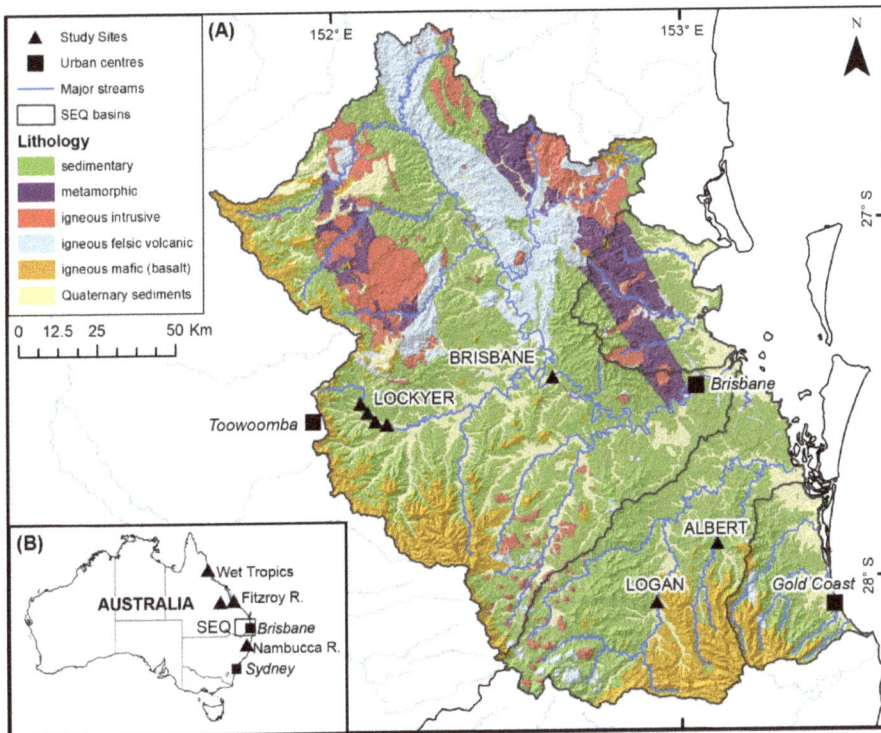

Figure 2. (**A**) Location map of study sites, major streams, and basins in SEQ, Australia. Underlying base map shows 1:1M Surface Geology (Geoscience Australia) (**B**) Inset shows the location of SEQ in Australia and additional site locations discussed in text.

Geomorphic mapping was undertaken by delineating units through hypsographic analysis, using a probability distribution function, of a detrended digital elevation model (DEM), derived from 1 m LiDAR data [61,62]. Alluvial valley floors were identified as relatively flat-lying (<2° slope) surfaces adjacent to the channels. Based on the presumption that alluvial units are generally flat-lying, breaks in the slope of the curve define the boundaries between adjacent landforms, which were classified based on their relative height above the channel. Peaks in the hypsographic curve are indicative of a higher proportion of the valley floor associated with a given elevation range. Breakpoints were assessed against slope class intervals and hillshade rasters derived from the DEM for consistency with user-identified geomorphic breaks.

Field observations were undertaken to characterize the geomorphic structure (Figure 1) and verify mapping at each site. Near-surface sediments on the adjacent terrace and inset units were targeted for geochronology to determine how recently surface sediments that form the macrochannel boundaries were deposited. The terraces are presumed to be fill terraces following chronostratigraphic descriptions in the Lockyer Creek basin [46]. In some circumstances, the Lockyer Creek terrace sits on an elevated bedrock strath surface, but preservation of older basal material within inset units suggests the underlying bedrock strath existed prior to the deposition of the alluvial fill [46]. However, as only short cores were collected in this study, the alluvial units investigated in this study are presumed to be equivalent to those of Lockyer Creek given the proximity of the sites and comparative forms of the valley floors.

The sampling approach assumed that ages from terrace surfaces reflected the last period of valley floor aggradation, vis-à-vis abandonment. Bank exposures were analyzed at two sites, and short cores up to 3 m long were extracted in units directly adjacent to channel banks using a hand-held percussion corer to date the abandonment age of various alluvial surfaces.

Table 1. Summary details of field sites.

Site Location	Lockyer Creek	Logan River	Albert River	Brisbane River
Gauge ID	143203	145008	145102	143001
Gauge Name	Helidon	Round Mountain	Bromfleet	Vernor
Drainage area (km)	357	1262	544	10,172
Distance upstream (km)	99	125	47	131
Elevation (m)	129	44	28	18
Slope (m·m^{-1})	0.002	0.001	0.001	0.0002
Stream order	6	6	6	8
Channel setting	partly confined	partly confined	partly confined	partly confined–unconfined
Bankfull AEP (a^{-1})	>200	20	40	48
Bed load	mixed sand-cobble	m/c sand	m/c sand	mixed sand–cobble
Primary geology [1]	Lithofeldspathic labile and feldspathic labile sandstone; quartzose sandstone, siltstone, shale, minor coal, ferruginous oolite marker	Lithofeldspathic labile and feldspathic labile sandstone; quartzose sandstone, siltstone, shale, minor coal, ferruginous oolite marker	Lithofeldspathic labile and feldspathic labile sandstone; quartzose sandstone, siltstone, shale, minor coal, ferruginous oolite marker	Lithofeldspathic labile and feldspathic labile sandstone
Secondary geology [1]	Quartz-lithic and quartzose sandstone, quartz-rich granule conglomerate, silty sandstone, siltstone, claystone; carbonaceous siltstone and claystone	Alkali-olivine basalt	Alkali-olivine basalt	Feldspathic and lithic meta-arenite, metasiltstone and conglomerate proximal turbidites, with structurally intercalated or stratigraphically underlying chert, jasper and basic meta-volcanics.
Soils [2]	Fine textured alluvium—hardsetting brown and grey loam surface soil over neutral to alkaline clay	Fine textured alluvium—black earth and grey clays	Fine textured alluvium—black earth and grey clays	Fine textured alluvium—black earth and grey clays

[1] Geological descriptions after 1:25,000 scale Geoscience Data, Queensland Mapping Data. [2] Soil descriptions after 1:2,000,000 scale Digital Atlas of Australian Soils.

2.2. Chronology

Cores were examined under subdued red-light (>590 nm) conditions and sampled for luminescence and particle size analysis. Grain size was described from field texture tests using grain size class intervals [58], to primarily distinguish between clay, loam, sand, and gravels. Particle size analysis using a Mastersizer (Malvern Instruments, Malvern, UK) was undertaken on representative samples to verify field textures and the relative change with depth. The degree of alteration or pedogenesis (e.g., iron staining, carbonates, or ped development) was described based on the presence of sedimentary structures and the nature of contact between unit boundaries.

A combination of optically stimulated luminescence (OSL) and radiocarbon dating was used to establish the chronology of the sites across the four rivers. In total, nine samples were analyzed for OSL analysis and two charcoal samples were analyzed for radiocarbon determinations. OSL samples were collected from the center of cores with particular care to avoid stratigraphic boundaries following standard procedures by [63,64]. Following the pre-treatment of samples to remove carbonates, organics, feldspars, and heavy minerals, single-grain (180–220 μm) quartz sands were etched in 48% hydrofluoric acid for 40 min to remove the outer 10 μm α-irradiated rind. Etched grains were analyzed using optical stimulation, and equivalent dose (D_e) values on individual grains were determined according to the modified single aliquot-regenerative dose (SAR) protocol [65] and Risø instrumentation described therein, using the acceptance criteria listed by [66]. Age modelling from measured D_e values followed the well-established approach of [67–70]. Equivalent dose determination of two samples with high overdispersion (σ_d) followed [66] by fitting a single Gaussian curve to the peak of a multi-Gaussian summed probability distribution. The final age determinations are summarized in Table 2.

Lithogenic radionuclide concentrations were analyzed from sediments adjacent to OSL samples using high-resolution gamma spectrometry for ^{238}U, ^{226}Ra, ^{210}Pb, ^{232}Th, and ^{40}K concentrations [71]. Dose rates were calculated using the conversion factors [72] and β-attenuation factors [73]. Cosmic dose rates were calculated from [74] using an alpha-efficiency "α" value of 0.04 ± 0.02 [75]. Water content was measured directly from each sample and assigned errors of ±5% to account for uncertainty. Water content was assumed to be representative of saturation levels over the full period of burial.

Radiocarbon analyses on charcoal fragments were conducted at Beta Analytic Inc. laboratory using the ^{14}C Accelerator Mass Spectrometry (AMS) technique, with age and error calculations following parameters outlined by [76]. Table 3 presents radiocarbon results as conventional ^{14}C ages and calibrated ages in cal. ka BP using the SHCAL13 database [77].

3. Results

3.1. Integrity of Chronology

Chronological data for OSL samples is presented in Table 2 and for radiocarbon in Table 3. All eight OSL samples exhibited good luminescence characteristics. Optical ages were all from single-grain determinations. For samples 16-0115-002, 16-0115-005, 16-0115-007, 16-0115-008, and 16-0115-009, D_es were calculated using the central age model (CAM) [69]. The low σ_d values in the range of 15–20% suggested complete bleaching and hence the use of the CAM (Figure 3A). For sample 16-0115-001, although the σ_d was 28%, no clear stratigraphic distinction could be made between this and the underlying sample (with a lower σ_d, sample 16-0115-002; Table 2). In this case, the CAM was accepted. Some samples had high σ_d (25–60%), and in these instances, age models were employed on a sample-by-sample basis.

In some examples, the difference between the D_e returned from the CAM and the minimum age model (MAM) had no impact on the final age determination, with the ages for each model within 1σ error (sample 16-0115-004 in Table 2; the D_es for both models are presented in Figure 3B). Sample 16-0115-0003 displayed a high degree of mixing and over-dispersion of 60%. For samples such as this (and 16-0115-004, Figure 2B), equivalent doses determined that using the MAM yields results were consistent with the dominant population of a finite mixture model (FMM) and the multi-Gaussian summed probability approach. As such, the MAM was selected for these two samples and it was assumed they were partially bleached. However, the low-dose components identified in some of the samples were likely to be a contamination signal due to bioturbation (e.g., grain instrusion).

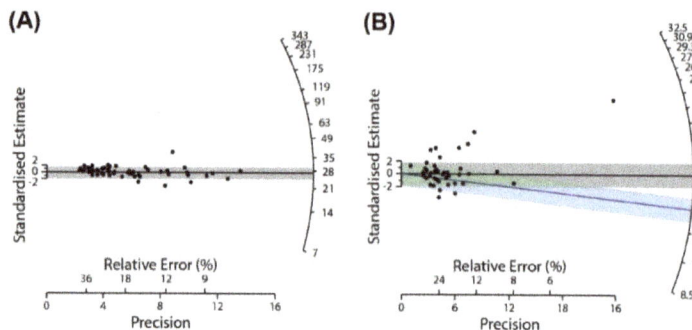

Figure 3. Radial plots for OSL samples displaying equivalent doses (De) and model predictions for (**A**) 16-0115-0002, and (**B**) 16-0115-0004. The black line is CAM and the blue line is MAM.

3.2. Valley Floor Characteristics

Valley floors across the four sites were 0.5–3 km wide and were characterized by dissected floodplain surfaces in partly confined settings. The Logan and Albert Rivers had similar bed slopes of 0.001 m·m^{-1} and basin areas varied between 540 and 1260 km^2 (Table 1). All four study sites featured the same primary underlying geology (Table 1). Three of the four basins shared the same geological units in their upper basins, draining predominantly basalts and feldspathic sedimentary lithologies (Figure 2). The Brisbane River site was in the partly confined mid-reaches of the basin and was located several kilometers downstream of its junction with Lockyer Creek. The reach features the same underlying geology found in the mid- to lower Lockyer, but upstream it was dominated by metasediments (Table 1). The stream abruptly changed course at the downstream extent of the study reach as it again abutted this lithological unit (Figure 4A). The Brisbane reach had a slope of 0.0002 m·m^{-1} and a basin area an order of magnitude larger than the other sites (Table 1). All streams featured comparative morphologies to the Lockyer Creek [46,49]. In all settings, infrequently inundated valley floors bound comparatively narrow macrochannels 50–100 m wide that featured an active channel and inset alluvial units, with up to four alluvial units across the valley floor; bars, benches/inset floodplains, and terraces. The channels were low-sinuosity, but featured sharp, structurally-controlled bends. However, two valley settings could be distinguished by the spatial extent of their floodplains.

Figure 4A shows the distribution of alluvial units along the Brisbane River at Vernor (mid-Brisbane). The wide floodplain setting was comparable to the Lockyer Creek at Helidon, where floodplain formation had been influenced by changes in variations in underlying lithology [46]. The valley floor on the Brisbane River had a complex morphology (Figure 4A). The site featured a number of alluvial units, with floodplains comprising 30–40% of the valley floor within the reach (Figure 4A). However, the floodplain had a recurrence interval for overtopping of 50 years under the current flow regime (annual maximum series), while the T1 terrace on the Brisbane River exceeded the 100-year recurrence interval flood and there were no records it had been inundated in historical records. In the mid-Brisbane reach, a higher terrace (T2 in Figure 4A) occurred as discontinuous remnants ≈4 m above the T1 terrace. However, it was unclear whether there were two terrace surfaces or whether the apparent height difference between the two terraces was simply a function of drainage development and terrace dissection. The T1 terrace and floodplain on the mid-Brisbane were heavily dissected, with numerous palaeochannels and meander cut-offs within inset units across the valley floor. While T2 was comparatively less dissected, it featured higher order drainage development along its margin with

T1 (Figure 4A). At this location, we have mapped the terraces as two separate units, but we do not have chronological or stratigraphic control that allowed us to verify this. This upper T2 terrace was also clearly identified along the Logan River, with similar drainage features (Figure 4C), but was not present along other streams and was not further investigated in this study.

In comparison to the Brisbane River, the valley floor morphology in the other locations was comparatively simple (Figure 4B). In these settings, the valley floor was dominated by a single terrace (T1) which occupied up to 90% of the valley floor. Floodplains were comparatively narrow (10–30 m wide) and are commonly referred to as alluvial benches within the Australian literature [40,78,79]. These were inset within the macrochannel and are inundated by 2–5-year annual exceedance probability (AEP) floods. Valley floors were up to 2 km wide and were dominated by the T1 terrace, which formed the boundary of the macrochannels, was longitudinally continuous down valley, and has inundation frequencies of between the 20 and the 50-year AEP. Occasional palaeochannels along T1 suggested comparative channel widths prior to abandonment (Figure 4B).

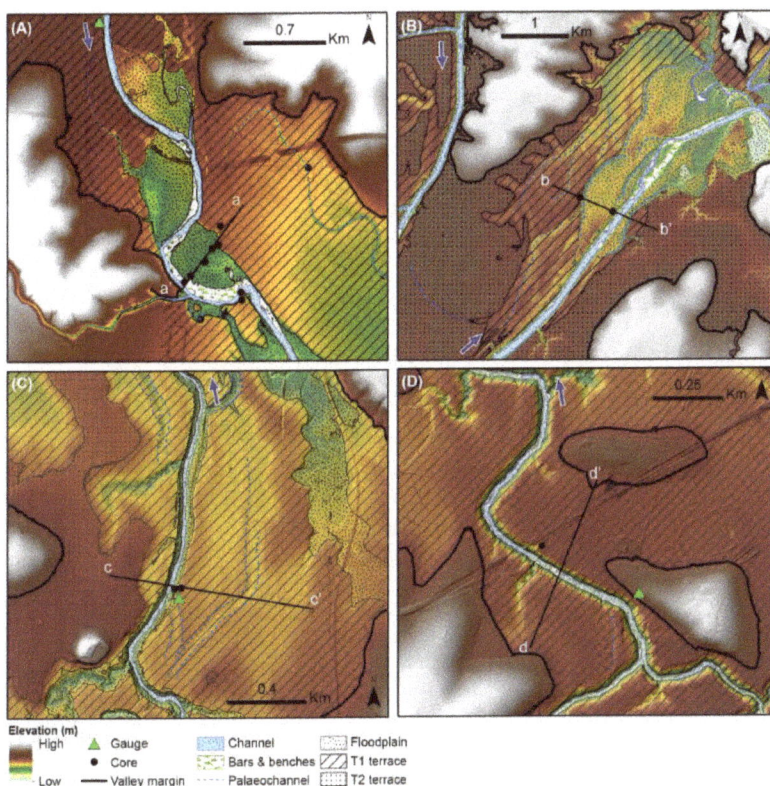

Figure 4. Representative geomorphic maps showing the distribution of alluvial units and the spatial extent of terraces along the valley floor for (**A**) Lockyer Creek at Site 3 (Helidon), (**B**) Brisbane River at Vernor, (**C**) Logan River at Round Mountain, (**D**) Albert River at Bromfleet. Also shown are the location of cores, stream monitoring gauges, and valley cross sections shown in Figure 5. Blue arrows indicate flow direction. Gauge locations for Brisbane River are <500 m beyond the mapping extent of (**B**).

3.3. Alluvial Sedimentology and Chronology

Short cores 1–4 m in length were collected from units adjacent to the channel at each site (Figure 5). Terraces sediments were consistent with overbank vertical accretion and were comprised of fining up, light brown, very fine–fine sandy silts, to a dense, black silty clay. Silt and clay content varied with depth, but was overall relatively homogenous. The stratigraphic relationships varied across sites, but overall expression of sediments was relatively consistent (Figure 5). At three sites, the upper facies of the terrace consisted of highly cohesive black clay (7.5 yr 2.5/1) and showed a gradual down-profile transition to sandier units. No mottling was apparent in the cores, but Fe concretions and detrital carbonates <2 mm diameter were present throughout the cores at 0.5–1.5 m depths.

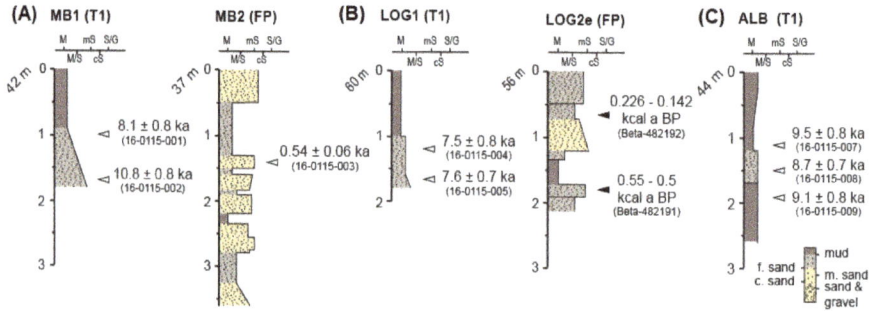

Figure 5. Core logs sedimentology and chronological data for (**A**) Brisbane River, (**B**) Logan River, and (**C**) Albert River. Geomorphic units are shown in brackets next to the log name; see Figures 1 and 4 for locality of sites and Figure 6 for transects.

Floodplain sediments were more stratigraphically variable and consisted of coarser-grained sediment than terraces, dominantly composed of unconsolidated clay loams to massive sands. The surficial units reflected overbank sediments on the floodplain levee comprised of vertically accreted, decimeter-thick beds. The loams consisted of light-brown to brown sandy clay loam to sandy clay with fine sands and small quantities of fragmented charcoal and minor ferruginous staining. Sharp boundaries occurred with massive beds of brown, fine–medium sand that featured increasing silt content down the profile. Sands were well sorted, unaltered, and had sharp contacts with adjacent units.

A preliminary terrace abandonment chronology was provided using OSL dating of near-surface sediments within terraces adjacent to terrace scarps within four basins, as well as basal floodplain material in Lockyer Creek (Table 2). OSL ages were obtained from 0.4–1.2 m to avoid soil disturbance associated with agriculture and have been used to infer the age of terrace abandonment. Sediments from the upper part of the T1 terrace along the mid-Brisbane River yielded age estimates ranging from 8.08 ± 0.77 ka at 1.0 m and 10.83 ± 0.84 ka at 1.7 m depth (Figure 6D). The ages were within 2σ error of each other and had a mean age of 9.46 ± 0.90 ka. While no basal ages were determined for the adjacent floodplain, a 0.4 m thick flood unit of fine sands in the floodplain levee returned a minimum age of 0.54 ± 0.06 ka at 1.4 m depth (MB2, Table 2). Such deposition on this floodplain has been recorded anecdotally with recent flooding depositing a unit of massive sands at the surface 0.5 m thick. These chronological results from the Brisbane terrace site were comparable to what is observed in Lockyer Creek (Figure 6A,B), where nine OSL ages from the upper parts of the terraces, as well as basal ages from adjacent floodplains (Table 2), constrained the age of terrace abandonment to 9.19 ± 1.2 ka (Figure 6).

Along the Logan River, sediments in the upper part of the T1 terrace have been dated at 7.51 ± 0.82 ka (1.2 m; Figure 6D) and 7.61 ± 0.68 ka (1.6 m; Figure 6D). The upper sample at 1.2 m yielded slightly older age estimates when a CAM was employed (8.88 ± 0.81 ka) but the estimates remained within 1σ of the underlying sample, and as such, the age estimates from the MAM were used as the depositional age. In the adjacent inset floodplain, radiocarbon determinations on charcoal yielded calibrated ages of 226–142 cal. years BP at 0.65 m and 550–500 cal. years BP at 1.8 m (Table 3). Despite different methods of age determination, these much more recent ages compared with the age of flood deposits in the inset floodplains along the Brisbane River and Lockyer Creek, although the present inundation frequency of these units differs by an order of magnitude (e.g., 5 year AEP at Logan River vs 50 year AEP at Brisbane River, Figure 6).

In the sediments along the Albert River, three terrace samples dated with OSL were all within error and indicated an age range of 9.51 ± 0.82 ka, 8.71 ± 0.66 ka, and 9.08 ± 0.81 ka (1.1, 1.5, and 1.9 m respectively; Figure 6E). The ages yielded a mean pooled age of 9.10 ± 0.87 ka, coeval with the age of abandonment in the Lockyer and Brisbane River (Figure 6C).

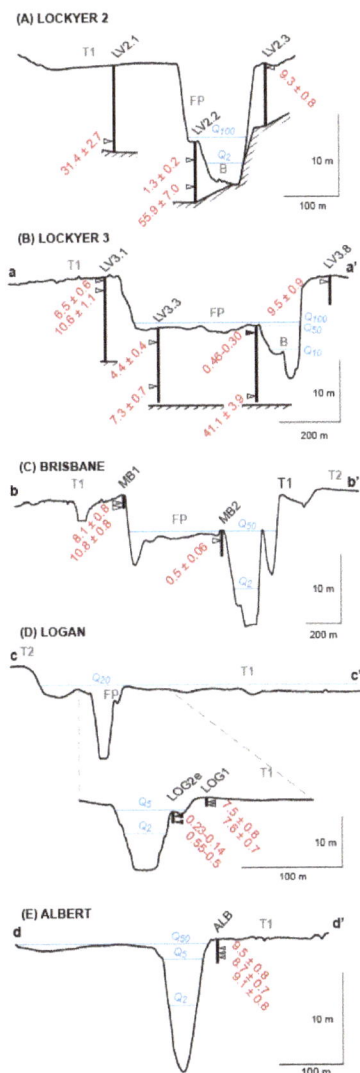

Figure 6. Representative valley cross sections and chronology for study sites (**A**) Lockyer Creek, Site 2, (**B**) Lockyer Creek, Site 3, (**C**) Brisbane River, (**D**) Logan River; and (**E**) Albert River. Location of cross sections are shown in planform in Figure 4. (**A**,**B**) Lockyer Creek sites and chronology after [46]. Flood frequency analyses and heights after [60].

Table 2. Summary of OSL data, description and ages (1σ errors). See Table S1 for complete OSL results.

Basin	Core ID	Lab Code	Depth (m)	Landform	Depositional Environment	Age (ka)	Method
Mid-Brisbane	MB1	16-0115-001	1.0	Terrace	Overbank	8.08 ± 0.77	CAM
					Overbank	5.86 ± 0.80	MAM
	MB1	16-0115-002	1.7	Terrace	Overbank	10.83 ± 0.84	CAM
	MB2	16-0115-003	1.4	Floodplain	Overbank	0.54 ± 0.06	MAM
Logan	LOG1	16-0115-004	1.2	Terrace	Overbank	8.88 ± 0.81	CAM
					Overbank	7.51 ± 0.82	MAM
	LOG1	16-0115-005	1.6	Terrace	Overbank	7.61 ± 0.68	CAM
Albert	ALB	16-0115-007	1.1	Terrace	Overbank	9.51 ± 0.82	CAM
	ALB	16-0115-008	1.5	Terrace	Overbank	8.71 ± 0.66	CAM
	ALB	16-0115-009	1.9	Terrace	Overbank	9.08 ± 0.81	CAM
Lockyer [46]	LV1.1	14-0528-001	0.4	Terrace	Overbank	11.51 ± 1.29	CAM
	LV1.2	14-0596-005	5.7	Floodplain	Basal	7.97 ± 1.02	MAM
	LV2.3	14-0596-015	0.2	Terrace	Overbank	9.25 ± 0.84	CAM
	LV3.1	14-0596-001	0.4	Terrace	Overbank	6.45 ± 0.57	MAM
	LV3.1	14-0528-010	2.1	Terrace	Overbank	10.60 ± 1.09	MAM
	LV3.3	14-0528-008	9.4	Floodplain	Basal	7.34 ± 0.67	MAM
	LV3.8	14-0528-003	0.4	Terrace	Overbank	9.50 ± 0.93	CAM
	LV4.3	14-0596-007	2.0	Terrace	Overbank	10.01 ± 1.23	MAM
	LV4.4	14-0596-012	10.3	Floodplain	Basal	10.08 ± 1.03	MAM

Table 3. Radiocarbon ages for the Logan River at Round Mountain.

Basin	Core #/Depth (m)	Lab Code	Material	pMc	δ13C (‰)	Conventional Age (a BP) ± 1σ Error	95.4% Calibrated Age (cal. BP) [%]
Logan	LOG2/0.65	BETA-482192	Charred material	97.18	−24.3	230 ± 30	226–142 [67.8%] 306–252 [27.2%] 79–74 [0.4%]
	LOG2/1.8	BETA-482191	Charred material	93.5	−25	540 ± 30	550–500

4. Discussion

4.1. Regional Correlation in Terrace Abandonment

Regional dating demonstrates that terraces ubiquitously bounded the macrochannels analyzed here. At all locations, valley fills were dominated by a terrace comprising 60–90% of the valley floor area. In most instances, these terrace surfaces were still inundated under the modern flow regime, but only by high-magnitude floods greater than a 20-year AEP (Figure 6). At all sites, the age determination of alluvium in the mid-reaches demonstrated an incisional episode between 7.5–10.8 ka (Figure 7), which was broadly synchronous with estimates of terrace abandonment in the mid-reaches of Lockyer Creek at 9.6 ± 0.98 ka. Estimates of terrace abandonment throughout the region were all within 1σ error and yielded a mean pooled age of 9.24 ± 0.93 ka. Surficial ages of inset floodplains in two of the streams (Brisbane and Logan Rivers) suggested these units have been forming at least at timescales at 10^2–10^3 years and were the locale of "active" flood deposition.

Contemporaneous early Holocene terrace abandonment was apparent in the regional correlation of terrace surface ages in north-eastern Australia (Figure 7). On the mid-north coast of New South Wales, early Holocene terrace abandonment was associated with laterally active river systems [18,21,22]. In the subtropical Fitzroy River basin, north of SEQ (Figure 2B), chronostratigraphic evidence indicates a peak in fluvial activity at 11 ka followed by a sharp decline through the remaining Holocene period [80]. However, the adjacent upper units of the terraces exhibited a mean pooled age of 12.2 ± 1.4 ka [80]. These ages were obtained from samples 3–6 m below the surface and therefore probably underestimated the true age of terrace abandonment. As such, terrace abandonment in the Fitzroy River basin was likely to be younger than that presented in Figure 7 and may well be in the range presented for other subtropical streams. The comparative entrenched channel form (e.g., enlarged macrochannel that was constrained between the 10- and 100-year AEP floods) in this basin suggested a further regional phenomenon.

In the wet tropics (Figure 2B), nine OSL ages from terrace sediments and adjacent basal deposits in four basins suggest terraces were abandoned slightly earlier at 10.3 ± 1.3 ka [23,24,81]. Basal OSL ages from the adjacent alluvial unit had a pooled mean age of 7.5 ± 0.9 ka [81], which was similar to that investigated on the Lockyer. This was further indicative of a possible regional correlation with the subtropical rivers of SEQ. The fact that terrace abandonment ages in six of the seven basins presented in Figure 7 overlapped within 1σ is a compelling argument for a regional climatic control and is discussed below.

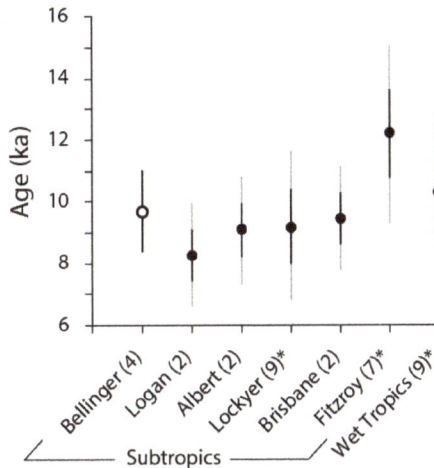

Figure 7. Mean terrace abandonment ages in north-eastern Australia. The number of samples per region is represented in brackets. Samples include OSL age determination from the upper terraces and, where noted by (*), from basal samples in adjacent inset units. Fitzroy and wet tropics OSL ages after [24,80,81], respectively. Black bars represent 1σ errors and grey bars represent 2σ errors. The Bellinger basin age (hollow circle) was derived from conventional and AMS ^{14}C ages [18,82], calibrated. Due to the differences in error calculations between OSL and ^{14}C methods, errors for the Bellinger ages were the standard deviation of the population.

4.2. Relationships to Late Quaternary Climatic Changes

The documented entrenchment/incision and the abandonment of previously active floodplains across SEQ closely aligned with widespread evidence of the onset of an early Holocene wet period in the subtropics and throughout eastern Australia between ca. 10–6 ka [26–28,37]. Channel incision during a glacial–interglacial transition is a commonly considered process in models of climatically-induced terrace development [83,84]. An increasingly warmer and wetter environment in SEQ may indeed have driven channel incision in the last glacial–interglacial transition and early Holocene [29,31]. Rapidly warming sea surface temperatures (SST) [85,86] may have presented a subtropical climate commensurate with the delivery of more extreme weather events [31].

Increased erosion in the wet tropics associated with the onset of wet conditions may have been more related to the strengthening of the summer monsoon [35,37,87,88]. Catchment instability, in the form of alluvial fan incision, slope instability, and changes to vegetation communities, were likely associated with higher humidity and rainfall events after 12 ka [88,89]. However, terrace abandonment ages within the wet tropics fall within 1σ of those presented in this study (Figure 7) and may be indicative of climatic forcing throughout eastern Australia across the latitudinal range of 16–27° S, despite differences in climatic regimes.

Despite broad, regional differences in present climatic regimes, the proposed concept of a regional climatic factor could be observed in the modern regime. Within the present climate, dominated by multi-decadal wet and dry phases influenced by both ENSO [58] and the Interdecadal Pacific oscillation (IPO), there was generally a dominant spatial synchronicity of years with above average rainfall and flooding (Figure 8) [57]. In years where a strong La Niña had occurred, much of eastern Australia experienced very high anomalous rainfall and catastrophic flooding, as occurred in 1974 and 2011 (Figure 8). Such external climatic forcing may have also been reflected in the late Holocene riverine responses, with three of the four sites investigated in this study featuring pronounced flood deposits on adjacent surfaces dating to c. 0.5 ka (Figure 6).

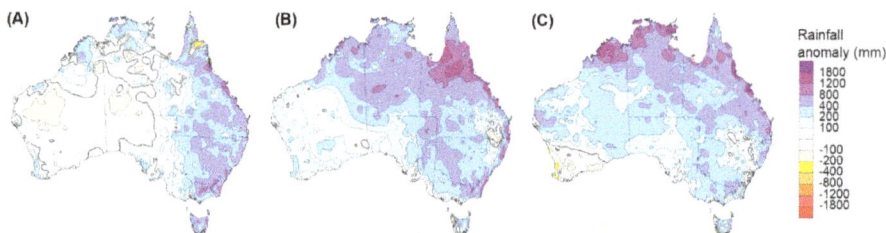

Figure 8. Maps of rainfall anomaly from long-term average conditions in Australia for water years September 1–August 31, (**A**) 1955, (**B**) 1974, and (**C**) 2011, demonstrating the tendency for broad continental trends in rainfall along the eastern margin. After [90].

4.3. Alternative Drivers of Terrace Abandonment

The correlation of terrace abandonment provided an intriguing opportunity to explore the geomorphic controls on fluvial activity. If such terraces were driven by intrinsic factors [41], it is unlikely there would be basin-wide coherence in terrace formation, let alone across the sub-tropical region. However, the regional synchronicity of terrace abandonment and inset basal ages invoked an extrinsic control. Whilst a regional landscape response suggested a climate driver of channel incision in the Australian sub-tropics, it was important to evaluate alternatives.

While the minimal rates of uplift throughout the late Cenozoic [14] ruled out the role of tectonic processes, alternative interpretations of broad cut-and-fill episodes in SEQ could implicate eustatic base-level changes [91] or prehistoric human–landscape interactions [92]. Base-level changes over the late Quaternary would undoubtedly have had an impact to coastal rivers as sea-level transgression and regression of 124 ± 4 m occurred along the eastern margin [93,94]. For such a control to be dominant on the timing of incision, as documented here, it would require a temporally consistent lag since lowstand-driven incision (assuming incision was initiated at, or prior to, the Last Glacial Maximum), was expressed across all basins. The incisional phase at 9.24 ± 0.93 ka that formed the macrochannels in SEQ (Figure 7) occurred as the sea level rapidly rose from −30 m below present mean sea level to +0.5–1.5 above modern at c.7 cal. ka BP [94]. As such, lowstand-driven incision could be discounted. This phase of incision counters an expected depositional response to sea-level changes [95,96]. The sites investigated in this study had elevations that ranged between 16–110 m and were above the inferred impacts of an immediate sea-level response [96].

Likewise, a changing environment throughout the late Quaternary most likely played a significant role in the population dynamics of Indigenous Australian people [97–99], who first inhabited the region >40 ka [100,101]. Indigenous land management practices would have been spatially localized [98,102]. Although impacts may have occurred, it is challenging to decouple these from climate forcing at the broad landscape level and rationalize a ca. 30 ka lag in riverine response to human–landscape interactions. Without any sound evidence on the role of human impacts in the early Holocene and given the regional synchronicity in terrace abandonment ages and their association with other palaeoenvironmental records, we favor the role of a climatically-driven broad-scale fluvial response in

SEQ. What awaits further validation is the ultimate cause of this period of terrace formation, and the quantitative evidence for changes to runoff in this late Pleistocene/early Holocene period.

4.4. Modern and Late Holocene Floodplain Dynamics

Following channel incision in the early Holocene, many streams in SEQ have been largely confined to their entrenched "macrochannel" form that remains preserved within the valley floor. While the macrochannels themselves may reflect a broadly similar entrenched morphology to arroyos in south-western United States [103,104], a lack of sediment supply limited the capacity for dynamic, cyclic responses commonly observed in arroyo valley fills [38,105]. As such, the modern floodplains in SEQ equated to narrow, depositional benches that formed within the macrochannel; in some cases, wider inset floodplains developed. The observed lack of a Holocene signal in the Fitzroy River basin may have been due to a lack of sediment preservation in a narrow (entrenched) fluvial corridor, or the dominance of in-channel bench formation [80]. Chronostratigraphic data of alluvial benches in Lockyer Creek indicated that these units had been the primary depositional sequences over the last 2000 years [78]. These inset units had been periodically stripped and reformed in relation to stochastic flood events and their formation may well be driven by intrinsic thresholds of channel capacity [41], relating to longitudinal distribution of flood power.

4.5. Comparison to Terrace Abandonment in Other Climatic Regions

Invoking climate forcing in the formation of terraces is not a new or novel idea. Indeed, it is understood to be a principle driver of Cenozoic terrace formation globally [9,13,84,96]. Particularly, in regions affected by glacial ice sheets, global syntheses suggest the cyclicity of terrace staircases that reflect glacial-interglacial cycles [9,106]. Cold to warm transitions have produced coarse-grained deposits during glacial periods, with terraces capped by early-interglacial, vertically-accreting fines prior to abandonment [84]. This is a significant factor in glacially-fed fluvial systems, where melt water and changes to both flow regime and sediment supply were significant between glacial and interglacial cycles. Likewise, other regions may have experienced similar temperature-driven cyclicity in a fluvial response due to changes in vegetation cover and associated effects on sediment supply [106–108].

In humid climates, there is greater uncertainty as to the drivers of terrace formation and whether temperature or moisture flux had a more significant role (although these may be difficult to decouple). The important role of precipitation (aridity-humidity cycles influenced by global climate patterns) is seen as the dominant control on fluvial processes [109,110]. In these regions, humidity had a similar impact on ecological communities as temperature in cooler regions. In the wet tropics, moisture-driven changes in vegetation cover during the Last Glacial–Interglacial Transition initiated processes of landscape instability [88,111]. A shift between rainforest and dry sclerophyll forests critically impacted slope stability and sediment supply [89], although lowland terraces in the wet tropics did not appear to have been abandoned until the early Holocene (Figure 7). In contrast, empirical evidence from subtropical Australia suggests a positive moisture balance during the last Glacial period, with pollen from rainforest communities persisting in both the headwaters and lower reaches of the coastal-draining streams [31,112,113]. Indeed, the distribution of precipitation may have been a more important factor, particularly with regard to extreme events [114]. A precipitation maximum driven by changes in SST may have been significant enough to alter weather systems and increased the intensity of rainfall throughout the subtropical region during the early Holocene. As such, precipitation-driven terrace abandonment in the subtropics may have been associated with extreme events and may well have occurred rapidly. However, further research is required to evaluate exactly what elements of the environment the fluvial systems were responding to in the late Glacial to early Holocene.

Interestingly, entrenched channels with compound forms have also been identified in other regions noted for a lack of late Cenozoic uplift, such as the Indian subcontinent and the Kaapvaal Craton in South Africa [115]. While it is challenging to draw a comparison between the exact mechanisms of terrace abandonment, particularly one of global climate forcing, crustal stability may be an important factor in the preservation of alluvial fills in such regions [9,115]. This tectonic stability may have ensured that terraces were abandoned slowly, had a polycyclic nature, and were retained in the landscape for extensive periods, exposing them to considerable weathering processes. Such factors would have increased the preservation potential of alluvial fills [116], and as such, allowed streams to retain entrenched channel forms over considerable periods, as with the macrochannels found in these regions [50,52].

5. Conclusions

This study presented a preliminary regional terrace chronology for subtropical SEQ, Australia. The results indicated a synchronous response of early Holocene channel incision and terrace abandonment that likely reflected large-scale climatic changes in eastern Australia. Chronostratigraphic interpretations from four basins within SEQ suggested a major phase of terrace abandonment between 7.5–10.8 ka with a mean age of 9.24 ± 0.93 ka. This age range closely aligned with a precipitation maximum identified from other palaeoenvironmental records in eastern Australia. While intrinsic factors have likely impacted local scale variations in channel morphology, particularly within in-channel alluvial units, external climatic factors appear to have been the principal drivers of major phases of valley-scale erosion and deposition in subtropical Australia. The correlation with fluvial records elsewhere in the subtropics, and more broadly across eastern Australia, are further indication that terraces have been driven by climatic forcing. Further clarity around late Quaternary riverine responses within this region would be significantly enhanced by the addition of floodplain and palaeochannel chronologies and quantitative palaeodischarge calculations.

Supplementary Materials: The following are available online at http://www.mdpi.com/2571-550X/1/3/23/s1.

Author Contributions: Conceptualization, J.S.D. and T.J.C.; Methodology, J.S.D. and T.J.C.; Validation, J.S.D. and T.J.C.; Formal Analysis, J.S.D.; Investigation, J.S.D.; Data Curation, J.S.D.; Writing-Original Draft Preparation, J.S.D.; Writing-Review & Editing, J.S.D. and T.J.C.; Visualization, J.S.D.; Supervision, T.J.C.; Project Administration, J.S.D.

Funding: This research was funded in part by an Australian Research Council Linkage project, LP120200093, and industry partners DSITI, Seqwater, and Lockyer Valley Regional Council. Additional funding support was provided by a University of Queensland (UQ) Australian Postgraduate Award Scholarship.

Acknowledgments: The authors thank Jacky Croke as the LP project leader and Ashneel Sharma for his assistance with OSL analyses. We thank the journal editors and two anonymous reviewers for their comments and constructive input on this paper.

Conflicts of Interest: The authors declare no conflict of interest.

References

1. Malatesta, L.C.; Prancevic, J.P.; Avouac, J.-P. Autogenic entrenchment patterns and terraces due to coupling with lateral erosion in incising alluvial channels. *J. Geophys. Res. Earth Surf.* **2017**, *122*, 335–355. [CrossRef]
2. Schumm, S.A. *The Fluvial System*; Wiley: New York, NY, USA, 1977.
3. Vandenberghe, J.; Cordier, S.; Bridgland, D.R. Extrinsic and intrinsic forcing of fluvial development: Understanding natural and anthropogenic influences. *Proc. Geol. Assoc.* **2010**, *121*, 107–112. [CrossRef]
4. Brierley, G.J.; Fryirs, K. *Geomorphology and River Management: Application of the Rivers Styles Framework*; Blackwell Publishing Ltd.: Malden, MA, USA, 2005.
5. Cooke, R.U.; Reeves, R.W. Arroyos and environmental change. *Oxford* **1976**, *328*, 520–557.
6. Dühnforth, M.; Anderson, R.S.; Ward, D.J.; Blum, A. Unsteady late pleistocene incision of streams bounding the colorado front range from measurements of meteoric and in situ ^{10}be. *J. Geophys. Res. Earth Surf.* **2012**, *117*. [CrossRef]
7. Malatesta, L.C.; Avouac, J.P.; Brown, N.D.; Breitenbach, S.F.M.; Pan, J.; Chevalier, M.L.; Rhodes, E.; Saint-Carlier, D.; Zhang, W.; Charreau, J.; et al. Lag and mixing during sediment transfer across the tian shan piedmont caused by climate-driven aggradation-incision cycles. *Basin Res.* **2017**. [CrossRef]

8. Maddy, D.; Bridgland, D.; Westaway, R. Uplift-driven incision and climate-controlled river terrace development in the Thames Valley, UK. *Quat. Int.* **2001**, *79*, 23–36. [CrossRef]

9. Bridgland, D.R.; Westaway, R. Climatically controlled river terrace staircases: A worldwide quaternary phenomenon. *Geomorphology* **2008**, *98*, 285–315. [CrossRef]

10. Hanson, P.R.; Mason, J.A.; Goble, R.J. Fluvial terrace formation along wyoming's laramie range as a response to increased late pleistocene flood magnitudes. *Geomorphology* **2006**, *76*, 12–25. [CrossRef]

11. Macklin, M.G.; Lewin, J.; Jones, A.F. River entrenchment and terrace formation in the uk holocene. *Quat. Sci. Rev.* **2013**, *76*, 194–206. [CrossRef]

12. Macklin, M.G.; Lewin, J.; Woodward, J.C. The fluvial record of climate change. *Philos. Trans. Ser. A Math. Phys. Eng. Sci.* **2012**, *370*, 2143–2172. [CrossRef] [PubMed]

13. Vandenberghe, J. The relation between climate and river processes, landforms and deposits. *Quat. Int.* **2002**, *91*, 17–23. [CrossRef]

14. Czarnota, K.; Roberts, G.G.; White, N.J.; Fishwick, S. Spatial and temporal patterns of australian dynamic topography from river profile modeling. *J. Geophys. Res. Solid Earth* **2014**, *119*, 1384–1424. [CrossRef]

15. Nott, J.; Young, R.; McDougall, I. Wearing down, wearing back, and gorge extension in the long-term denudation of a highland mass: Quantitative evidence from the Shoalhaven catchment, Southeast Australia. *J. Geol.* **1996**, *104*, 224–232. [CrossRef]

16. Peel, M.C.; McMahon, T.A.; Finlayson, B.L. Continental differences in the variability of annual runoff-update and reassessment. *J. Hydrol.* **2004**, *295*, 185–197. [CrossRef]

17. Rustomji, P.; Bennett, N.; Chiew, F. Flood variability east of australia's great dividing range. *J. Hydrol.* **2009**, *374*, 196–208. [CrossRef]

18. Cohen, T.J.; Nanson, G.C. Topographically associated but chronologically disjunct late quaternary floodplains and terraces in a partly confined valley, south-eastern australia. *Earth Surf. Process. Landf.* **2008**, *33*, 424–443. [CrossRef]

19. Kermode, S.J.; Cohen, T.J.; Reinfelds, I.V.; Nanson, G.C.; Pietsch, T.J. Alluvium of antiquity: Polycyclic terraces in a confined bedrock valley. *Geomorphology* **2012**, *139–140*, 471–483. [CrossRef]

20. Cheetham, M.D.; Keene, A.F.; Erskine, W.D.; Bush, R.T.; Fitzsimmons, K.; Jacobsen, G.E.; Fallon, S.J. Resolving the holocene alluvial record in southeastern australia using luminescence and radiocarbon techniques. *J. Quat. Sci.* **2010**, *25*, 1160–1168. [CrossRef]

21. Nanson, G.C.; Cohen, T.J.; Doyle, C.J.; Price, D.M. Alluvial evidence of major late-quaternary climate and flow-regime changes on the coastal rivers of New South Wales, Australia. In *Palaeohydrology: Understanding Global Change*; Gregory, K.J., Benito, G., Eds.; John Wiley & Sons, Ltd.: Chichester, UK, 2003; pp. 233–258.

22. Cohen, T.J.; Nanson, G.C. Mind the gap: An absence of valley-fill deposits identifying the holocene hypsithermal period of enhanced flow regime in southeastern australia. *Holocene* **2007**, *17*, 411–418. [CrossRef]

23. Hughes, K.; Croke, J.C.; Bartley, R.; Thompson, C.J.; Sharma, A. Alluvial terrace preservation in the wet tropics, northeast queensland, australia. *Geomorphology* **2015**, *248*, 311–326. [CrossRef]

24. Leonard, S.; Nott, J. Rapid cycles of episodic adjustment: Understanding the holocene fluvial archive of the daintree river of northeastern australia. *Holocene* **2015**, *25*, 1208–1219. [CrossRef]

25. Nott, J.; Hayne, M. High frequency of 'super-cyclones' along the great barrier reef over the past 5000 years. *Nature* **2001**, *413*, 508–512. [CrossRef] [PubMed]

26. Kershaw, A.P. Environmental change in greater australia. *Antiquity* **1995**, *69*, 656–675. [CrossRef]

27. Kershaw, A.P.; D'Costa, D.M.; McEwen Mason, J.R.C.; Wagstaff, B.E. Palynological evidence for quaternary vegetation and environments of mainland southeastern australia. *Quat. Sci. Rev.* **1991**, *10*, 391–404. [CrossRef]

28. Petherick, L.; Bostock, H.; Cohen, T.J.J.; Fitzsimmons, K.; Tibby, J.; Fletcher, M.S.S.; Moss, P.; Reeves, J.; Mooney, S.; Barrows, T.; et al. Climatic records over the past 30ka from temperate australia—A synthesis from the oz-intimate workgroup. *Quat. Sci. Rev.* **2013**, *74*, 58–77. [CrossRef]

29. Reeves, J.M.; Barrows, T.T.; Cohen, T.J.; Kiem, A.S.; Bostock, H.C.; Fitzsimmons, K.E.; Jansen, J.D.; Kemp, J.; Krause, C.; Petherick, L.; et al. Climate variability over the last 35,000 years recorded in marine and terrestrial archives in the australian region: An oz-intimate compilation. *Quat. Sci. Rev.* **2013**, *74*, 21–34. [CrossRef]

30. Kershaw, A.P.; McKenzie, G.M.; Porch, N.; Roberts, R.G.; Brown, J.; Heijnis, H.; Orr, M.L.; Jacobsen, G.; Newall, P.R. A high-resolution record of vegetation and climate through the last glacial cycle from Caledonia Fen, Southeastern Highlands of Australia. *J. Quat. Sci.* **2007**, *22*, 481–500. [CrossRef]

31. Moss, P.T.; Tibby, J.; Petherick, L.; McGowan, H.A.; Barr, C. Late quaternary vegetation history of North Stradbroke Island, Queensland, Eastern Australia. *Quat. Sci. Rev.* **2013**, *74*, 257–272. [CrossRef]

32. Turney, C.S.M.; Kershaw, A.P.; Clemens, S.C.; Branch, N.; Moss, P.T.; Fifield, L.K. Millennial and orbital variations of el niño/southern oscillation and high-latitude climate in the last glacial period. *Nature* **2004**, *428*, 306–310. [CrossRef] [PubMed]

33. Jones, R.; Bowler, J.; McMahon, T. A high resolution holocene record of p/e ratio from Closed Lakes, Western Victoria. *Palaeoclimates* **1998**, *3*, 51–82.

34. Wilkins, D.; Gouramanis, C.; De Deckker, P.; Fifield, L.K.; Olley, J. Holocene lake-level fluctuations in Lakes Keilambete and Gnotuk, Southwestern Victoria, Australia. *Holocene* **2013**, *23*, 784–795. [CrossRef]

35. Wyrwoll, K.-H.H.; Miller, G.H. Initiation of the australian summer monsoon 14,000 years ago. *Quat. Int.* **2001**, *83–85*, 119–128. [CrossRef]

36. Woodward, C.; Shulmeister, J.; Bell, D.; Haworth, R.; Jacobsen, G.; Zawadzki, A. A holocene record of climate and hydrological changes from Little Llangothlin Lagoon, South Eastern Australia. *Holocene* **2014**, *24*, 1–10. [CrossRef]

37. Reeves, J.M.; Bostock, H.C.; Ayliffe, L.K.; Barrows, T.T.; De Deckker, P.; Devriendt, L.S.; Dunbar, G.B.; Drysdale, R.N.; Fitzsimmons, K.E.; Gagan, M.K.; et al. Palaeoenvironmental change in tropical australasia over the last 30,000 years—A synthesis by the oz-intimate group. *Quat. Sci. Rev.* **2013**, *74*, 97–114. [CrossRef]

38. Harvey, J.E.; Pederson, J.L.; Rittenour, T.M. Exploring relations between arroyo cycles and canyon paleoflood records in buckskin wash, utah: Reconciling scientific paradigms. *Bull. Geol. Soc. Am.* **2011**, *123*, 2266–2276. [CrossRef]

39. Erskine, W.D.; Livingstone, E.A. In-channel benches: The role of floods in their formation and destruction on bedrock- confined rivers. In *Varieties of Fluvial Form*; Miller, A.J., Gupta, A., Eds.; John Wiley & Sons, Ltd.: Chichester, UK, 1999; pp. 445–476.

40. Nanson, G.C. Episodes of vertical accretion and catastrophic stripping: A model of disequilibrium flood-plain development. *Geol. Soc. Am. Bull.* **1986**, *97*, 1467–1475. [CrossRef]

41. Thompson, C.J.; Croke, J.C.; Fryirs, K.; Grove, J.R. A channel evolution model for subtropical macrochannel systems. *CATENA* **2016**, *139*, 199–213. [CrossRef]

42. Erskine, W.D.; Saynor, M. Effects of catastrophic floods on sediment yields in Southeastern Australia. In *Erosion and Sediment Yield: Global and Regional Perspectives, Proceedings of the International Symposium, Exeter, UK, 15–19 July 1996*; IAHS: London, UK; p. 381.

43. Baker, V.R. Stream-channel responses to floods, with examples from central texas. *Geol. Soc. Am. Bull.* **1977**, *88*, 1057–1071. [CrossRef]

44. Croke, J.C.; Fryirs, K.; Thompson, C.J. Channel-floodplain connectivity during an extreme flood event: Implications for sediment erosion, deposition, and delivery. *Earth Surf. Process. Landf.* **2013**, *38*, 1444–1456. [CrossRef]

45. Croke, J.C.; Reinfelds, I.; Thompson, C.J.; Roper, E. Macrochannels and their significance for flood-risk minimisation: Examples from Southeast Queensland and New South Wales, Australia. *Stoch. Environ. Res. Risk Assess.* **2014**, *28*, 99–112. [CrossRef]

46. Daley, J.S.; Croke, J.; Thompson, C.; Cohen, T.; Macklin, M.; Sharma, A. Late quaternary channel and floodplain formation in a partly confined Subtropical River, Eastern Australia. *J. Quat. Sci.* **2017**, *32*, 729–743. [CrossRef]

47. Grove, J.R.; Croke, J.C.; Thompson, C.J. Quantifying different riverbank erosion processes during an extreme flood event. *Earth Surf. Process. Landf.* **2013**. [CrossRef]

48. Thompson, C.J.; Croke, J.C. Geomorphic effects, flood power, and channel competence of a catastrophic flood in confined and unconfined reaches of the Upper Lockyer Valley, Southeast Queensland, Australia. *Geomorphology* **2013**, *197*, 156–169. [CrossRef]

49. Croke, J.C.; Todd, P.; Thompson, C.J.; Watson, F.; Denham, R.; Khanal, G. The use of multi temporal lidar to assess basin-scale erosion and deposition following the catastrophic January 2011 Lockyer Flood, Se Queensland, Australia. *Geomorphology* **2013**, *184*, 111–126. [CrossRef]

50. Heritage, G.L.L.; Broadhurst, L.J.J.; Birkhead, A.L.L. The influence of contemporary flow regime on the geomorphology of the Sabie River, South Africa. *Geomorphology* **2001**, *38*, 197–211. [CrossRef]

51. Van Niekerk, A.W.; Heritage, G.L.; Broadhurst, L.J.; Moon, B.P. Bedrock anastomosing channel systems: Morphology and dynamics in the Sabie River, Mpumalanga Province, South Africa. In *Varieties of Fluvial Form*; Miller, A.J., Gupta, A., Eds.; John Wiley & Sons, Ltd.: Chichester, UK, 1999; pp. 33–52.

52. Gupta, A. Magnitude, frequency, and special factors affecting channel form and processes in the seasonal tropics. In *Natural and Anthropogenic Influences in Fluvial Geomorphology*; Costa, J.E., Miller, A.J., Potter, K.W., Wilcock, P.R., Eds.; American Geophysical Union: Washington, DC, USA, 1995; pp. 125–136.

53. Hoyle, J.; Brooks, A.P.; Brierley, G.J.; Fryirs, K.; Lander, J. Spatial variability in the timing, nature and extent of channel response to typical human disturbance along the Upper Hunter River, New South Wales, Australia. *Earth Surf. Process. Landf.* **2008**, *33*, 868–889. [CrossRef]

54. Woodyer, K.D.D. Bankfull frequency in rivers. *J. Hydrol.* **1968**, *6*, 114–142. [CrossRef]

55. Harden, T.; Macklin, M.G.; Baker, V.R. Holocene flood histories in South-Western USA. *Earth Surf. Process. Landf.* **2010**, *35*, 707–716. [CrossRef]

56. Crimp, S.J.; Day, K.A. Evaluation of multi-decadal variability in rainfall in queensland using indices of el nino-southern oscillation and inter-decadal variability. In *Science for Drought, Proceedings of the National Drought Forum'*; Stone, R.C., Partridge, I., Eds.; National Drought Forum: Brisbane, Australia, 2003; pp. 106–115.

57. Kiem, A.S. Multi-decadal variability of flood risk. *Geophys. Res. Lett.* **2003**, *30*, 1035–1035. [CrossRef]

58. Power, S.; Casey, T.; Folland, C.; Colman, A.; Mehta, V. Inter-decadal modulation of the impact of enso on australia. *Clim. Dyn.* **1999**, *15*, 319–324. [CrossRef]

59. Risbey, J.; Karoly, D.; Reynolds, A.; Braganza, K. Drought and climate change. In Proceedings of the National Drought Forum, Brisbane, Australia, 15–16 April 2003; Stone, R.C., Partridge, I., Eds.; National Drought Forum: Brisbane, Australia, 2003; pp. 8–11.

60. Daley, J.S. *Late Quaternary Stream Channel Adjustment in Hydrologically Variable Catchments, Subtropical Australia*; University of Queensland: Brisbane, Australia, 2018.

61. Pasternack, G.B. *Lower Yuba River Corridor Inundation Zones. Prepared for The Yuba County Water Agency*; University of California: Davis, CA, USA, 2017.

62. Weber, M.D.; Pasternack, G.B. *2014 Topographic Mapping of the Lower Yuba River. Prepared for the Yuba Accord River Management Team*; University of California: Davis, CA, USA, 2016.

63. Aitken, M.J. *An Introduction to Optical Dating: The Dating of Quaternary Sediments by the Use of Photon-Stimulated Luminescence*; Oxford Science Publications: Oxford, UK, 1998.

64. Wallinga, J. Optically stimulated luminescence dating of fluvial deposits: A review. *Boreas* **2002**, *31*, 303–322. [CrossRef]

65. Olley, J.M.; Pietsch, T.J.; Roberts, R.G. Optical dating of holocene sediments from a variety of geomorphic settings using single grains of quartz. *Geomorphology* **2004**, *60*, 337–358. [CrossRef]

66. Pietsch, T.J. Optically stimulated luminescence dating of young (<500 years old) sediments: Testing estimates of burial dose. *Quat. Geochronol.* **2009**, *4*, 406–422.

67. Galbraith, R.F. Graphical display of estimates having differing standard errors. *Technometrics* **1988**, *30*, 271–281. [CrossRef]

68. Galbraith, R.F.; Laslett, G.M. Statistical models for mixed fission track ages. *Nucl. Tracks Radiat. Meas.* **1993**, *21*, 459–470. [CrossRef]

69. Galbraith, R.F.; Roberts, R.G.; Laslett, G.M.; Yoshida, H.; Olley, J.M. Optical dating of single and multiple grains of quartz from Jinmium Rock Shelter, Northern Australia, part 2, results and implications. *Archaeometry* **1999**, *41*, 339–364. [CrossRef]

70. Roberts, H.M.; Wintle, A.G. Equivalent dose determinations for polymineralic fine-grains using the sar protocol: Application to a holocene sequence of the chinese loess plateau. *Quat. Sci. Rev.* **2001**, *20*, 859–863. [CrossRef]

71. Murray, A.; Marten, R.; Johnston, A.; Martin, P. Analysis for naturally occuring radionuclides at environmental concentrations by gamma spectrometry. *J. Radioanal. Nucl. Chem.* **1987**, *115*, 263–288. [CrossRef]

72. Stokes, S.; Ingram, S.; Aitken, M.J.; Sirocko, F.; Anderson, R.; Leuschner, D. Alternative chronologies for late quaternary (last interglacial–holocene) deep sea sediments via optical dating of silt-sized quartz. *Quat. Sci. Rev.* **2003**, *22*, 925–941. [CrossRef]

73. Mejdahl, V. Thermoluminescence dating: Beta-dose attenuation in quartz grains. *Archaeometry* **1979**, *21*, 61–72. [CrossRef]

74. Prescott, J.R.; Hutton, J.T. Cosmic ray contributions to dose rates for luminescence and ESR dating: Large depths and long-term time variations. *Radiat. Meas.* **1994**, *23*, 497–500. [CrossRef]

75. Bowler, J.M.; Johnston, H.; Olley, J.M.; Prescott, J.R.; Roberts, R.G.; Shawcross, W.; Spooner, N.A. New ages for human occupation and climatic change at Lake Mungo, Australia. *Nature* **2003**, *421*, 837–840. [CrossRef] [PubMed]

76. Stuiver, M.; Polach, H.A. Discussion; reporting of c-14 data. *Radiocarbon* **1977**, *19*, 355–363. [CrossRef]

77. Reimer, P.J.; Bard, E.; Bayliss, A.; Beck, J.W.; Blackwell, P.G.; Bronk Ramsey, C.; Buck, C.E.; Cheng, H.; Edwards, R.L.; Friedrich, M.; et al. Intcal13 and marine13 radiocarbon age calibration curves 0–50,000 years cal BP. *Radiocarbon* **2013**, *55*, 1869–1887. [CrossRef]

78. Croke, J.; Thompson, C.; Denham, R.; Haines, H.; Sharma, A.; Pietsch, T. Reconstructing a millennial-scale record of flooding in a single valley setting: The 2011 flood-affected Lockyer Valley, South-East Queensland, Australia. *J. Quat. Sci.* **2016**, *31*, e2919. [CrossRef]

79. Kermode, S.J.; Cohen, T.J.; Reinfelds, I.V.; Jones, B.G. Modern depositional processes in a confined, flood-prone setting: Benches on the Shoalhaven River, NSW, Australia. *Geomorphology* **2015**, *228*, 470–485. [CrossRef]

80. Croke, J.C.; Jansen, J.D.; Amos, K.; Pietsch, T.J. A 100 ka record of fluvial activity in the Fitzroy River Basin, Tropical Northeastern Australia. *Quat. Sci. Rev.* **2011**, *30*, 1681–1695. [CrossRef]

81. Hughes, K.; Croke, J.C. How did rivers in the wet tropics (Ne Queensland, Australia) respond to climate changes over the past 30 000 years? *J. Quat. Sci.* **2017**, *32*, 744–759. [CrossRef]

82. Warner, R.F. Radio-carbon dates for some fluvial and colluvial deposits in the Bellinger Valleys, New South Wales. *Aust. J. Sci.* **1970**, *32*, 368–369.

83. Gibbard, P.L.; Lewin, J. River incision and terrace formation in the Late Cenozoic of Europe. *Tectonophysics* **2009**, *474*, 41–55. [CrossRef]

84. Vandenberghe, J. River terraces as a response to climatic forcing: Formation processes, sedimentary characteristics and sites for human occupation. *Quat. Int.* **2015**, *370*, 3–11. [CrossRef]

85. Chang, J.C.; Shulmeister, J.; Woodward, C.; Steinberger, L.; Tibby, J.; Barr, C. A chironomid-inferred summer temperature reconstruction from subtropical australia during the last glacial maximum (LGM) and the last deglaciation. *Quat. Sci. Rev.* **2015**, *122*, 282–292. [CrossRef]

86. Woltering, M.; Atahan, P.; Grice, K.; Heijnis, H.; Taffs, K.; Dodson, J. Glacial and holocene terrestrial temperature variability in subtropical east australia as inferred from branched gdgt distributions in a sediment core from lake mckenzie. *Quat. Res.* **2014**, *82*, 132–145. [CrossRef]

87. Field, E.; Tyler, J.; Gadd, P.S.; Moss, P.; McGowan, H.; Marx, S. Coherent patterns of environmental change at multiple organic spring sites in northwest australia: Evidence of indonesian-australian summer monsoon variability over the last 14,500 years. *Quat. Sci. Rev.* **2018**, *196*, 193–216. [CrossRef]

88. Thomas, M.F. Understanding the impacts of late quaternary climate change in tropical and sub-tropical regions. *Geomorphology* **2008**, *101*, 146–158. [CrossRef]

89. Thomas, M.F.; Nott, J.; Murray, A.S.; Price, D.M. Fluvial response to late quaternary climate change in Ne Queensland, Australia. *Palaeogeogr. Palaeoclim. Palaeoecol.* **2007**, *251*, 119–136. [CrossRef]

90. Bureau of Meteorology. *Rainfall Anomalies (mm)—Product of the National Climate Centre*; Australian Government, Bureau of Meteorology: Melbourne, Australia, 2012.

91. Beckmann, G.G. *The Post-Tertiary History of brisbane And Surroundings*; University of Queensland: Brisbane, Australia, 1959.

92. Bowman, D.M.J.S. The impact of aboriginal landscape burning on the australian biota. *New Phytol.* **1998**, *140*, 385–410. [CrossRef]

93. Chivas, A.R.; García, A.; van der Kaars, S.; Couapel, M.J.J.; Holt, S.; Reeves, J.M.; Wheeler, D.J.; Switzer, A.D.; Murray-Wallace, C.V.; Banerjee, D.; et al. Sea-level and environmental changes since the last interglacial in the gulf of carpentaria, australia: An overview. *Quat. Int.* **2001**, *83–85*, 19–46. [CrossRef]

94. Lewis, S.E.; Sloss, C.R.; Murray-Wallace, C.V.; Woodroffe, C.D.; Smithers, S.G. Post-glacial sea-level changes around the australian margin: A review. *Quat. Sci. Rev.* **2013**, *74*, 115–138. [CrossRef]

95. Blum, M.D.; Aslan, A. Signatures of climate vs. Sea-level change within incised valley-fill successions: Quaternary examples from the texas gulf coast. *Sediment. Geol.* **2006**, *190*, 177–211. [CrossRef]

96. Blum, M.D.; Törnqvist, T.E. Fluvial responses to climate and sea-level change: A review and look forward. *J. Int. Assoc. Sedimentol.* **2000**, *47*, 2–48. [CrossRef]

97. Turney, C.S.M.; Hobbs, D. Enso influence on holocene aboriginal populations in queensland, australia. *J. Archaeol. Sci.* **2006**, *33*, 1744–1748. [CrossRef]

98. Williams, A.N.; Mooney, S.D.; Sisson, S.A.; Marlon, J. Exploring the relationship between aboriginal population indices and fire in australia over the last 20,000 years. *Palaeogeogr. Palaeoclim. Palaeoecol.* **2015**, *432*, 49–57. [CrossRef]
99. Williams, A.N.; Ulm, S.; Goodwin, I.D.; Smith, M. Hunter-gatherer response to late holocene climatic variability in Northern and Central Australia. *J. Quat. Sci.* **2010**, *25*, 831–838. [CrossRef]
100. Allen, J.; O'Connell, J. Both half right:Updating the evidence for dating first human arrivals in Sahul. *Aust. Archaeol.* **2014**, *79*, 86–108. [CrossRef]
101. Neal, R.; Stock, E. Pleistocene occupation in the South-East Queensland Coastal Region. *Nature* **1986**, *323*, 618. [CrossRef]
102. Bird, R.B.; Bird, D.W.; Codding, B.F.; Parker, C.H.; Jones, J.H. The "fire stick farming" hypothesis: Australian aboriginal foraging strategies, biodiversity, and anthropogenic fire mosaics. *Proc. Natl. Acad. Sci. USA* **2008**, *105*, 14796–14801. [CrossRef] [PubMed]
103. Bull, W.B. Discontinuous ephemeral streams. *Geomorphology* **1997**, *19*, 227–276. [CrossRef]
104. Elliott, J.G.; Gellis, A.C.; Aby, S.B. Evolution of arroyos: Incised channels of the southwestern united states. In *Incised River Channels: Processes, Forms, Engineering and Management*; Darby, S.E., Simon, A., Eds.; Wiley: Chichester, UK, 1999; pp. 153–186.
105. Waters, M.R. Alluvial chronologies and archaeology of the Gila River Drainage Basin, Arizona. *Geomorphology* **2008**, *101*, 332–341. [CrossRef]
106. Macklin, M.G.; Fuller, I.C.; Lewin, J.; Maas, G.S.; Passmore, D.G.; Rose, J.; Woodward, J.C.; Black, S.; Hamlin, R.H.B.; Rowan, J.S. Correlation of fluvial sequences in the mediterranean basin over the last 200 ka and their relationship to climate change. *Quat. Sci. Rev.* **2002**, *21*, 1633–1641. [CrossRef]
107. Macklin, M.G.; Lewin, J. Alluvial responses to the changing earth system. *Earth Surf. Process. Landf.* **2008**, *33*, 1374–1395. [CrossRef]
108. Doğan, U. Fluvial response to climate change during and after the last glacial maximum in Central Anatolia, Turkey. *Quat. Int.* **2010**, *222*, 221–229. [CrossRef]
109. Jain, M.; Tandon, S.K. Fluvial response to late quaternary climate changes, Western India. *Quat. Sci. Rev.* **2003**, *22*, 2223–2235. [CrossRef]
110. Latrubesse, E.M.; Bocquentin, J.; Santos, J.C.R.; Ramonell, C.G. Paleoenvironmental model for the late cenozoic of southwestern amazonia: Paleontology and geology. *Acta Amazon.* **1997**, *27*, 103–117. [CrossRef]
111. Kershaw, A.P. Climatic change and aboriginal burning in North-East Australia during the last two glacial/interglacial cycles. *Nature* **1986**, *322*, 47. [CrossRef]
112. Ellerton, D.; Shulmeister, J.; Woodward, C.; Moss, P. Last glacial maximum and last glacial–interglacial transition pollen record from Northern NSW, Australia: Evidence for a humid late last glacial maximum and dry deglaciation in parts of Eastern Australia. *J. Quat. Sci.* **2017**, *32*, 717–728. [CrossRef]
113. Petherick, L.M.; Moss, P.T.; McGowan, H.A. An extended last glacial maximum in Subtropical Australia. *Quat. Int.* **2016**, *432*, 1–12. [CrossRef]
114. Vandenberghe, J. Climate forcing of fluvial system development: An evolution of ideas. *Quat. Sci. Rev.* **2003**, *22*, 2053–2060. [CrossRef]
115. Westaway, R.; Bridgland, D.; Mishra, S. Rheological differences between archaean and younger crust can determine rates of quaternary vertical motions revealed by fluvial geomorphology. *Terra Nova* **2003**, *15*, 287–298. [CrossRef]
116. Lewin, J.; Macklin, M.G. Preservation potential for late quaternary river alluvium. *J. Quat. Sci.* **2003**, *18*, 107–120. [CrossRef]

quaternary

MDPI

Article

River Systems and the Anthropocene: A Late Pleistocene and Holocene Timeline for Human Influence

Martin R. Gibling

Department of Earth Sciences, Dalhousie University, Halifax, NS B3H 4R2, Canada; mgibling@dal.ca

Academic Editors: David R. Bridgland, Jef Vandenberghe and Xianyan Wang
Received: 15 June 2018; Accepted: 28 September 2018; Published: 4 October 2018

Abstract: Rivers are central to debate about the Anthropocene because many human activities from antiquity focused on channels and floodplains. A literature compilation for the onset of human modification of rivers identifies six stages that represent key innovations focused in the Near East and adjoining areas: (1) *minimal effects* before about 15,000 cal yr BP, with the use of fire and gathering of plants and aquatic resources; (2) *minor effects* from increased cultivation after about 15,000 cal yr BP, with plant and animal domestication after about 10,700 cal yr BP; (3) *agricultural era* after about 9800 cal yr BP, with legacy sediments, widespread fire use, the first dams and irrigation, and mud-brick manufacture; (4) *irrigation era* from about 6500 cal yr BP, with large-scale irrigation, major cities, the first large dam, urban water supplies, expanded groundwater use, river fleets, and alluvial mining; (5) *engineering era* with embankments, dams, and watermills after about 3000 cal yr BP, especially in the Chinese and Roman empires; and (6) *technological era* after about 1800 CE. Anthropogenic river effects were more varied and intense than commonly has been recognised, and they should be considered routinely in interpreting Late Pleistocene and Holocene fluvial archives.

Keywords: agriculture; Anthropocene; archaeology; dams; deforestation; dikes; domestication; fire; legacy sediments; river engineering

---

1. Introduction

Humans exert a geomorphic force that now rivals that of the natural Earth [1,2]. The period of human dominance has been termed the Anthropocene, and several dates have been proposed for its onset. Many researchers have emphasised the dramatic changes associated with the Industrial Revolution in Europe after about 1750 CE (Common Era) and the Great Acceleration in technology at about 1950 CE [3–6]. However, a detectable human imprint on the environment extends back for thousands of years [7–10], and an emphasis on recent changes minimises the enormous landscape transformation caused by humans in antiquity [11]. Important earlier human effects with significant environmental consequences include megafaunal extinctions between 14,000 and 10,500 cal yr BP [12]; domestication of plants and animals close to the start of the Holocene at 11,700 cal yr BP; agricultural practices and deforestation at 10,000 to 5000 cal yr BP; and widespread generation of anthropogenic soils at about 2000 cal yr BP [6,13–16].

Rivers are central components of the terrestrial realm, and historically many human settlements have been located along rivers. As used here, *river systems* include the channels themselves, the riparian zone, floodplains and terraces, adjoining uplands dissected by lower order channels, and deltas. Anthropogenic activities are often considered as discrete elements, including the use of fire, domestication of plants and animals, soil development, cropland expansion, the establishment of settlements, and irrigation. However, they are all affected by, and contribute to, the dynamics of nearby rivers, and it is important to consider them in a holistic environmental and geomorphic context.

Key evidence of anthropogenic activity is encoded in early fluvial successions [17,18], long predating anthropogenic effects that have intensified over the past centuries and led to the modern worldwide river crisis [19,20].

During the Late Pleistocene and Holocene, rivers around the world were affected by deglaciation, sea-level rise, and climatic instability on timescales from millennia to decades. Effects included damming by ice and glacial sediment, pulses of meltwater, runoff over permafrost, and major monsoonal and other climatic fluctuations as seen in alluvial-plain and terrace successions [21–26]. Geoscientists commonly interpret fluvial successions with reference to such powerful forcing factors. However, these factors may have masked early anthropogenic effects, contributing to a limited awareness of human influence on rivers through time. Assessing the balance between anthropogenic and natural processes, which vary greatly on temporal and spatial scales, is a major challenge, and cultural and geomorphic records must both be evaluated in order to assess human impacts [18,27].

The present paper is an exploratory attempt to identify and investigate some of the key processes through which humans have modified rivers from antiquity, including processes that are less obvious but potentially important [18]. Because many anthropogenic effects took place in fluvial settings, any comprehensive assessment of land-use change needs to highlight rivers explicitly [18]. Drawing on a compilation of literature from several disciplines, we set out a provisional timeline and stages for the onset and intensity of human modification of river systems worldwide, from the Late Pleistocene to the present, covering a wide geographic and climatic range but with a focus on the Near East and adjoining areas. Such an assessment of key processes and their onset is a necessary precursor for constructing realistic timelines for river modification in certain areas; timelines for some regions and quantitative approaches to assessing the intensity of modification are discussed briefly. Comparisons are drawn with available quantitative estimates of population and land use from 12,000 years cal yr BP to the modern era.

2. Major Anthropogenic Influences and Their Effects

Table 1 summarises key anthropogenic influences on modern river landscapes. Some are illustrated in Figures 1–3 with reference to low-technology settings that have some application to the Late Pleistocene and Holocene. The influences may have direct or indirect effects on rivers, commonly altering the boundary conditions of processes such as sediment erosion and transport [18]. They are substantiated in the table with references that demonstrate their impact in modern and submodern settings.

A central set of influences concerns agriculture. Fire is used for hunting and deforestation (Figure 1A) and, along with other anthropogenic vegetation changes, converts forests and other undisturbed ecosystems to cropland and grazing land, commonly with terrace construction (Figure 1B–D). *Cropland* is the sum of land under permanent crops, with the addition of arable land for temporary crops, meadows for mowing or pasture, market and kitchen gardens, and temporarily fallow land [8,9]; it covers rain-fed and irrigated crops. *Grazing land* includes pasture (permanent meadows and pastures and land used for growing cultivated or wild crops for forage) and wider rangelands. Animal husbandry involves grazing, with additional use of animals for ploughing and transport (Figure 1E,F). Collectively, these influences reduce regolith strength and enhance erosion, increasing sediment supply to rivers.

Table 1. Effects of some anthropogenic activities on modern and submodern channel and floodplain systems. Some of the cited literature refers to human activity at a technological level much greater than that of the earlier Quaternary. Italics denote processes not substantiated here by citations.

Human Activity	Effects on Modern River Systems	References
Fire use	Vegetation loss promotes rapid runoff, soil erosion, and enhanced flux of sediment and charcoal to rivers. Change in soil properties enhances erosion. Fire may trigger vegetation change.	[28–30]
Agriculture and deforestation	Reduced resistance of river banks and hillslopes where crops with shallow roots replace natural vegetation. Widening of channels and sediment coarsening. Slope failure, and increased sediment flux to fluvial, deltaic, and eolian systems. Change in palynomorph associations.	[17,31–33]
Animals used for food, ploughing, and transport	Herds reduce vegetation cover and enhance soil erosion, gullying, and sediment flux to rivers. Trampling breaks down river banks, widens channels, and increases suspended load. Ploughing intensifies use of floodplains and hillslopes and, along with animal transport trails, enhances erosion.	[34–40]
Embankments along channels	Embankments narrow channels, reduce their migration, increase flow velocity, funnel sediment to deltas, and reduce channel siltation; complex upstream feedbacks. Sediment trapping within embanked floodplains reduces inundation capacity. Embankments raise the channel base, promoting catastrophic avulsion.	[41–43]
Dams and irrigation systems	Dams alter river flow regime, cause deposition in millponds and reservoirs, and increase downstream erosion. Irrigation reduces river discharge, and the use of river and groundwater promote soil waterlogging and salination.	[44–47]
Navigation and bank structures	*Riverside construction (wharves, steps, access roads, bridges) affects banks and channels.* Dredging, riverbed scour, and removal of wood snags and jams to aid navigation alters river morphology. Reduced number of delta distributary channels aids year-round navigation.	[48,49]
City water supplies	Remove river and groundwater from the hydrological system, with water pollution from sewage and waste.	[50]
Warfare	River diversions during warfare causes catastrophic floods and floodplain aggradation. Deliberate 1938 breach of the Yellow River dikes caused death toll of >800,000.	[51,52]
Extraction of channel and floodplain materials	Alluvial mining of channels and terraces for gold and other minerals increases aggradation and erosion rates locally. Pits on floodplains for bricks, tiles, pottery, and ochre, and on terraces for laterite remove fertile soil and aquifer media, cause soil erosion, increase suspended load, lower the water table, and cause waterlogging.	[43,53–55]
Extraction of aquatic materials	Reeds, papyrus and other in-channel plants influence flow dynamics and sedimentation. *Fisheries and aquatic harvesting enhance human activity along river banks.*	[56]
Cultural events	Water festivals involve large populations, with infrastructure on river banks and sand bars. Khumbha Mela festival at Allahabad, India, had 120 million attendees in 2 months in 2013.	[57]

Anthropogenic effects mediated through a change in vegetation, however, are unlikely to have resulted in a simple linear change from forest to open terrain. Prior to anthropogenic modification, postglacial lowland Europe and other areas may have been a park-like landscape of forests, grasslands and shrublands, where keystone species that included large herbivorous mammals and birds influenced the vegetation succession [58,59].

More direct effects on rivers involve the modification of channels and changes to the connectivity of channels and floodplains, including the transfer of large volumes of water to floodplains for irrigation and other human uses. *Irrigation* is defined as applying water, in addition to natural rainfall, to the soil to enhance crop yield [60]. Irrigation systems are commonly linked to dams and barrages, with the use of specialised irrigation technology (Figure 2A,B). Settlements and agricultural land are protected by embankments along channel margins (Figure 2C). The transport of materials by boats, rafts, and barges (here termed *navigation*) requires infrastructure that includes wharves and access roads along river banks (Figure 2D), as well as dredging of channels and the removal from the channel of snags such as fallen trees and large logs. Linked with these activities are riverside settlements and cities, which require large-scale water supplies. Bridges and river crossings are a common focus for human activity (Figure 2E). The deliberate breaching of embankments during warfare has been responsible for catastrophic avulsions and widespread floodplain aggradation.

A third set of influences involves resource extraction from floodplains and channels, commonly yielding technofossils [61]. The influences include placer mining in channels for alluvial gold and other minerals, and the extraction of floodplain clay for bricks, tiles, pottery, and ochre (Figure 2F). In some regions, terraces are capped by indurated regolith that includes laterite, extracted on a large-scale for city construction. The removal of aquatic plants such as reeds and papyrus potentially alters river dynamics, and fisheries (Figure 2G) may involve considerable traffic and bank infrastructure.

Figure 1. Anthropogenic effects on modern river landscapes due to fire, agriculture, and animal husbandry. Examples are from settings with modest technological influence. See Table 1 for some modern results of these effects. (**A**) Firing of forests for tapioca plantations, southeast Thailand. (**B**) Hillslope terracing for barley cultivation above tributary of Kali Gandaki River, Himalayan foothills, Nepal. (**C**) Ricefields on floodplain of Chao Phraya tributary, Thailand. Note well sweep for water supply at centre-left. (**D**) Crop growing on sand bars during seasonal low-flow stage, with temporary causeway in foreground, Sabarmati River, India. (**E**) Olive grove at oasis along creek, Sinai Desert, Egypt. (**F**) Ploughing of floodplain of Marsyandi River, Nepal. (**G**) Camel-cart carrying brushwood for fires, and water buffalo, Sabarmati River, India. (**H**) Yak caravan crossing mountain pass by Kali Gandaki River, Nepal.

Figure 2. Anthropogenic effects on modern river landscapes due to irrigation, embankments, navigation, floodplain workings, fishing, and cultural events. Examples are mainly from settings with modest technological influence. See Table 1 for some modern results of these effects. (**A**) Irrigation canal from barrage on Sutlej River (in distance), Ropar, India. (**B**) Waterwheel for irrigation from Yellow River, Lanzhou, China. (**C**) Embankment for flood protection, Baghmati River, India. (**D**) Rice barges and riverside structures, Chao Phraya tributary, Phitsanulok, Thailand. (**E**) Low-season bridge, stilt buildings, and access path, Mae Moei River, Thailand. (**F**) Brickpit, Ganges River plains, India. (**G**) Fishing in Ganges River, India, with temple ruins at left. (**H**) Festival on sand flats at confluence of the Ganges and Yamuna rivers, Allahabad, India.

Major cultural events take place at rivers around the world. For example, tens of millions of people attend the festival of Khumba Mela, held every twelve years at the confluence of the Ganges and Yamuna rivers in India (Figure 2H).

Subdivided in a different way, the processes identified above form two inter-related groups that have varied in intensity in space and time: (1) a group linked to the development of agriculture in association with fire, irrigation systems, and dams; and (2) a group associated with cities and strings of settlements that are concentrated along rivers, with effects from urban water use, boat use and navigation infrastructure, fish and vegetation use, bridges, brick and pottery manufacture on floodplains, and water-related cultural events. The first group of processes has received considerable attention in the fluvial literature, affecting extensive areas of river plains, adjoining uplands, and channel reaches. The second group of processes has received much less attention, but it is the contention of the author that these processes had a considerable and underestimated effect on rivers.

Alluvial mining and the effects of warfare have been locally important.

Figure 3. Schematic diagram to show range of anthropogenic influences on river systems. Illustrated examples are mainly from settings with modest technological influence, especially in the period of about 10,000 to 4000 cal yr BP. Figure designed and drafted by Meredith Sadler.

3. Methods

Fluvial scientists have drawn largely on sedimentological evidence to assess anthropogenic influence on rivers, linked especially to developments in agriculture. However, this approach yields only a partial view of river modification: humans have interacted with rivers over the past millennia in ways that are enormously varied, culturally and climatically influenced, and in ways that are often obscure to modern technological societies. To make a realistic evaluation of anthropogenic effects, it was necessary to consider information from a wide range of disciplines that included archaeology, anthropology, genetics, ancient history, engineering, geomorphology, palaeoclimatology, and material science (Tables 2–5). Each discipline has its own approach to age assessment. The literature examples cited here are selections, and they represent only a small sample of a voluminous literature. They variously include the oldest known instances of certain processes, times when certain processes became prominent, and good examples from later times, where distinguishable from natural climatogenic and autogenic processes.

The evaluation required a careful analysis of each cited age assessment. Radiocarbon dates were mainly reported in the cited literature as calibrated years BP, with a few as radiocarbon years BP (uncalibrated), linked to a 1950 datum. Calibrated dates variously represented age models that were updated periodically over more than two decades. A few studies used optically stimulated luminescence (OSL), thermoluminescence (TL), and U-Th dates, reported in years BP. The ages and uncertainties were commonly reported without indication of confidence limits, which could be either 2 or 1 sigma, even for the same chronometric technique. In some studies, calibrated radiocarbon dates and OSL, TL, or U-Th dates were used collectively to provide a general assessment. Some studies provided a detailed account of methods, including supplementary data files, whereas other studies gave only brief information. Studies of plant and animal domestication report dates linked to a genetic assessment of selected characters, some of which are controversial as indicators of human influence.

The selected studies include dates from individual sites as well as summaries of dates for numerous sites, some of which reported a range of methods applied over several decades. Direct dates from pottery and parts of domesticated plants and animals were provided in some studies. In other studies, indirect dates on charcoal, wood, and other materials constrained the timing of key events, from the stratum in question or from adjoining strata. These dates are commonly subject to some uncertainty due to reworking and mixing of materials.

Archaeological studies commonly provided dates based on regional comparisons. These include widely accepted designations such as Pre-Pottery Neolithic B (not necessarily dated at the site in question), the duration of a dynasty in Egypt or China, or the reign of a ruler for which historical information is available. Some assessments were constrained only to part of a millennium BCE (Before Common Era), and some studies provided dates in BCE without any indication of the data sources.

A usable compilation requires a representation of all age dates in a comparable manner. In view of the wide variation in type of information, methods used, and the reliability of sites and materials, no attempt was made to apply a uniform calibration to the dates. Many publications did not provide the measured uncalibrated radiocarbon dates and uncertainty, and it was not possible to recalibrate with a single more current calibration. The great majority of age assessments used here is based on calibrated radiocarbon dates, and in these cases the dates are given as calibrated years BP (cal yr BP). A few dates, noted in the tables, were provided as uncalibrated (uncal) years BP. A minority of studies used OSL, TL, and U-Th dates along with radiocarbon dates; although these dates are in years BP, they are considered as supporting the radiocarbon dates in cal yr BP. Dates within the past two thousand years are reported with respect to CE. A broad archaeological assessment such as "in the fifth millennium BCE" was converted to "greater than 6000 years ago", taking the duration of the Common Era as 2000 years. The variable datums of 1950, 2000, and years BP for different sources yield a slight discrepancy, but error ranges are unavailable for most dates and the effects of this discrepancy are modest. Formal subdivisions for the Holocene have recently been ratified, with the Early, Middle, and Late Holocene termed the Greenlandian, Northgrippian, and Meghalayan with basal dates of 11,700 yr b2k (years before 2000 CE), 8326 yr b2k, and 4250 yr b2k, respectively [62].

Tables 2–5 document the basis of age assessment for each study cited, and dates in the text note the units recorded in the original studies. The author acknowledges the limitations of this approach to a complex problem of age assessment. However, the purpose of the study is to provide a general overview of human effects on rivers and their approximate timing, rather than to reconcile dates for specific sites. The reader is referred to the original references if precise age assessment is required.

4. Early Hominins and Fire

The earliest hominins date to >7 Ma in Chad [63], and early hominins used stone tools along Ethiopian rivers by at least 2.6 Ma [64]. The use of stone tools was widespread along European rivers after the introduction of handaxe technology at about 600 ka [65], and continued through the Neolithic (Figure 4A,B). However, there is little evidence that early hominins modified rivers.

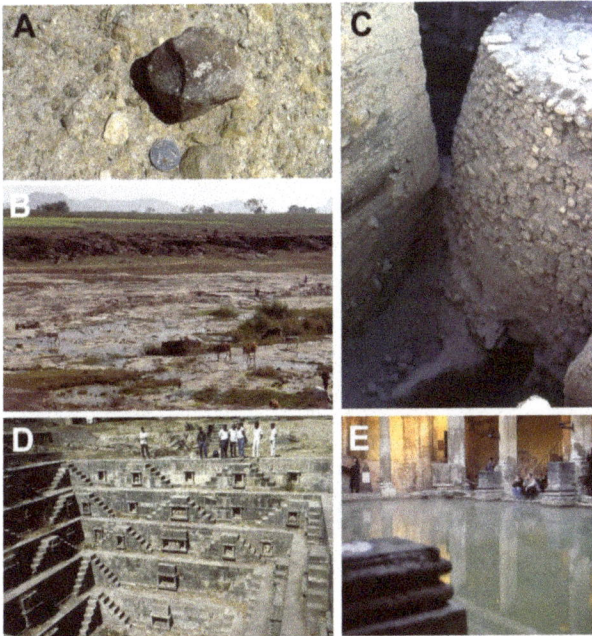

Figure 4. Archaeological sites associated with rivers and springs. (**A**) Palaeolithic quartzite corestone extracted from cemented river gravels, Rajasthan, India. Coin 2.5 cm in diameter. (**B**) Palaeolithic to Mesolithic site above bedrock river bank in middle distance, Belan River, India [66]. The site had a small settlement where stone tools were worked. Stone extraction in progress in river bed. (**C**) Structure dated to ca. 10,300 to 9800 years ago at Jericho, Israel, possibly a watchtower for flash floods [67]. (**D**) Ancient well during dry season, near Sabarmati River, India. (**E**) Roman bath fed by hot springs, Bath, UK.

In contrast, nonhuman animals greatly influenced rivers worldwide prior to European settlement, and they continue to do so intensively in some regions. In North America, some 60 to 400 million beaver ponds trapped hundreds of billions of cubic metres of sediment; huge bison herds widened stream channels at crossings and increased the streambed silt fraction; and colonies of black-tailed prairie dogs moved large volumes of earth [68,69]. In Africa, hippo trails across levees promote avulsion and the formation of new channel systems [70]. Other examples include earth movement by worms [71] and disturbance of river gravel by fish and crayfish [72].

The use of fire may be the first detectable anthropogenic influence on river landscapes, perhaps as early as 1.6 Ma in Africa [73]. Fire was used for warmth, cooking, and hafting of stone artefacts after about 300 to 400 ka in Europe and the Near East [74–76]. Indigenous populations had entered Australia by 65,000 years ago [77], and they may have altered vegetation through controlled burns to manipulate resources, although charcoal records also reflect climatic fluctuations [78–80]. Human use of fire was identified in New Guinea at 28,000 cal yr BP based on charcoal in slopewash deposits at a site with artefacts [81]. As noted below, human use of fire intensified as agriculture developed. However, a close linkage exists between climate and fire on a global scale [82], and the attribution of charcoal to human activity requires careful analysis.

5. Domestication of Plants

The Neolithic rise of agriculture [50,83] had a major impact on river systems around the world. Floodplains, terraces, and uplands were deforested, and croplands were irrigated from rivers and springs. Crops are commonly grown on modern river bars during seasonal low-stage flow (Figure 1D), a tradition probably inherited from antiquity. Many major crops were domesticated and came under production from about 10,700 to 7000 cal yr BP at localities around the world (Table 2).

Food procurement initially involved *gathering wild plants*, with progressive *cultivation* that required local clearance of vegetation and modest tillage, although wild plants remained the predominant food source [84]. Subsequent *domestication* involved genetic changes through selection pressures imposed by cultivation, with increased labour to maintain the crops, and cultivation continued to accompany domestication [85]. Plant resources on river plains were increasingly managed by manipulating growth conditions and the environment to increase their relative abundance and predictability [86].

Table 2. Estimates of time and place of domestication of some plants and animals. Evidence for domestication is based on a genetic selection process that leads to organisms adapted to cultivation and utilisation by humans [86]. See Section 3 for the approach used in presenting age dates. Dates shown are in calibrated years BP from the cited papers, with some general age assessments (*) and dates based on genetic analysis (**). cal: calibrated, uncal: uncalibrated.

Crop or Animal	Data Sources	Approximate Date (years BP)	Location	References
	Plant Domestication			
Barley, emmer, and einkorn wheat	Varied sources, cal	c. 12,000–10,000	Near East	[87]
	Approximate uncal date for domestic spikelets in einkorn wheat	9250	Near East	[85]
Flax	Varied sources, cal	10,700–10,200	Near East	[88]
	Varied sources	9000	Near East	[89]
Date palm	Varied cal and uncal sources	>6000	Arabia	[90]
Millet	Varied sources	8000–7000	China	[91]
Rice	Varied cal sources for initiation of rice cultivation	9000	China	[92]
	Varied sources for domesticated rice, after 6000 cal BCE	8000	China	[84]
	Varied sources	9000–8000	China	[93]
Squash	Cal date for charred or desiccated squash seed	10,400–10,160	Peru	[94]
	Cal dates on plant and other materials	>8700	Mexico	[95]
Maize	Cal dates on plant and other materials	>8700	Mexico	[95]
Peanut	Cal dates for charred peanut hull	8640–8440	Peru	[94]
Cotton	Cal dates for cotton boll	6280–5950	Peru	[94]
Banana	Cal dates on materials that include banana phytoliths	6950–6440	New Guinea	[96]
Grape and wine culture	Cal dates on pottery sherds, ca. 7000–6600 BCE	9000	China	[97]
	Cal dates on pottery sherds and soil, ca. 5900–5750 BCE	8000	Georgia	[98]

Table 2. *Cont.*

Crop or Animal	Data Sources	Approximate Date (years BP)	Location	References
	Animal Domestication			
Cattle	Varied sources, cal date for Early Pre-Pottery Neolithic site, Syria	10,650–10,250	Near East	[99]
Sheep, goats	Cal dates for Early Pre-Pottery Neolithic at Near East sites	10,250–9500	Near East	[100,101]
Pigs	Intensification of relationship between humans and pigs, second half of 9th millennium BCE, sites in Turkey and Cyprus	>10,000	Near East	[102]
	Assessment of zooarcheological data	9000	Near and Far East	[103]
Water buffalo	Varied sources suggest 3rd to 5th millennium BCE in parts of Asia and Near East; authors support origin in Indian Subcontinent	5000	Indian Subcontinent	[104] *
Horses	Faunal remains, bridles and milk suggest mid-4th millennium BCE, using cal dates	6000–5000	Eurasian steppes	[105,106] *
Dromedary	Varied archeological sources suggest late 2nd millennium BCE	>3000	Arabia	[107] *
Donkey	Varied archeological sources suggest late 5th to first half of 4th millennium BCE in Egypt	7000–6000	Africa	[108,109] *
Yak	Genome model for domestication, confidence interval of 7914 to 7227 BP	7300	Tibetan Plateau	[110] **
Llama, Alpaca	Archaeozoological data at dated sites and genome models	7000–6000	Peruvian Andes	[111] *, [112] **

As demonstrated from starch grains on grinding tools, hunter-gatherers processed cereals and probably produced flour by 32,000 cal yr BP in Europe [113,114]. Food processing also took place along the Tigris, Euphrates and Jordan rivers and adjoining uplands in the Near East, with dates from 23,000 cal yr BP [115,116]. The emergence of the Natufian culture in the Near East approximately 14,500 years ago involved cultural innovations that made later agricultural developments possible, although there is no evidence that plants were domesticated at that time [115,117]. By 11,700 to 10,700 cal yr BP and prior to domestication, cultivation of wild cereal species, legumes, goatgrass, fruits, and nuts was widespread across the Near East [88].

From 10,700 to 10,200 cal yr BP, some Near East sites show increased proportions of domesticated cereal chaff that includes barley, emmer and einkorn wheat (Table 2). *Agriculture*, "a provisioning system based primarily on the production and consumption of domesticated resources" [86], was in evidence by about 9800 cal yr BP [88]. Domesticated traits gradually emerged through to about 8500 cal yr BP, and over the next few thousand years most Near East villagers were farmers with key crops playing a crucial role in establishing the region's civilisations [50,87,118]. By approximately 7000 years ago, cereal production had spread across most of Europe, with migration of some farmers from the Near East [119], and wheat and barley reached northern China by 3000 to 4000 cal yr BP [120].

In Asia, millet was cultivated in northern China 7000 to 8000 years ago. Rice was gathered by 11,000 to 12,000 cal yr BP in the Yangtze area, with a mixture of collection and cultivation by 8000 to 10,000 cal yr BP [92]. Genome analysis of wild and cultivated rice suggests that domestication took place between 8000 and 9000 years ago along the Pearl river in southern China, probably with a single domestication event [93,121]. Domesticated rice spread to the Yellow River area by 7500 cal yr BP and to Southeast Asia, Korea, Japan, and the Philippines over the next few thousand years, reaching the Indus Valley and other parts of the Indian Subcontinent by 4000 years ago and Mesopotamia by 2000 years ago [84,122–124]. Grape and fruit cultivation for wine dates to 9000 cal yr BP in China.

In Central and South America, squash was domesticated before 10,000 cal yr BP in Peru and maize by about 9000 cal yr BP in Mexico (Table 2). Sustainable agriculture has been practised along the Amazon for at least 2500 years, as indicated by widespread dark anthropogenic soils [125]. In eastern North America, maize-based agriculture increased valley sedimentation from about 1000 to 1300 CE, some 500 years prior to major European settlement [126]. In the Pacific region, the banana was domesticated nearly 7000 years ago in New Guinea, and anthropogenic soils in Australia may date to 3500 years ago [127]. Crops widely used for textiles, including cotton and flax, were domesticated early in several parts of the world.

Agricultural terraces are a major modifier of alluvial landscapes, serving to retain water and soil, reduce erosion, improve ploughing, and promote irrigation [128]. However, abandoned terraces may increase gully erosion and ultimately fail, liberating sediment [128,129]. Terracing is believed to have originated in Asia, and terraces date to about 6000 years ago (ca. 3500–4000 cal yr BCE) in Yemen and to about 5000 years ago (ca. 3000 cal yr BCE) in the Mediterranean and NW Europe during the Bronze and Iron Ages [18,130].

6. Domestication of Animals

The domestication of many animals took place between about 11,000 and 3000 years ago (Table 2). Products included meat, wool and milk, the latter used before 8000 years ago (by the seventh millennium BCE) in the Near East [131]. Nomadic herdsmen exploited extensive pasturelands from early in the Holocene [8]. Grazing and watering of cattle, sheep, goats, pigs, yaks, llamas, and alpacas would have caused land degradation, sediment runoff, and trampling of river banks and sediments (Table 1), especially near settlements, although such effects may have been limited largely to parts of the Near East and Asia [40]. Cattle and water buffalo would have aided ploughing of floodplain soils, enhancing erosion. The earliest known image of a plough dates to ca. 5000 years ago (3000 BCE) in Mesopotamia [129].

Domestication of the horse, dromedary, camel, and yak revolutionised transport and communication, aiding the establishment of trading routes such as the Silk Roads across Asia, many of them along river systems. The early domestication of the donkey promoted food transport and distribution early in the development of the Egyptian state [108].

Sites in Europe dating from 7900 to 4400 years ago (ca. 5900 to 2400 cal yr BCE) show that Neolithic farmers used livestock manure and water management to enhance crop yields, linking plant cultivation and animal herding through a strong investment in the land [132]. The production of fodder plants such as goatgrass and legumes may have been linked to caprine management in parts of the Near East [88].

7. River and Floodplain Modification for Surface Water and Groundwater

Croplands are largely rain-fed, but agriculture in many seasonal settings requires irrigation (Table 3). The earliest archaeological evidence for irrigated farming comes from the Near East and Egypt approximately 8000 years ago, developing across the Near East and Mediterranean over the following millennia and emerging independently in the Indian Subcontinent and China [9,60,129]. Irrigation led to the development of hydraulic technology [133,134] and may be the most important environmental modification practised intentionally by humans [60]. In the geological record, identification of irrigation channels requires analysis of conduit form, sediments, and artefacts [134–137].

A settlement existed by 10,000 years ago at Jericho [67] (Figure 4C), where early farmers probably practised irrigation at least 8000 years ago [129] from springs along the margin of the Jordan Valley [138]. On the northeastern margin of Mesopotamia in Iraq and Iran, indirect evidence supports irrigation on alluvial fans approximately 7000 to 8000 years ago (during the sixth millennium BCE), mainly following natural water courses that yield ceramic fragments [135]. By 7000 years ago (5000 BCE) in parts of this area, wide lateral canals followed hillside contours, and river flows were manipulated for agriculture.

On the southern plains of Mesopotamia, agriculture was underway by about 8000 to 7500 years ago (6000 to 5500 BCE) [135]. Although evidence for irrigation is indirect, farmers probably took advantage of levee breaks and crevasse splay formation to grow irrigated crops, under a seasonal climate with unstable westerlies [139]. By 7000 years ago, croplands of barley and wheat supported tens of thousands of city dwellers [140]. After about 6500 years ago (in the second half of the fifth millennium BCE), many irrigation canals traversed the Tigris and Euphrates plains, with a progressively increased scale of canals and the construction of thousands of brick sluice gates along the main water courses over the following millennia [135,141]. Irrigation progressively raised the water table on the plains and, as early as 4400 to 3700 years ago (2400 to 1700 BCE), salt precipitation in the soils led to a major decline in productivity, causing farmers to replace salt-sensitive wheat with barley [129,141,142].

In the Nile Valley, long irrigation canals were impracticable, and farmers used artificial levees and short canals to trap floodwaters [129,143]. The falling water table following flooding allowed recession agriculture and leached salt back into the river. A ceremonial macehead from approximately 5100 years ago shows irrigated fields and the Scorpion King cutting an irrigation ditch [129,142].

Table 3. Some early events in the development of water technology and river navigation. See Section 3 for the approach used in presenting age dates. Dates in original papers are based on assessments of archaeological evidence and historical events, with a few dates in calibrated years BP (*). cal: calibrated.

Location	Activity	Approximate Date (Years BP)	References
Irrigation and Drainage Systems			
Mesopotamia and uplands	Irrigation systems in existence in SW Iran by 5000 BCE, maybe by 6000 BCE, with large organised systems on Mesopotamian plains by 4500 BCE	~8000–7000	[129,135]
Near East	Water raising systems: shaduf (well sweep) by 2300 BCE, with later appearance of saqia (water wheel) and Archimedes screw in Ptolemaic times	~4300	[133]
New Guinea	Drainage ditches and channels, from cal dates on charcoal in ditch fills	4350–3980	[96] *
China	Large-scale engineering of embankments and canals on Yellow River for flood control and irrigation, with large dike projects in first half of 7th century BCE [144] to the 1800s	~3000	[41,144,145]
Peru	Irrigation canals dated at 5000 to 3750 cal BCE	~7000	[146]
U.S.A.	Gila River, Arizona: large areas under irrigation by 900 BCE, and earth dam 5 km long and 7 m high with irrigation canals by 1000 CE [147]; individual irrigation systems of 10,000 hectares (100 km^2), with large developments in 850–1450 CE [148]; distinction of canals from natural channels [136]	~2900	[136,147,148]
Dams			
Near East	Weirs of brushwood, stones and earth divert water into irrigation canals [135], estimated at 8000 years ago on E flank of Mesopotamia [149]	~8000	[135,149]
Mesopotamia	Nimrud Dam, probably an earth dam, diverted Tigris; dam later failed and returned the Tigris to its former course	>4000	[133,150]
Egypt	Sadd-el-Kafara Dam (Dam of the Pagans), 14 m high, >100 m long, with 17,000 cut stone blocks; built to aid quarrying ca. 2650 BCE	4650	[133]
Near East and Europe	45 dams built under the Roman Empire in the Near East, with irrigation canals, waterwheels, aqueducts, and qanat systems, from 63 BCE to 636 CE [151]. Earthen dams in Spain: Prosperina 12 m high and 427 m long, Cornalvo 20 m high and 194 m long with dam crest 8 m wide [133].	2700–1400	[133,151]
China	Dam at Anfeng Tan, east China, ca. 600 BCE	2600	[152]
Groundwater Systems			
Iran	Qanat systems of sloping tunnels that bring water from water table or springs at a high elevation to fields downslope, dated to ca. 3000 years ago in Iran, introduced from Indus to Nile under Persian rule, 550–331 BCE	~3000	[133]
Israel	Chamber excavated at Hazor, where water rises up the Dead Sea Fault; access to the underground water table by staircases; construction coeval with stratum dated at 9th century BCE	2800	[153]
Israel	Tunnel at Siloam, Jerusalem, routes spring within walls, dated by U-Th dating of speleothems, cal radiocarbon dating, and historically to ca. 700 BCE	2700	[154]

Table 3. *Cont.*

Location	Activity	Approximate Date (Years BP)	References
	River Navigation		
Egypt	Models of reed and log rafts at ca. 5500 to 4000 BCE [155]. Boats preserved to >5000 years ago, with a wooden boat 43.6 m long preserved from 4th dynasty, ~2500 BCE; barge with two 350-ton obelisks towed by 27 boats with 810 oarsmen, ca. 1472 to 1458 BCE [156].	>7500	[155,156]
Mesopotamia	Fleets for transport and warfare on Tigris, Euphrates, and canals; request for 600 vessels to bring grain to Ur in 3rd Dynasty [156]; channel straightening for boats, 3rd to 1st millennium BCE [135].	~4000	[135,156]
China	Grand Canal, 1800 km long, linked pre-existing canals, rivers, and lakes; connected southern and northern plains, locks across Yellow River; built by 5 million workers	609 CE	[157]

Technical innovations aided irrigation across the Near East [129,133,142]. By about 4300 years ago, the well sweep or *shaduf* allowed a bucket to be swung round to irrigate the fields (Figure 1C). Later inventions included the waterwheel or *saqia* with an animal-powered treadmill, typically raising water from a well or standing water; the *noria* or water-powered wheel to draw river water (Figure 2B), operative in Egypt in the first millennium BCE; and the Archimedes screw, an inclined tube with a spiral fin. In the Negev Desert, the Nabateans directed rare torrential rains down cleared hill slopes into underground cisterns, and fed runoff into dry river beds for cultivation, supporting tens of thousands of people in the early centuries CE [129,158].

Irrigation channels were used in Peru by about 7000 years ago (Table 3), with sophisticated pre-Columbian dryland terrace systems and water-control structures [159,160]. From approximately 700 to 1130 CE, the Hohokam culture of Arizona built hundreds of kilometres of canals with a carefully engineered gradient, with large areas under irrigation as early as 3000 years ago. Drainage ditches were excavated in New Guinea prior to 4000 cal yr BP. In China, major dike and canal systems for water supply and flood control date back at least 3000 years, with large public irrigation projects along the Yellow River over the past 2000 years. Over the past thousand years up to 1855, Chinese emperors organized large-scale dike and channel management on the Yellow River, employing hundreds of thousands of workers and in some years using ten percent of the imperial budget [41,161].

The earliest dams to supply irrigation canals were probably constructed about 8000 years ago in the Zagros uplands bordering Mesopotamia (Table 3). Rock and earth dams some four metres high and 80 m long supplied water in Jordan about 5000 years ago [149], and the Nimrud Dam in Mesopotamia diverted the Tigris River before 4000 years ago. In Egypt, the large Sadd-el-Kafara Dam was constructed more than 4600 years ago, faced with cut stone blocks and designed to store water for quarrying [133].

Using Egyptian and Mesopotamian infrastructure, the Romans built dams in many parts of the Mediterranean and Near East from about 2700 years ago to 400 CE, including 45 large dams on rivers in the Near East, with irrigation canals, waterwheels, and aqueducts. Some Roman dams in Spain are more than 400 m long and 20 m high. The earliest known dam in China dates to approximately 2600 years ago (600 BCE), and the Mogollon culture built a large earth dam in Arizona at about 1000 CE [147].

Groundwater was widely used for agriculture from wells dug under alluvial plains (Figure 4D), and by 3000 years ago farmers in the Near East had a good understanding of the water table [133]. The *qanat* systems of Iran used a gently sloping tunnel from the water table or springs at a high elevation to fields as much as 70 km downflow, with many vertical shafts to provide ventilation. At Jerusalem and Hazor in Israel, engineers exploited chambers and tunnels at the water table to provide water from 2700–2800 years ago (700 to 800 BCE).

8. Riverine Cities and Water Supplies

Most early settlements were built along rivers [135,162], and some were large enough for a considerable environmental footprint (Table 4). Hunter-gatherers built monumental structures at Göbekli Tepe in Turkey before 11,000 years ago (second half of the tenth and ninth millennia cal BCE) [163], and structures at Jericho date to at least 10,000 years ago [67]. Çatalhöyük in Anatolia may have been the world's first major settlement, occupied between about 9400 and 7600 years ago (ca. 7400 to 5600 BCE) [164]. In the Fertile Crescent, settlement density increased dramatically after about 6500 years ago, with high points at about 4400 BCE and 400 CE [165]. Settlements large enough to be considered "cities" date back to the Uruk period in Mesopotamia some 6000 years ago, with urban areas as large as 400 hectares (4 km^2) by about 4700 years ago. Early Dynastic cities in Egypt date to about 5000 years ago. By about 2000 years ago, organised societies had generated anthropogenic soils over large areas [14].

Table 4. Estimates of time and place of early use of channel, floodplain and aquatic resources, and the development of large settlements on floodplains. See Section 3 for the approach used in presenting age dates. Most dates are based on assessments of archaeological evidence and historical events. Some dates (*) are based on calibrated (cal) dates, uncalibrated (uncal) dates, OSL, and TL dates.

	Activity and Location	Approximate Date (Years BP)	References
	Large settlements and cities		
	Mesopotamia, Uruk period ca. 4000–3000 BCE (uncal) [166]; area of 400 ha by Early Dynastic II period, ca. 2700 BCE [167]	>6000	[166,167]
	Indus Civilisation, first cities in the region before 3500 BCE; >500 ha of large settlements in aggregate by late Mature phase; peak populations of 40,000 in Harappa and Mohenjodaro, and several million inhabitants in the region in the Mature phase	5500	[168]
	Egypt, city of Memphis in Early Dynastic and Old Kingdom times, ca. 3000 to 2165 BCE	5000	[167]
	China, Longshan times, ca. 2600 to 2000 BCE; Anyang covered 15 km^2 along 6 km of the Yellow River by 1200 BCE	4600	[167]
	Mesoamerican cities, lowland Maya cities from 900 to 300 BCE	2900	[167]
	Urban water systems		
Indian Subcontinent	Indus Civilisation, with early urban phase ca. 2800 to 2600 BCE [167,169]; extensive public and private systems: wells, baths, street drains, and sewage pits during 3rd millennium BCE [133].	~4800	[133,167,169]
Crete	Minoan culture, with wells, cisterns, fountains, and aqueducts from rivers and springs, in Early Minoan period ca. 3500 to 2150 BCE	5500–4150	[133]
	Resources from channel and floodplain sediments		
Alluvial gold	Working of placer deposits in rivers in China (Xia, Shang, and Zhou dynasties, ca. 2100 to 256 BCE), with advances in dredging and hydraulic methods	4100	[170]
Mud bricks (sun dried)	Earliest known bricks at Jericho ca. 7500 BCE, with later Indus cultural examples in Baluchistan ca. 7000 BCE	~9500	[168]
	Structures preserved in Egypt with bricks of Nile mud and straw; buildings with reed layers, ca. 3050 to 2687 BCE	~5050–4690	[171]
Baked bricks	Indus Civilisation, widely used in Mature Phase from 2800 BCE	4800	[168]
Earliest pottery	China, with other Late Pleistocene occurrences in Asia, based on calibrated dates on associated bone and charcoal [172] and assessment of dates from varied calibrated sources [173]	~20,000	[172,173] *
Continuous pottery record	At sites in Mesopotamia and Anatolia, with initial pottery levels dated to ca. 7000 BCE	~9000 on	[174]
Earliest pottery	At sites along the Amazon in Brazil, based on cal dates and a TL date	8000–7000	[175] *
Ochre	Poland, excavation of 25 acres of red floodplain clay to >1 m depth; occupied from ca. 15,500 to 11,500 years ago [176]	15,500	[176,177]

Table 4. *Cont.*

	Activity and Location	Approximate Date (Years BP)	References
	Aquatic resources		
Reeds	Large reed buildings in Uruk period of Mesopotamia, >5000 years ago [178]; reeds used in Egyptian tombs of Old Kingdom, ca. 2686 to 2160 BCE [179]	>5000	[178,179]
Papyrus	Used in Egypt since at least 3000 BCE, with large-scale production probably controlled by the state; Greco-Roman factories in the Nile Delta [180]. Oldest inscribed at Wadi al-Jarf with hundreds of fragments from reign of Khufu (2589–2566 BCE) [181]	>5000, ~4570	[180,181]
Fish and other aquatic organisms	Known from Middle Palaeolithic sites on the Nile. Prominent Nile fisheries from > ca. 14,000 years ago, with salting for storage from ca. 9000 years ago; along Amazon in Brazil from ca. 11,200 years ago, Yangtze in China from ca. 10,000 years ago, and Columbia in USA from ca. 9300 years ago. Sites include molluscs, turtles, and amphibians. Based on cal and uncal dates and OSL dates [182], cal dates [183], dates from varied sources with calibration not reported [184], and calibrated dates [185]	>14,000	[182] *, [183] *, [184] *, [185] *, [186], [187]

Climate change influenced settlements across the Near East and North Africa, leading to the migration of irrigation-based farmers and periods of societal collapse during droughts, especially during major events at 8200, 5200, and 4200 cal yr BP [101]. By about 5000 years ago, declining precipitation rendered large areas of North Africa uninhabitable and settlement focused along the Nile [188,189]. In Mesopotamia over the past 7000 years, a dynamic interaction between climate, anthropogenic effects (including settlement, irrigation, and warfare), and river dynamics makes it difficult to ascribe avulsions to natural causes alone [162].

Settlements in the mature stage of the Indus Civilisation were present across 1 million km^2 of the river plains, with thousands of individual sites along rivers and an early urban phase from about 4800 years ago. Five cities with populations of tens of thousands were each more than 100 hectares (1 km^2) in area. In China, precincts with earthen walls date back to Longshan times at about 4600 years ago (2600 to 2000 BCE), and the city of Anyang extended widely along the Yellow River by about 3200 years ago. Large urban centres were present in Mexico nearly 3000 years ago (by 900 BCE). Spanish adventurers in the sixteenth century CE encountered a continuous line of large settlements along the Amazon river bluffs [190].

Several cultures developed innovative urban water systems (Table 4). Mohenjodaro in the Indus Civilisation had many wells, large public baths, street drains and sewage pits, and houses with bathrooms and terracotta conduits for waste water—the most sophisticated drainage system of any city of the time. The Minoan culture of Crete tapped rivers and springs for extensive public water systems prior to 5000 years ago.

Roman cities used river and spring water intensively [133,191,192], with aqueducts nearly 100 km long from springs along rivers to supply Rome. By the end of the first century CE, the city's population may have numbered half a million, using nearly one million m^3 of water daily, and waste water was flushed into a major sewage system, the *Cloaca Maxima*, with an outlet to the Tiber. Public baths across the empire (Figure 4E) used hot springs and water heated by wood-fired furnaces, requiring considerable woodland resources [193]. By the fourth century CE, Rome boasted 11 large baths, 856 public baths, 15 monumental fountains, and 1352 other fountains and basins [191].

9. Navigation and Trade Routes

Navigation would have required riverbank structure such as wharves and boatyards, as well as channel maintenance through dredging of sediment and clearance of woody debris (Table 1). In Egypt, preserved models of reed and log rafts date back about 7500 years, with large Nile barges capable of transporting heavy obelisks in the ensuing millennia (Table 3). In Mesopotamia, large fleets carried grain and other commodities on rivers and canals about 4000 years ago. In China, an especially remarkable construction was the Grand Canal, almost 1800 km long and completed in 609 CE, with incorporation of earlier waterways to link agricultural plains in the northern and southern parts of the empire.

Trade routes have followed river banks for thousands of years. Trade in obsidian from Turkey dates back about 12,000 years at sites on the Euphrates [164]. During the Uruk period and Early Bronze Age in Mesopotamia (after about 6000 years ago), many overland trade routes followed the Tigris and Euphrates and their tributaries, using ox- and donkey-wagons and pack donkeys to connect with river transport [194]. Wagons came into use in the Near East after the wheel appeared during the Uruk period before about 5000 years ago [194].

Before 6000 years ago (late fifth millennium BCE), boats transported flint along the Vistula and Oder rivers in Europe [195]. A wide range of goods was transported along the Silk Roads from China to the Mediterranean as early as 4000 years ago [196].

Transport routes would have required bridges and seasonal causeways across rivers. Ancient bridges of wood or stone are rarely preserved, but stone piers indicate that a bridge crossed the Euphrates at Babylon [178]. Wooden bridge pilings from the Bronze and Iron Ages, the earliest dating back more than 3000 years (1530 to 1210 cal yr BCE), are preserved in the Thames Valley, UK at a long-term crossing point where the bridge structures influenced local sedimentation and fluvial competency [197]. River beds would have served as routes at low stage in dry seasons, as in many parts of the world today. Coastal reaches were dredged to improve navigation on the Tiber river near Rome in the second century CE [198].

10. Resources of Channels and Floodplains

Utilisation of channel and floodplain sediments and vegetation has a direct impact on rivers. Modern examples indicate that these activities cause intense local disturbance (Table 1), although the cited examples are commonly on a larger scale than those of the earlier Quaternary.

Bedrock mining for copper and other commodities was underway by 7000 years ago and probably much earlier in the Near East and elsewhere [199–201], requiring local water supplies for ore separation and settlements. In China, lake records indicate increased copper concentrations approximately 4000 years ago, with widespread sediment pollution related to the intensity of smelting activities [202]. Placer mining for alluvial gold was active by 4100 years ago (Table 4), and improved methods of dredging and hydraulic extraction were developed in the first millennium BCE [170].

Construction of early settlements and later cities required excavation of floodplain mud to generate large volumes of sun-dried mud bricks, baked bricks, and tiles (Table 4). The first use of mud bricks worldwide was at Jericho about 9500 years ago (7500 BCE). In the Indus cultural tradition, mud bricks date to about 9000 years ago and early mud-brick villages to 8500 years ago, with larger mud-brick settlements by 5500 years ago [168]. Baked bricks first appear at about 4800 years ago (2800 BCE) and were widely used for baths, drainage systems, and flood-protection structures that required water-resistant materials. Baking of bricks required sufficient firewood for heating to >500 °C for several hours, although 200 hectares of riverine forest may have sufficed to supply Mohenjodaro with a century of baked bricks [168]. Mud-brick structures from about 5000 years ago are preserved in Egypt. Brick pits at Uruk in Mesopotamia must have covered a considerable area, perhaps >2.5 km^2, but they would be difficult to identify when filled with water- or wind-blown sediment and garbage [135].

City populations used large amounts of pottery, commonly made from floodplain clay. Currently the earliest known pottery dates to about 20,000 cal yr BP in China, and many sites in the Near East exhibit a continuous pottery record after about 9000 years ago [174]. In the Americas, pottery dates to before 7000 cal yr BP. Pottery kilns in southern Poland constituted Europe's largest production area during Roman times (second to fifth centuries CE), using river clays and Miocene clays exposed in river banks, and enough firewood to prevent regeneration of oak forests [176].

Ochre obtained from floodplains and bedrock was used as far back as 100,000 years ago for colouring many materials and for cave art [203]. At Rydno in Poland, ochre mining by 15,500 years ago took out red floodplain clay and hematite-rich gravel over 25 acres and down to more than 1 m depth [176,177].

Other river products widely used in antiquity include reeds, papyrus, hay, and (among many nutritional resources) fish (Table 4). The use of papyrus for writing dates back at least 5000 years in Egypt, with large-scale production under state control, and large reed buildings were constructed in Mesopotamia more than 5000 years ago. In the Thames Valley, the use of coppiced oak for bridge structures indicates Bronze and Iron Age woodland management along the river, and hay was grown on floodplains in Roman times [197]. Brushwood for fires and hedges would have been widely gathered along rivers in antiquity, as today.

Fishing was a major livelihood on rivers around the world. On the Nile, fish are preserved at Middle Palaeolithic sites [186], with examples of large-scale fishing by 14,000 years ago (Table 4). Palaeolithic fishers initially caught Nile fish that spawned in shallow water, but after about 10,000 years ago, fishers also used deeper water species, implying improved technology and probably more stable vessels [187]. In the Sudan at about 9000 cal yr BP, fish were salted and stored in pottery containers [185]. Fish are prominent at archaeological sites along rivers in China and the Americas, along with molluscs, turtles, and amphibians at Amazon sites.

Communities have used water for many other purposes historically. Modern cultural festivals may bring together millions of people along river banks (Figure 2H).

11. Early Legacy Sediments and Geomorphic Change (Middle to Early Late Holocene)

Anthropogenic changes in land use, especially agricultural development and deforestation, commonly result in *legacy sediments* [17,204] (Table 5). Direct evidence for anthropogenic influence in these deposits comes from sediment type, aggradation rate, charcoal, plant materials and pollen, artefacts, and associated paleosols that were anthropogenically overprinted and eroded [205–208]. Sediments dating to 7350 to 2450 cal yr BP in Iran yield charcoal, bones, burned byproducts of brick and ceramic making, collapsed plinth masonry, plaster fragments, possible slag, and phosphatic coprolites that may have been used as fuel [209]. Domestic animal tracks have been identified in some floodplain deposits [210,211]. Some floodplain events, especially episodes of aggradation and incision, may reflect climate effects with an anthropogenic overprint [205,208].

Legacy sediments may be best documented from Quaternary successions in small river valleys and tributaries [17,18,96,208] where shifts in the equilibrium profile of the channels promoted substantial aggradation and incision. The identification of anthropogenic signatures is more problematic for large rivers where regional climatic effects influence sediment flux in large catchments [21], earthquakes liberate large sediment volumes [212], and catastrophic glacial floods sweep river reaches [26]. However, large river systems may capture evidence for both human activities and climatic and other environmental effects [165].

Early legacy sediments are found in New Guinea (Table 5), where human impacts at 7800 cal yr BP include valley alluviation, a more open herbaceous cover, and increased charcoal. Banana cultivation and managed planting commenced shortly after 7000 cal yr BP, and within about 500 years crops were growing in a largely anthropogenic landscape. Drainage ditches were excavated by 4000 cal yr BP.

In the Yellow River area of China, alluvium dating to about 7700 cal yr BP indicates the onset of slash-and-burn methods [213], and successions predating 7000 years ago contain evidence for rice-paddy farming in the form of phytoliths, charcoal, and pottery [122]. As valleys filled with sediment, wet floodplains suitable for rice cultivation increased in area until, approximately 4000 years ago, incision narrowed the floodplains and reduced wetland areas. After 3000 cal yr BP, the sediment load of the Yellow River increased steadily, linked to progressive cultivation and deforestation on the Loess Plateau. Changes in stream regime through this period largely represent human impacts rather than climatic forcing, and human and climatic effects are so intertwined on the Loess Plateau as to be inseparable by about 2000 years ago [214], with such drylands especially prone to both climatic and anthropogenic effects [18]. In southwest China [215], changing pollen spectra mark the expansion of irrigated rice cultivation after 2200 cal yr BP, with gully erosion by about 1400 cal yr BP and 54,000 hectares (540 km^2) of land under irrigation during the Tang Dynasty (618–907 CE).

Table 5. Some examples of anthropogenic sediment and effects in river systems. Middle and Late Holocene dates are in calibrated years BP, and dates for the past millennium are based on historical evidence.

Location	Legacy Sediments	References
	Middle Holocene to Early Late Holocene	
New Guinea	From 7800 cal yr BP on, clearance of forest, expansion of herbaceous vegetation, and increased charcoal from use of fire. Aggradation in *Baliem River* valley. Banana and other crops in an anthropogenic landscape by 6950 to 6440 cal yr BP. Drainage ditches and channels dated at ~4350–3980 cal yr BP. Based on calibrated dates.	[81,96]
China	Agriculture and population growth after ~7700 cal yr BP in *Yellow River* system, based on charcoal, pollen and sediment records, with onset of slash-and-burn [213]. Widespread effects by 7000 cal yr BP [214]. Valley alluviation and hillslope erosion from ~5300/5010 BCE through 2130/1870 BCE, in *Yiluo River* system; pottery, bones, charcoal, and fossil evidence for rice farming [122]. Deforestation of Loess Plateau increased Yellow River sediment load at ~3000 cal yr BP [216]. Based on OSL and cal dates.	[122,213,214,216]
Spain	Aggradation and incision episodes in the *Ebro Basin*, with valley-floor sedimentation from ~8000 cal yr BP (6000 BCE), intensifying during the Bronze and Iron Ages and through the Roman period. Associated charcoal and pottery. Based on cal dates and artefacts.	[205]
UK	Clays dated at 4440–3560 cal yr BP, during change from woodland to scrubland with cereal crops. Overlying silty sand at ~4100 cal yr BP in *Frome River* system. Linked to deforestation and expanded arable farming, with increased sedimentation rate. Based on cal and OSL dates.	[17]
Italy	Original vegetation lost by Bronze Age ~3300 cal yr BP in *Po River* plain. Drainage ditches and irrigation channels since Early Iron Age. Widespread Roman deforestation, intensification of farming, roads and ditches after ~2100 cal yr BP. River courses embanked and prevented from silting. Soil erosion caused Po delta to prograde further at ca. 300 CE than in previous centuries. Based on cal dates.	[217,218]
Mexico	Agriculture and deforestation enhanced erosion by 3450 cal yr BP, intensifying with urban development after 2350 cal yr BP. *Río Verde* shows increased sediment load, aggradation, and change from meandering to braided. Spread of overbank fines expanded agricultural areas. Coastal morphology modified. Based on cal dates.	[219]
	Past Millennium	
NW Europe	Salmon stocks declined by 90% from Early Middle Ages (~450–900 CE) to ~1600 CE. Decline matches watermill expansion; weirs and millponds blocked access to upstream spawning grounds. River gravel covered with silt and mill waste.	[220]
USA	Buildup of slackwater sediment from ~1700 CE, behind tens of thousands of 17–19th century milldams in northeastern *rivers*. Underlying soils and gravel represent forested wetlands with anabranching channels, now mainly lost.	[204]
China	After ~1800 CE in *SW China*, intense cultivation and population pressure caused soil erosion and hillslope collapse. Fluvial aggradation of 2 m/yr, channel blockages.	[221]
U.S.A.	Release of 1.3 billion m^3 from hydraulic gold mining in *Sacramento River* system, California after 1849 gold rush. Aggradation of 60 m of alluvium, followed in a few decades by incision. Rapid sedimentation of San Francisco Bay.	[53]
Global	Currently >50,000 large dams >15 m high. 17 million dammed reservoirs of all sizes, with a volume >8000 km^3. Severe effects on sediment transport, flow regimes, and biodiversity of freshwater ecosystems.	[20,222]
Global	Rivers swept by waves of millions of cubic metres of toxic waste and cyanide from mine-dam failures and pipeline breaks.	[223,224]
China	South-North Diversion Project, with capacity to deliver 25 billion m^3 from Yangtze to northern China along two routes, each >1000 km long. Partially constructed.	[152]

In the Ebro basin of Spain, human occupation has been intense since the Neolithic about 10,000 to 8000 years ago. Rapid sedimentation was attributed to anthropogenic influence after about 8000 years ago, intensifying during the Bronze and Iron Ages and Roman times [205]. For sites across central Europe, no major anthropogenic impact was noted prior to 7500 cal yr BP, with a changing pollen spectrum but little effect on sedimentation from 7500 to 6300 cal yr BP [225]. After 6300 cal yr BP, an increased thickness of colluvium and flood sediments, increased charcoal, and strongly altered pollen spectra suggest vigorous human activity, which may have amplified drought effects. For the Frome River in UK, accelerated silt buildup prior to 4000 cal yr BP was accompanied by a decline in oak-elm pollen and a rise in scrub and hedge taxa and cereal pollen, testifying to deforestation and agricultural development.

In Italy, the central Po Plain had lost its original vegetation cover by 3300 cal yr BP (Table 5), but deforestation and soil erosion had intensified by 2100 years ago under the Roman Empire, with control of embanked rivers and irrigation channels, reclamation of swamps, and the establishment of agricultural plots for veterans. Some creeks aggraded whereas others were dredged. As the empire declined, fields were abandoned and fluvial aggradation set in under cooler and wetter conditions, with a rapid advance of the Po Delta by 400 CE related to anthropogenic activity. By late Roman times, deforestation had been sufficiently intense to promote climatic alteration around the Mediterranean [226]. Intensive agricultural development is evident elsewhere in the Roman Empire [197,217].

In Mexico, deforestation linked in part to maize cultivation triggered landscape erosion by 3450 cal yr BP, although terracing reduced soil loss [219]. Key effects included increased sediment load, a shift from meandering to braided rivers, and coastal geomorphic changes.

Fire played an important role in generating legacy sediments. For southern Europe, anthropogenic fire effects have been important since 8000 to 7000 cal yr BP in Neolithic times, with increased frequency during the Bronze Age (ca. 3800–3600 cal yr BP) and significant vegetational change as shrublands replaced forests [28,29]. In China, biomass burning for land reclamation intensified after 3100 years ago, based on OSL and radiocarbon dates [227].

12. Pre-Industrial and Industrial Changes (Past Millennium)

Anthropogenic influences have changed rivers enormously within the past centuries. A few effects are highlighted here (Table 5).

From the Early Middle Ages (about 450–900 CE) to Early Modern Times (about 1600 CE), the construction of weirs and millponds on European rivers blocked fish migration routes and covered river gravel with fine sediment and mill wastes, leading to a steep decline in fish populations [220]. In England, the 1086 Domesday Book listed 5624 watermills along rivers, one for every two hundred and fifty people [149], and by the early eighteenth century a river in France had twenty mills per linear kilometre [220]. In the sixteenth century, debris from mining was pouring into the rivers of southwestern UK, silting up Plymouth Harbour [17].

Over the past centuries, many multiple-channel reaches were converted to single channels as anabranches were taken over for agriculture, mill channels, and canals [228,229]. By 1840 more than 65,000 watermills with dams a few meters high were operating in the northeastern United States alone, each ponding the river for some kilometres upstream and converting multichannel wetlands into single channels with aggradational floodplains [204]. More than 1000 snags per kilometre, in the form of drifted trees and logs, were removed from rivers in the central United States over a few decades in the late 1800s [48], and many rivers were extensively dredged [230]. Blasted out in the 1830s to improve navigation, the Great Raft of Louisiana was a log jam 250 km long that had formed over hundreds of years [231].

Rapid alluviation affected many rivers in recent centuries through agriculture and mining (Table 5). In southwest China, the wholesale collapse of upland slopes as agriculture intensified caused extreme aggradation and channel blockage. Hydraulic mining of gold in the Sacramento River system of California following the 1849 gold rush choked valleys with 60 m of gravel, agricultural land was flooded with tailings, and a debris wave reached San Francisco Bay. After legislation ended hydraulic mining in 1884, the channels readjusted and cut through the gravel, leaving flights of young terraces. Overall, more than a cubic kilometre of tailings (1.3 billion m^3) was released [53]. In the Fraser River of Canada, 58 million m^3 of gold tailings were dumped in the channel, with a debris wave 500 km long that is advancing at a few kilometres a year, and tin mining in Tasmania added 40 million m^3 of mine waste to a local creek [232,233].

The construction of tens of thousands of large dams has seriously affected most large rivers worldwide, as have failures of mining dams and pipelines that released waves of toxic waste (Table 5). Sand and gravel mining in many river channels is now a serious issue [234]. Large water transfers between river systems include the South-North Diversion Project in China. Many of the world's largest rivers are at risk through pollution, dams, water overextraction, navigation, climate change, invasive species, and overfishing [19], and wilderness is effectively gone [235].

Deliberate diversion of rivers and breaching of dikes during warfare goes back to ancient times. Herodotus records that, in 539 BCE, the Persian king Cyrus diverted the Euphrates upstream of Babylon into a canal, allowing his troops to storm the city along the river bed (sections 1.190–191 in [236]). A recent example is the 1938 breaking of dikes on the Yellow River by Chinese Nationalist forces to impede the Japanese advance, inundating an enormous area and causing a huge loss of life (Table 1).

13. Stages in River Modification

Figure 5 contains some of the evidence set out above for the timing of key processes by which humans influenced rivers. The analysis emphasises many kinds of events (Figure 3) along numerous large rivers, including the Nile, Tigris, Euphrates, Jordan, and Orontes (Near East and North Africa), the Indus and Ganges (Northwestern Indian Subcontinent), the Yellow, Yangtze, and Pearl (China), the Amazon and Gila (Americas), and rivers in parts of Europe. This broad compilation suggests that humans had the capability to influence rivers by the latest Pleistocene in many parts of the world, with a progressively strong influence in places from the Early Holocene onwards.

Six provisional stages highlight the onset of key innovations, which represent important changes in the ways that humans viewed and used river systems. The durations of the stages are typically a few thousand years and their boundaries are generalised, denoting broad developments rather than historical events. These developments were strongly diachronous globally, depending principally on the trajectory of agricultural development in different regions and continents [8,16–18], and the stages apply most closely to the Near East and adjoining areas. However, once innovations appeared, they tended to spread through human migration and the transmission of ideas, especially along trade routes. Examples include the spread of rice, wheat, and barley across Asia [84,120] and the spread of irrigation and groundwater technology across North Africa and the Near East [133].

Stage 1 (Minimal River Effects) covers the appearance of the earliest hominins to about 15,000 years ago in the Late Pleistocene. Activities late in the stage include fishing along the Nile in the Middle Palaeolithic, cereal processing in Europe (32,000 cal yr BP) and the Near East (23,000 cal yr BP), and the use of fire in New Guinea (28,000 cal yr BP). Although hunter-gatherers left traces along many rivers, the evidence for effects on rivers is largely circumstantial and the cumulative effect was probably minimal.

Stage 2 (Minor River Effects) extends from about 15,000 to 9800 years ago. The stage spans a period of widespread gathering and cultivation of plants, especially by the Natufian culture and early settlements in the Near East, and it includes the initial domestication of plants and animals after about 10,700 cal yr BP through to the emergence of formal agricultural organisation. Human activity along rivers included fisheries, ochre mining, and construction of early settlements.

Stage 3 (Agricultural Era) extends from about 9800 to 6500 years ago. The stage covers the onset of organised agriculture, domestication of additional crops and animals, widespread use of fire for land clearance, and more extensive settlements with mud-bricks and pottery made from floodplain clay. Early dams and irrigation features appear, with spring-fed irrigation at Jericho (Figure 4C) by about 8000 years ago. The first well-documented legacy sediments appear in several regions during this stage.

Stage 4 (Irrigation Era) extends from about 6500 to 3000 years ago. The stage covers the rise of organised irrigation systems in Mesopotamia, Egypt, and the Indus region, the first large dam in Egypt, and urban water supplies. Rulers and administrators deployed large labour forces and utilised resources for construction projects. Other activities included the use of fleets of boats with implications for riverside construction, the harvesting of aquatic resources such as reeds and papyrus, alluvial mining, and the increased use of groundwater on floodplains.

Stage 5 (Engineering Era) extends from about 3000 years ago to about 1750 CE. The stage marks the onset of large-scale river engineering under Chinese and Roman emperors, with management of extensive embankments, canals, dams, and urban water supplies. Agriculture and sediment flux intensified in both empires, with effects on river deltas. From the Early Middle Ages onwards in Europe, watermills partitioned many rivers and freshwater fish stocks were decimated. Navigation became increasingly important on many rivers.

Stage 6 (Industrial Era) extends from about 1750 CE at the start of the Industrial Revolution to the present, marking the widespread use of rivers for waterpower, factories, and transport. The modern Big Dam Era from 1882 onwards ushered in technological river modification and regulation such that most rivers on Earth now experience major anthropogenic influence.

Stages 2–6 encompass less than 10,000 years of river history. The human river footprint has expanded with great rapidity: such a duration is commonly represented in the geological record by a few metres of sediment on an alluvial plain, a few avulsions of a trunk channel, or part of a rhythm of sea-level rise.

Key anthropogenic events that influenced rivers differ markedly in time and in space over short distances. River modification commenced relatively recently in some regions, stages were bypassed entirely in places, and events such as periods of soil erosion [237] influenced a region's subsequent history, irrespective of the passage of time. In parts of the polar regions, river modification has yet to commence. The brief regional analysis below covers areas for which considerable information is available in the current compilation (Figure 5, Tables 2–5), drawing also on surveys of land use changes and lake records over the past few thousand years [8,9,238]. A full analysis for these regions is beyond the scope of the present paper and would require consideration of local cultural factors, climatic, hydrological and sea-level history, as well as a full resolution of available age data [27]. However, such an analysis would need to take into account the many processes of anthropogenic modification set out here. The technological era of Stage 6 affected all areas sometime after about 1750 CE.

The six stages of Figure 5 accord most closely with the Near East, for which abundant archaeological evidence is available from Mesopotamia and adjoining areas through Anatolia and the Fertile Crescent to Egypt and adjoining parts of North Africa. Within this area, Stage 2 featured intensified cultivation and the early domestication of many plants and animals. Stage 3 features the early onset of agriculture, as well as the rise of irrigation and dams, the first cities, and brick manufacture. By the end of Stage 3, most villagers in the region were farmers. Extensive irrigation systems in Mesopotamia mark the onset of Stage 4, with the progressive use of irrigation technology and groundwater, as well as the deployment of shipping on the Nile, Tigris, and Euphrates. River engineering in Stage 5 was prominent under local rulers, culminating in large-scale engineering under the Roman Empire.

In Europe, cereal production spread across much of the continent by 7000 years ago, with widespread legacy sediments and the use of fire. By 2000 years ago the Roman Empire was conducting major river engineering (Stage 3) in many areas.

China also experienced early river modification, with Stage 3 marked by the early domestication of rice and millet and legacy sediments that date back about eight thousand years. Engineering of the Yellow River (Stage 5) included the construction under a centralised economy of dikes and canals after approximately 3000 years ago. Rice-based agriculture spread across Asia over several millennia [87], and brick-built cities with sophisticated urban water supplies were present in the Indus Basin by 5500 years ago.

In the Americas, Stages 1 and 2 commenced with the first crossing of the Bering Strata after about 16,500 years ago, although earlier occupations have been suggested [239]. Trajectories of river modification varied greatly in local areas. A relatively early onset of Stages 3 and 4 is suggested for Central and South America by the early domestication of squash and maize, with irrigation systems in place locally by 7000 cal yr BP. Large cities developed in some areas shortly after 3000 years ago.

In Oceania, a relatively early onset of Stage 3 in New Guinea is indicated by legacy sediments that predate 8000 cal yr BP and the domestication of the banana shortly after 7000 years ago. Humans arrived in Australia about 65,000 years ago [77], and the use of fire may have been the main effect on rivers until European settlement in the late 1700s. Stage 1 commenced after Polynesians reached New Zealand 700–800 years ago, with severe fires that began to transform the vegetation and enhanced erosion [240].

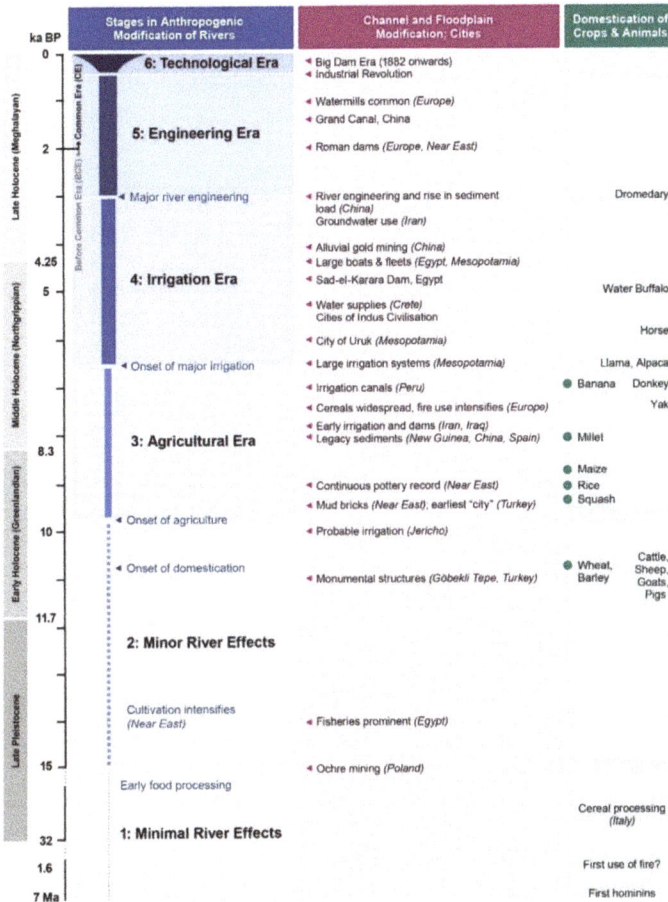

Figure 5. Timeline and stages for the anthropogenic modification of river systems. See Tables 2–5 for sources of the dates shown in the figure, with dates for early food processing [113,114] and intensification of cultivation and onset of domestication and agriculture [88]. Suggested stages in modification are based largely on information from the Near East, North Africa, Northwestern Indian Subcontinent, China, and parts of Europe. Note that the timing and intensity of anthropogenic influence vary greatly regionally [17,18]. Dates for Early, Middle, and Late Holocene (E, M, L) from [62].

14. Comparison of Stages with Models for Population Growth and Land Use

Human population growth is regarded as the main driving force of global change [7]. Because many communities depend on rivers, global population estimates may serve as a general proxy for anthropogenic influence on rivers. The HYDE 3.2 model uses historical population data and land-use allocation algorithms [8,9]. Model results have high uncertainty levels, especially for the earlier Holocene, because the quality of historical population data varies in time and space and because the models depend on assumptions about the relationship between population and land use and sociocultural factors [241]. Land-use data are reasonably reliable for rice-based civilisations of Asia and hydraulic civilisations of North Africa and the Mediterranean, but the area of grazing land is poorly constrained.

A central estimate for global population is 4.4 million at 12,000 years ago, near the start of the Holocene [8,9] (Figure 6). Population rose four-fold to 19 million by 7000 years ago and thereafter by more than an order of magnitude to 232 million by 1 CE. Population was clustered locally, reaching a few million during the Mature phase of the Indus Civilisation from 4600 to 3900 years ago [168]. The global population reached 461 million by 1500 CE and 7.26 billion by 2015.

The area of global cropland (mainly rain-fed) increased from 6 million hectares at 7000 years ago to 146 million hectares at 1 CE and 209 million hectares at 1400 CE. The area of grazing land increased from 30 million hectares at 7000 years ago to 199 million hectares at 1 CE and 483 million hectares at 1400 CE (Figure 6). Based on these figures, the combined area of cropland and grazing land rose by an order of magnitude between 7000 years ago and 1 CE, further doubling between 1 CE and 1400 CE. Irrigated cropland was a small proportion of the total, with 0.2 million hectares by 7000 years ago and 2.6 million hectares by 1 CE, with a rise 6000 years ago due to large irrigated areas in Egypt and the Near East. Soil erosion rates increased strongly after about 4000 years ago (ca. 2000 BCE), linked overwhelmingly to changes in land use rather than climate [242].

Stages in river modification from Figure 5 correlate broadly with global population, land-use, and river-sediment trends shown in Figure 6, supporting a trend of intensified anthropogenic influence on rivers through the Holocene. Stages 3 to 5 are in accord with a progressive increase in population and the area of cropland and grazing land, with initial irrigation prior to 6500 years ago. Although the area of irrigated land remained low, the land use per capita was higher in earlier times, especially during the early stages of hydraulic civilisation in North Africa and the Near East [7], and irrigation may have had a relatively large effect on rivers. After the onset of Stage 5 at about 3000 years ago with major river engineering in China and elsewhere, population, cropland and grazing land increased at a progressively higher rate globally, and the sediment load in the Yellow River increased as deforestation in the Loess Plateau promoted soil erosion. Rice fields initially covered a modest area but began to exceed irrigated land after 3000 years ago, with important effects on floodplains.

More direct proxy records typically focus on rates of erosion and sediment transport. They include measurements of sediment flux to the world oceans [49,243], local erosion rates under natural and modified vegetation [18,244], soil-erosion models [242], basinwide denudation rates derived from cosmogenic isotopes [33,245], and human impacts in lake records [202,238]. However, the timing and rates of events derived from different proxies commonly vary because they measure different effects, and proxy records are rarely truly independent variables and are subject to complex feedbacks [18,130]. Some large river systems transfer sediment to the ocean with little delay, but in other cases sediment storage in valley fills, terraces, and mass-wasting deposits leads to considerable time lags in sediment transfer [244–248].

Stages in Anthropogenic Modification of Rivers

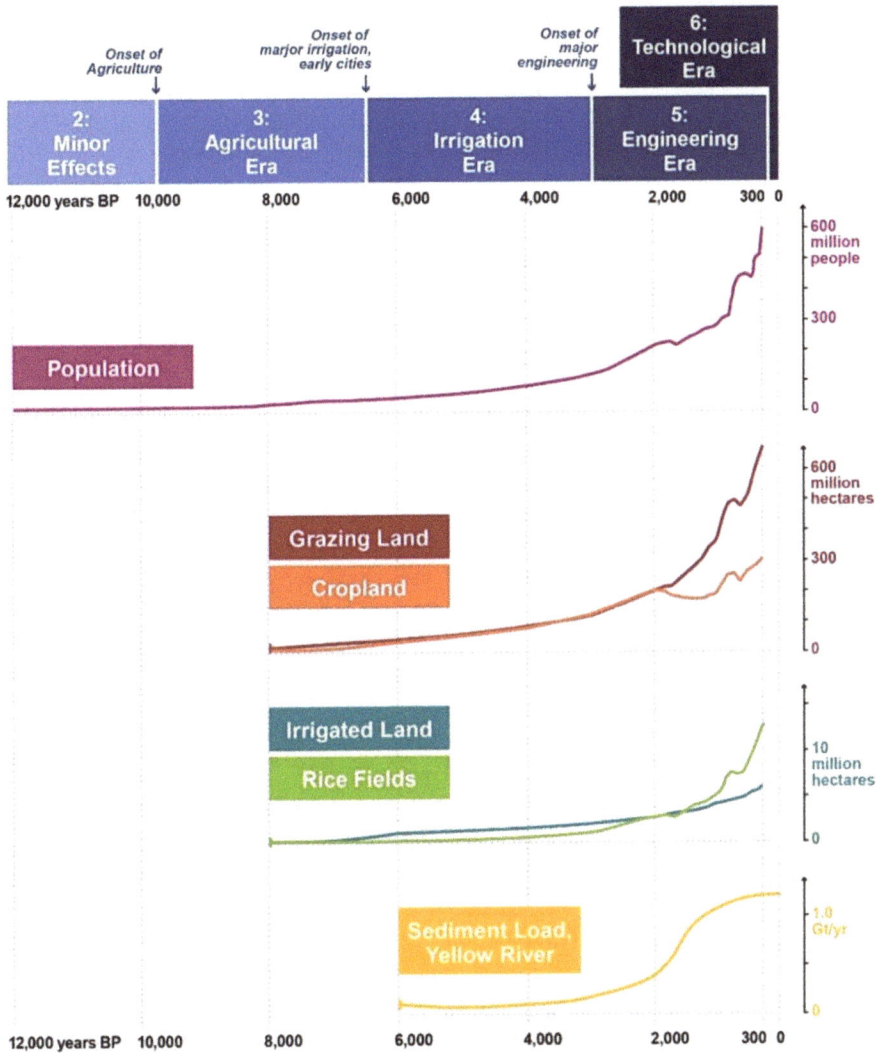

Figure 6. Comparison of stages in anthropogenic modification of rivers (Figure 5) with datasets for population and land use [8,9] and Yellow River sediment load [216]. Original graphs plotted in years before present.

15. Implications for the Concept of the Anthropocene

The present analysis highlights the great variety of human activities along rivers worldwide from ancient times, many of which probably have been undervalued as geomorphic processes. Collectively, the data suggest that humans were capable of considerable local modification of river systems during the Late Pleistocene, although more reliable case studies are needed for this period. River modification intensified in the Early Holocene (11,700 to 8300 years ago) through numerous processes and was widespread in many regions worldwide from the Middle Holocene (8300 years ago) onwards, as documented by a wealth of direct evidence.

As noted in the Introduction, strong allogenic effects are likely to have masked early anthropogenic processes, but they also would have affected human societies. Aridification caused changes of hydrological regime, collapse of civilisations, abandonment of settlements, and migration of farmers in the Nile Valley, the eastern Mediterranean, the Near East, and the Indian Subcontinent [27,101,140,189,249]. However, human responses to such changes were complex and, locally, provided opportunities for farmers [250]. Climate change probably played a role in the decline of the Roman Empire, leading to reafforestation, drainage changes, and aggradation [217]. Such climatic changes may produce local discontinuities in response to rhythms of aggradation and degradation [22], and these surfaces may correspond to widespread punctuated steps in river modification. A landslide dam outburst flood on the Yellow River nearly 4000 years ago (ca. 1920 BCE) may have led to the establishment of the Xia Dynasty at the Neolithic–Bronze Age transition [251].

Holocene sea-level rise displaced populations and promoted cultural change [252]. After about 8000 years ago, a decreased rate of sea-level rise caused the formation of deltas by large sediment-charged rivers, promoting occupation and the emergence of complex societies on the Nile, Yangtze, and other deltas [253–256]. In estuaries and bays of the Texas Gulf Coast, rapid Holocene sea-level rise adversely affected estuarine productivity but stillstands promoted aquatic resources, influencing human occupation [257].

Complex positive and negative feedback systems between climate, vegetation, and human activities complicate the assessment of anthropogenic influence [130]. As land-use changed in China, sediment flux into river valleys caused aggradation and increased the area of floodplains, promoting additional agriculture until later incision narrowed the cultivable area [122]. Fire and ploughing would have promoted soil erosion and nutrient loss, causing declines in crop production [237], and may have encouraged cropland expansion, as may waterlogging and salination of irrigated soils. Land-use changes may have exacerbated the effects of droughts [225].

The onset of river legacy sediments is diachronous worldwide [17]. No evidence adduced here supports the identification of a key event that could be used to demarcate the onset of the Anthropocene, which would additionally require careful discrimination of human from natural processes [18]. Nevertheless, the analysis supports the view [11] that researchers studying the Anthropocene need to acknowledge how early and profoundly humans began to modify the terrestrial environment. Anthropogenic effects commonly mimic natural river dynamics, making discrimination difficult. Such effects include the buildup of (artificial) levees, levee breaks used for irrigation, avulsion where embankments were breached by floods or warfare, rhythms of aggradation and incision, downstream migration of sediment slugs, conversion from multichannel to single channel and from meandering to braided planforms, and delta extension.

16. Conclusions

Humans have lived along rivers since hominins first appeared, engaging in a wide range of ingenious activities (Figure 3) that potentially influence river dynamics and sediments. Legacy sediments linked to the rise of agriculture provide direct evidence for anthropogenic influence on rivers, but the evidence for many other important activities with geomorphic consequences remains largely circumstantial. Humans have influenced rivers from the Late Pleistocene onwards, with increasingly prominent effects through the Early and Middle Holocene (Greenlandian and Northgrippian).

Six provisional stages for river modification (Figure 5), centred on the Near East, represent key innovations in human activity and use of rivers. They include the increased use and cultivation of crops in the Late Pleistocene and the domestication of many plants and animals in the Early Holocene; the onset of agricultural systems at about 9800 cal yr BP; the rise of major irrigation systems under centralised administrations after about 6500 cal yr BP; the development of major river engineering after about 3000 cal yr BP; and the industrial era after about 1750 CE. A trend of intensified river modification coincides broadly with available models that document a rapid increase in population and land use through the Holocene (Figure 6).

The evidence presented here supports diachronous anthropogenic change in river systems worldwide, and no indication was found of processes or deposits that could provide a reliable early boundary for defining the Anthropocene, although regional markers may be present. Geoscientists have commonly interpreted fluvial archives in terms of natural forcing functions such as climate and sediment flux, but the analysis presented here suggests that the human impact on rivers has been underestimated. Anthropogenic effects should be considered routinely when interpreting fluvial archives from the Late Pleistocene onwards.

The present study is a preliminary overview of a topic that requires much fuller development. Recommendations for future fluvial research include documentation of some human activities for which the evidence is currently circumstantial, as well as the further identification and correlation of quantitative proxy records. In particular, it is important to develop timelines for human influence in local areas, beyond the generalised global record set out here.

Funding: This research was funded by the Natural Sciences and Engineering Research Council of Canada, Discovery Grant number 03877. The funding sponsors had no role in the design of the study; in the collection, analyses, or interpretation of data; in the writing of the manuscript, and in the decision to publish the results.

Acknowledgments: I thank many colleagues, especially members of the Fluvial Archives Group (FLAG) and John Gosse, for helpful discussions about the concepts represented here. Thoughtful comments from three anonymous reviewers and the editors resulted in considerable improvements to the manuscript. I thank Meredith Sadler for her artistic skill in drafting Figures 3, 5 and 6.

Conflicts of Interest: The author declares no conflict of interest.

References

1. Hooke, R.L. On the history of humans as geomorphic agents. *Geology* **2000**, *28*, 843–846. [CrossRef]
2. Wilkinson, B.H. Humans as geologic agents: A deep-time perspective. *Geology* **2005**, *33*, 161–164. [CrossRef]
3. Crutzen, P.J. Geology of mankind. *Nature* **2002**, *415*, 23–24. [CrossRef] [PubMed]
4. Steffen, W.; Crutzen, P.J.; McNeill, J.R. The Anthropocene: Are humans now overwhelming the great forces of nature? *Ambio* **2007**, *36*, 614–621. [CrossRef]
5. Steffen, W.; Broadgate, W.; Deutsch, L.; Gaffney, O.; Ludwig, C. The trajectory of the Anthropocene: The Great Acceleration. *Anthr. Rev.* **2015**, *2*, 81–98. [CrossRef]
6. Waters, C.N.; Zalasiewicz, J.; Summerhayes, C.; Barnosky, A.D.; Poirier, C.; Galuszka, A.; Cearreta, A.; Edgeworth, M.; Ellis, E.C.; Ellis, M.; et al. The Anthropocene is functionally and stratigraphically distinct from the Holocene. *Science* **2016**, *351*. [CrossRef] [PubMed]
7. Klein Goldewijk, K.; Beusen, A.; van Drecht, G.; de Vos, M. The HYDE 3.1 spatially explicit database of human-induced global land-use change over the past 12,000 years. *Glob. Ecol. Biogeogr.* **2011**, *20*, 73–86. [CrossRef]
8. Klein Goldewijk, K.; Dekker, S.C.; van Zanden, J.L. Per-capita estimations of long-term historical land use and the consequences for global change research. *J. Land Use Sci.* **2017**, *12*, 313–337. [CrossRef]
9. Klein Goldewijk, K.; Beusen, A.; Doelman, J.; Stehfest, E. Anthropogenic land use estimates for the Holocene—HYDE 3.2. *Earth Syst. Sci. Data* **2017**, *9*, 927–953. [CrossRef]
10. Ellis, E.C.; Fuller, D.Q.; Kaplan, J.O.; Lutters, W.G. Dating the Anthropocene: Towards an empirical global history of human transformation of the terrestrial biosphere. *Elementa Sci. Anthr.* **2013**. [CrossRef]
11. Ruddiman, W.F.; Ellis, E.C.; Kaplan, J.O.; Fuller, D.Q. Defining the epoch we live in. *Science* **2015**, *348*, 38–39. [CrossRef] [PubMed]
12. Braje, T.J.; Erlandson, J.M. Human acceleration of animal and plant extinctions: A Late Pleistocene, Holocene, and Anthropocene continuum. *Anthropocene* **2013**, *4*, 14–23. [CrossRef]
13. Ruddiman, W.F. The anthropogenic greenhouse era began thousands of years ago. *Clim. Chang.* **2003**, *61*, 261–293. [CrossRef]

14. Certini, G.; Scalenghe, R. Anthropogenic soils are the golden spikes for the Anthropocene. *Holocene* **2011**, *21*, 1269–1274. [CrossRef]
15. Ruddiman, W.F.; Vavrus, S.; Kutzbach, J.; He, F. Does pre-industrial warming double the anthropogenic total? *Anthr. Rev.* **2014**, *1*, 147–153. [CrossRef]
16. Edgeworth, M.; Richter, D.D.; Waters, C.N.; Haff, P.; Neal, C.; Price, S.J. Diachronous beginnings of the Anthropocene: The lower bounding surface of anthropogenic deposits. *Anthr. Rev.* **2015**, *2*, 33–58. [CrossRef]
17. Brown, A.; Toms, P.; Carey, C.; Rhodes, E. Geomorphology of the Anthropocene: Time-transgressive discontinuities of human-induced alluviation. *Anthropocene* **2013**, *1*, 3–13. [CrossRef]
18. Brown, A.G.; Tooth, S.; Bullard, J.E.; Thomas, D.S.G.; Chiverrell, R.C.; Plater, A.J.; Murton, J.; Thorndycraft, V.R.; Tarolli, P.; Rose, J.; et al. The geomorphology of the Anthropocene: Emergence, status and implications. *Earth Surf. Process. Landf.* **2017**, *42*, 71–90. [CrossRef]
19. Wong, C.M.; Williams, C.E.; Pittock, J.; Collier, U.; Schelle, P. *World's Top 10 Rivers at Risk*; WWF (World Wildlife Fund) International: Gland, Switzerland, 2007; 53p.
20. Vörösmarty, C.J.; McIntyre, P.B.; Gessner, M.O.; Dudgeon, D.; Prusevich, A.; Green, P.; Glidden, S.; Bunn, S.E.; Sullivan, C.A.; Liermann, C.R.; et al. Global threats to human water security and river biodiversity. *Nature* **2010**, *467*, 555–561. [CrossRef] [PubMed]
21. Goodbred, S.L.; Kuehl, S.A. Enormous Ganges-Brahmaputra sediment discharge during strengthened early Holocene monsoon. *Geology* **2000**, *28*, 1083–1086. [CrossRef]
22. Gibling, M.R.; Tandon, S.K.; Sinha, R.; Jain, M. Discontinuity-bounded alluvial sequences of the southern Gangetic Plains, India: Aggradation and degradation in response to monsoonal strength. *J. Sediment. Res.* **2005**, *75*, 369–385. [CrossRef]
23. Rittenour, T.M.; Blum, M.D.; Goble, R.J. Fluvial evolution of the lower Mississippi River valley during the last 100 k.y. glacial cycle: Response to glaciation and sea-level change. *Geol. Soc. Am. Bull.* **2007**, *119*, 586–608. [CrossRef]
24. Bridgland, D.R.; Westaway, R. Climatically controlled river terrace staircases: A worldwide Quaternary phenomenon. *Geomorphology* **2008**, *98*, 288–315. [CrossRef]
25. Vandenberghe, J. River terraces as a response to climatic forcing: Formation processes, sedimentary characteristics and sites for human occupation. *Quat. Int.* **2015**, *370*, 3–11. [CrossRef]
26. Cordier, S.; Adamson, K.; Delmas, M.; Calvet, M.; Harmand, D. Of ice and water: Quaternary fluvial response to glacial forcing. *Quat. Sci. Rev.* **2017**, *166*, 57–73. [CrossRef]
27. Macklin, M.G.; Toonen, W.H.J.; Woodward, J.C.; Williams, M.A.J.; Flaux, C.; Marriner, N.; Nicoll, K.; Verstraeten, G.; Spencer, N.; Welsby, D. A new model of river dynamics, hydroclimatic change and human settlement in the Nile Valley derived from meta-analysis of the Holocene fluvial archive. *Quat. Sci. Rev.* **2015**, *130*, 109–123. [CrossRef]
28. Vannière, B.; Colombaroli, D.; Chapron, E.; Leroux, A.; Tinner, W.; Magny, M. Climate versus human-driven fire regimes in Mediterranean landscapes: The Holocene record of Lago dell'Accesa (Tuscany, Italy). *Quat. Sci. Rev.* **2008**, *27*, 1181–1196. [CrossRef]
29. Vannière, B.; Blarquez, O.; Rius, D.; Doyen, E.; Brücher, T.; Colombaroli, D.; Connor, S.; Feurdean, A.; Hickler, T.; Kaltenrieder, P.; et al. 7000-year human legacy of elevation-dependent European fire regimes. *Quat. Sci. Rev.* **2016**, *132*, 206–212. [CrossRef]
30. Shakesby, R.A. Post-wildfire soil erosion in the Mediterranean: Review and future research directions. *Earth-Sci. Rev.* **2011**, *105*, 71–100. [CrossRef]
31. Abernethy, B.; Rutherfurd, I.D. The effect of riparian tree roots on the mass-stability of riverbanks. *Earth Surf. Process. Landf.* **2000**, *25*, 921–937. [CrossRef]
32. Brooks, A.P.; Brierley, G.J.; Millar, R.G. The long-term control of vegetation and woody debris on channel and flood-plain evolution: Insights from a paired catchment study in southeastern Australia. *Geomorphology* **2003**, *51*, 7–29. [CrossRef]
33. Vanacker, V.; Bellin, N.; Molina, A.; Kubik, P.W. Erosion regulation as a function of human disturbances to vegetation cover: A conceptual model. *Landsc. Ecol.* **2013**, *29*, 293–309. [CrossRef]
34. Trimble, S.W. Erosional effects of cattle on streambanks in Tennessee, U.S.A. *Earth Surf. Process. Landf.* **1994**, *19*, 451–464. [CrossRef]
35. Trimble, S.W.; Mendel, A.C. The cow as a geomorphic agent—A critical review. *Geomorphology* **1995**, *13*, 233–253. [CrossRef]

36. Mwendera, E.J.; Saleem, M.A.M. Hydrologic response to cattle grazing in the Ethiopian highlands. *Agric. Ecosyst. Environ.* **1997**, *64*, 33–41. [CrossRef]

37. Deluca, T.H.; Patterson, W.A.I.; Freimund, W.A.; Cole, D.N. Influence of llamas, horses, and hikers on soil erosion from established recreation trails in western Montana, USA. *Environ. Manag.* **1998**, *22*, 255–262. [CrossRef]

38. Lefrancois, J.; Grimaldi, C.; Gascuel-Odoux, C.; Gilliet, N. Suspended sediment and discharge relationships to identify bank degradation as a main sediment source on small agricultural catchments. *Hydrol. Process.* **2007**, *21*, 2923–2933. [CrossRef]

39. Dunne, T.; Western, D.; Dietrich, W.E. Effects of cattle trampling on vegetation, infiltration, and erosion in a tropical rangeland. *J. Arid Environ.* **2011**, *75*, 58–69. [CrossRef]

40. Butler, D.R. Zoogeomorphology in the Anthropocene. *Geomorphology* **2018**, *303*, 146–154. [CrossRef]

41. Dodgen, R.A. *Controlling the Dragon: Confucian Engineers and the Yellow River in Late Imperial China*; University of Hawai'i Press: Honolulu, HI, USA, 2001; 243p.

42. Xu, J. Sedimentation rates in the lower Yellow River over the past 2300 years as influenced by human activities and climate change. *Hydrol. Process.* **2003**, *17*, 3359–3371. [CrossRef]

43. Hudson, P.F.; Middelkoop, H.; Stouthamer, E. Flood management along the Lower Mississippi and Rhine Rivers (The Netherlands) and the continuum of geomorphic adjustment. *Geomorphology* **2008**, *101*, 209–236. [CrossRef]

44. Kondolf, G.M. Hungry water: Effects of dams and gravel mining on river channels. *Environ. Manag.* **1997**, *21*, 533–551. [CrossRef]

45. Rengasamy, P. World salinization with emphasis on Australia. *J. Exp. Bot.* **2006**, *57*, 1017–1023. [CrossRef] [PubMed]

46. Ran, L.; Lu, X.X.; Xin, Z.; Yang, X. Cumulative sediment trapping by reservoirs in large river basins: A case study of the Yellow River basin. *Glob. Planet. Chang.* **2013**, *100*, 308–319. [CrossRef]

47. Smith, N.D.; Morozova, G.S.; Perez-Arlucea, M.; Gibling, M.R. Dam-induced and natural channel changes in the Saskatchewan River below the E.B. Campbell Dam, Canada. *Geomorphology* **2016**, *269*, 186–202. [CrossRef]

48. Sedell, J.R.; Reeves, G.H.; Hauer, F.R.; Stanford, J.A.; Hawkings, C.P. Role of refugia in recovery from disturbances: Modern fragmented and disconnected river systems. *Environ. Manag.* **1990**, *14*, 711–724. [CrossRef]

49. Syvitski, J.; Kettner, A. Sediment flux and the Anthropocene. *Philos. Trans. R. Soc. A* **2011**, *369*, 957–975. [CrossRef] [PubMed]

50. Brown, T.A.; Jones, M.K.; Powell, W.; Allaby, R.G. The complex origins of domesticated crops in the Fertile Crescent. *Trends Ecol. Evol.* **2009**, *24*, 103–109. [CrossRef] [PubMed]

51. Lary, D. Drowned earth: The strategic breaching of the Yellow River Dyke, 1938. *War Hist.* **2001**, *8*, 191–207. [CrossRef]

52. Dutch, S.I. The largest act of environmental warfare in history. *Environ. Eng. Geosci.* **2009**, *15*, 287–297. [CrossRef]

53. Gilbert, G.K. *Hydraulic-Mining Debris in the Sierra Nevada*; Professional Paper 105; United States Geological Survey: Reston, VA, USA, 1917; 154p.

54. Anbazhagan, S.; Dash, P. Environmental case study of Cauvery river floodplain. *GIS Dev.* **2003**, *7*, 30–35.

55. Santhosh, V.; Padmalal, D.; Baijulal, B.; Maya, K. Brick and tile clay mining from the paddy lands of Central Kerala (southwest coast of India) and emerging environmental issues. *Environ. Earth Sci.* **2013**, *68*, 2111–2121. [CrossRef]

56. Leonard, L.A.; Wren, P.A.; Beavers, R.L. Flow dynamics and sedimentation in *Spartina alterniflora* and *Phragmites australis* marshes of the Chesapeake Bay. *Wetlands* **2002**, *22*, 415–424. [CrossRef]

57. Khaleej Times. 11 March 2013. Available online: https://www.khaleejtimes.com/article/20130311/ARTICLE/303119878/1028 (accessed on 15 June 2018).

58. Vera, F.W.M. *Grazing Ecology and Forest History*; CAB International: Wallingford, UK, 2000; 506p.

59. Bakker, E.S.; Gill, J.L.; Johnson, C.N.; Vera, F.W.M.; Sandom, C.J.; Asner, G.P.; Svenning, J.-C. Combining paleo-data and modern exclosure experiments to assess the impact of megafauna extinctions on woody vegetation. *Proc. Natl. Acad. Sci. USA* **2016**, *113*, 847–855. [CrossRef] [PubMed]

60. Sojka, R.E.; Bjorneberg, D.L.; Entry, J.A. Irrigation: An historical perspective. In *Encyclopedia of Soil Science*; Lal, R., Ed.; Marcel Dekker: New York, NY, USA, 2002; pp. 745–749.

61. Zalasiewicz, J.; Williams, M.; Waters, C.N.; Barnosky, A.D.; Haff, P. The technofossil record of humans. *Anthr. Rev.* **2014**, *1*, 34–43. [CrossRef]

62. Gibbard, P. Formal Subdivision of the Holocene Series/Epoch. Available online: https://www.qpg.geog. cam.ac.uk/news/formalsubdivisionoftheholoceneseriesgeogr18.pdf (accessed on 16 September 2018).

63. Brunet, M.; Guy, F.; Pilbeam, D.; Lieberman, D.E.; Likius, A.; Mackaye, H.T.; Ponce de Leon, M.S.; Zollikofer, C.P.E.; Vignaud, P. New material of the earliest hominid from the Upper Miocene of Chad. *Nature* **2005**, *434*, 752–755. [CrossRef] [PubMed]

64. Semaw, S.; Rogers, M.J.; Quade, J.; Renne, P.R.; Butler, R.F.; Domínguez-Rodrigo, M.; Stout, D.; Hart, W.S.; Pickering, T.; Simpson, S.W. 2.6-million-year-old stone tools and associated bones from OGS-6 and OGS-7, Gona, Afar, Ethiopia. *J. Hum. Evol.* **2003**, *45*, 169–177. [CrossRef]

65. Bridgland, D.R.; Antoine, P.; Limondin-Lozouet, N.; Santisteban, J.K.; Westaway, R.; White, M.J. The Palaeolithic occupation of Europe as revealed by evidence from the rivers: Data from IGCP 449. *J. Quat. Sci.* **2006**, *21*, 437–455. [CrossRef]

66. Gibling, M.R.; Sinha, R.; Roy, N.G.; Tandon, S.K.; Jain, M. Quaternary fluvial and eolian deposits on the Belan River, India: Paleoclimatic setting of Paleolithic to Neolithic archeological sites over the past 85,000 years. *Quat. Sci. Rev.* **2008**, *27*, 391–410. [CrossRef]

67. Barkai, R.; Liran, R. Midsummer sunset at Neolithic Jericho. *Time Mind J. Archaeol. Conscious. Cult.* **2008**, *1*, 273–284. [CrossRef]

68. Butler, D.R.; Malanson, G.P. The geomorphic influences of beaver dams and failures of beaver dams. *Geomorphology* **2005**, *71*, 48–60. [CrossRef]

69. Butler, D.R. Human-induced changes in animal populations and distributions, and the subsequent effects on fluvial systems. *Geomorphology* **2006**, *79*, 448–459. [CrossRef]

70. McCarthy, T.S.; Ellery, W.N.; Bloem, A. Some observations on the geomorphological impact of hippopotamus (*Hippopotamus amphibilus* L.) in the Okavango Delta, Botswana. *Afr. J. Ecol.* **1998**, *36*, 44–56. [CrossRef]

71. Darwin, C. *The Formation of Vegetated Mould through the Action of Worms with Observation of Their Habitats*; Murray: London, UK, 1881; 328p.

72. Johnson, M.F.; Rice, S.P.; Reid, I. Topographic disturbance of subaqueous gravel substrates by signal crayfish (Pacifastacus leniusculus). *Geomorphology* **2010**, *123*, 269–278. [CrossRef]

73. Clark, J.D.; Harris, J.W.K. Fire and its roles in early hominid lifeways. *Afr. Archaeol. Rev.* **1985**, *3*, 3–27. [CrossRef]

74. Roebroeks, W.; Villa, P. On the earliest evidence for habitual use of fire in Europe. *Proc. Natl. Acad. Sci. USA* **2011**, *108*, 5209–5214. [CrossRef] [PubMed]

75. Shahack-Gross, R.; Berna, F.; Karkanas, P.; Lemorini, C.; Gopher, A.; Barkai, R. Evidence for the repeated use of a central hearth at Middle Pleistocene (300 ky ago) Qesem Cave, Israel. *J. Archaeol. Sci.* **2014**, *44*, 12–21. [CrossRef]

76. Shimelmitz, R.; Kuhn, S.L.; Jelinek, A.J.; Ronen, A.; Clark, A.E.; Weinstein-Evron, M. 'Fire at will': The emergence of habitual fire use 350,000 years ago. *J. Hum. Evol.* **2014**, *77*, 196–203. [CrossRef] [PubMed]

77. Clarkson, C.; Jacobs, Z.; Marwick, B.; Fullagar, R.; Wallis, L.; Smith, M.; Roberts, R.G.; Hayes, E.; Lowe, K.; Carah, X.; et al. Human occupation of northern Australia by 65,000 years ago. *Nature* **2017**, *547*, 306–310. [CrossRef] [PubMed]

78. Head, L. Prehistoric aboriginal impacts on Australian vegetation: An assessment of the evidence. *Aust. Geogr.* **1989**, *20*, 37–46. [CrossRef]

79. Turney, C.S.M.; Kershaw, A.P.; Moss, P.; Bird, M.I.; Fifield, L.K.; Cresswell, R.G.; Santos, G.M.; Di Tada, M.L.; Hausladen, P.A.; Zhou, Y. Redating the onset of burning at Lynch's Crater (North Queensland): Implications for human settlement in Australia. *J. Quat. Sci.* **2001**, *16*, 767–771. [CrossRef]

80. Black, M.P.; Mooney, S.D.; Haberle, S.G. The fire, human and climate nexus in the Sydney Basin, eastern Australia. *Holocene* **2007**, *17*, 469–480. [CrossRef]

81. Haberle, S.G.; Hope, G.S.; DeFretes, Y. Environmental change in the Baliem Valley, montane Irian Jaya, Republic of Indonesia. *J. Biogeogr.* **1991**, *18*, 25–40. [CrossRef]

82. Whitlock, C.; Higuera, P.E.; McWethy, D.B.; Briles, C.E. Paleoecological perspectives on fire ecology: Revisiting the fire-regime concept. *Open Ecol. J.* **2010**, *3*, 6–23. [CrossRef]

83. Balter, M. Seeking agriculture's ancient roots. *Science* **2007**, *316*, 1830–1835. [CrossRef] [PubMed]

84. Fuller, D.Q.; Yo-Ichiro, S.; Castillo, C.; Qin, L.; Weisskopf, A.R.; Kingwell-Banham, E.J.; Song, J.; Ahn, S.-M.; van Etten, J. Consilience of genetics and archaeobotany in the entangled history of rice. *Archaeol. Anthropol. Sci.* **2010**, *2*, 115–131. [CrossRef]

85. Tanno, K.; Willcox, G. How fast was wild wheat domesticated? *Science* **2006**, *311*, 1886. [CrossRef] [PubMed]

86. Zeder, M.A. Core questions in domestication research. *Proc. Natl. Acad. Sci. USA* **2015**, *112*, 3191–3198. [CrossRef] [PubMed]

87. Özkan, H.; Willcox, G.; Graner, A.; Salamini, F.; Kilian, B. Geographic distribution and domestication of wild emmer wheat (*Triticum dicoccoides*). *Genet. Resour. Crop Evol.* **2010**, *58*, 11–53. [CrossRef]

88. Arranz-Otaegui, A.; Colledge, S.; Zapata, L.; Teira-Mayolini, L.C.; Ibanez, J.J. Regional diversity on the timing for the initial appearance of cereal cultivation and domestication in southwest Asia. *Proc. Natl. Acad. Sci. USA* **2016**, *113*, 14001–14006. [CrossRef] [PubMed]

89. Karg, S. New research on the cultural history of the useful plant *Linum usitatissimum* L. (flax), a resource for food and textiles for 8,000 years. *Veg. Hist. Archaeobot.* **2011**, *20*, 507–508. [CrossRef]

90. Tengberg, M. Beginnings and early history of date palm garden cultivation in the Middle East. *J. Arid Environ.* **2012**, *86*, 139–147. [CrossRef]

91. Huang, C.C.; Pang, J.; Zhou, Q.; Chen, S. Holocene pedogenic change and the emergence and decline of rain-fed cereal agriculture on the Chinese Loess Plateau. *Quat. Sci. Rev.* **2004**, *23*, 2525–2535. [CrossRef]

92. Liu, L.; Lee, G.-A.; Jiang, L.; Zhang, J. Evidence for the early beginning (c. 9000 cal. BP) of rice domestication in China: A response. *Holocene* **2007**, *17*, 1059–1068. [CrossRef]

93. Molina, J.; Sikora, M.; Garud, N.; Flowers, J.M.; Rubinstein, S.; Reynolds, A.; Huang, P.; Jackson, S.; Schaal, B.A.; Bustamante, C.D.; et al. Molecular evidence for a single evolutionary origin of domesticated rice. *Proc. Natl. Acad. Sci. USA* **2011**, *108*, 8351–8356. [CrossRef] [PubMed]

94. Dillehay, T.D.; Rossen, J.; Andres, T.C.; Williams, D.E. Preceramic adoption of peanut, squash, and cotton in northern Peru. *Science* **2007**, *316*, 1890–1893. [CrossRef] [PubMed]

95. Ranere, A.J.; Piperno, D.R.; Holst, I.; Dickau, R.; Iriarte, J. The cultural and chronological context of early Holocene maize and squash domestication in the Central Balsas River Valley, Mexico. *Proc. Natl. Acad. Sci. USA* **2009**, *106*, 5014–5018. [CrossRef] [PubMed]

96. Denham, T.P.; Haberle, S.G.; Lentfer, C.; Fullagar, R.; Field, J.; Therin, M.; Porch, N.; Winsborough, B. Origins of agriculture at Kuk Swamp in the Highlands of New Guinea. *Science* **2003**, *301*, 189–193. [CrossRef] [PubMed]

97. McGovern, P.E.; Zhang, J.; Tang, J.; Zhang, Z.; Hall, G.R.; Moreau, R.A.; Nunez, A.; Butrym, E.D.; Richards, M.P.; Wang, C.; et al. Fermented beverages of pre- and proto-historic China. *Proc. Natl. Acad. Sci. USA* **2004**, *101*, 17593–17598. [CrossRef] [PubMed]

98. McGovern, P.; Jalabadze, M.; Batiuk, S.; Callahan, M.P.; Smith, K.E.; Hall, G.R.; Kvavadze, E.; Maghradze, D.; Rusishvili, N.; Bouby, L.; et al. Early Neolithic wine of Georgia in the South Caucasus. *Proc. Natl. Acad. Sci. USA* **2017**, E10309–E10318. [CrossRef] [PubMed]

99. Edwards, C.J.; Bollongino, R.; Scheu, A.; Chamberlain, A.; Tresset, A.; Vigne, J.-D.; Baird, J.F.; Larson, G.; Ho, S.Y.W.; Heupink, T.H.; et al. Mitochondrial DNA analysis shows a Near Eastern Neolithic origin for domestic cattle and no indication of domestication of European aurochs. *Proc. R. Soc. Lond. B* **2007**, *274*, 1377–1385. [CrossRef] [PubMed]

100. Harris, D. Development of the agro-pastoral economy in the Fertile Crescent during the Pre-Pottery Neolithic period. In *The Dawn of Farming in the Near East*; Cappers, R., Bottema, S., Eds.; Ex Oriente: Berlin, Germany, 2002; pp. 67–83.

101. Staubwasser, M.; Weiss, H. Holocene climate and cultural evolution in late prehistoric—Early historic West Asia. *Quat. Res.* **2006**, *66*, 372–387. [CrossRef]

102. Larson, G.; Albarella, U.; Dobney, K.; Rowley-Conwy, P.; Schibler, J.; Tresset, A.; Vigne, J.-D.; Edwards, C.J.; Schlumbaum, A.; Dinu, A.; et al. Ancient DNA, pig domestication, and the spread of the Neolithic into Europe. *Proc. Natl. Acad. Sci. USA* **2007**, *104*, 15276–15281. [CrossRef] [PubMed]

103. Ramos-Onsins, S.E.; Burgos-Paz, W.; Manunza, A.; Amills, M. Mining the pig genome to investigate the domestication process. *Heredity* **2014**, *113*, 471–484. [CrossRef] [PubMed]

104. Kierstein, G.; Vallinoto, M.; Silva, A.; Schneider, M.P.; Iannuzzi, L.; Brenig, B. Analysis of mitochondrial D-loop region casts new light on domestic water buffalo (*Bubalus bubalis*) phylogeny. *Mol. Phylogenet. Evol.* **2004**, *30*, 308–324. [CrossRef]

105. Outram, A.K.; Stear, N.A.; Bendrey, R.; Olsen, S.; Kasparov, A.; Zaibert, V.; Thorpe, N.; Evershed, R.P. The earliest horse harnessing and milking. *Science* **2009**, *323*, 1332–1335. [CrossRef] [PubMed]
106. Petersen, J.L.; Mickelson, J.R.; Cothran, E.G.; Andersson, L.S.; Axelsson, J.; Bailey, E.; Bannasch, D.; Binns, M.M.; Borges, A.S.; Brama, P.; et al. Genetic diversity in the modern horse illustrated from genome-wide SNP data. *PLoS ONE* **2013**, *8*, e54997. [CrossRef] [PubMed]
107. Almathen, F.; Charruau, P.; Mohandesan, E.; Mwacharo, J.M.; Orozco-terWengel, P.; Pitt, D.; Abdussamad, A.M.; Uerpmann, M.; Uerpmann, H.-P.; De Cupere, B.; et al. Ancient and modern DNA reveal dynamics of domestication and cross-continental dispersal of the dromedary. *Proc. Natl. Acad. Sci. USA* **2016**, *113*, 6707–6712. [CrossRef] [PubMed]
108. Rossel, S.; Marshall, F.; Peters, J.; Pilgram, T.; Adams, M.D.; O'Connor, D. Domestication of the donkey: Timing, processes, and indicators. *Proc. Natl. Acad. Sci. USA* **2008**, *105*, 3715–3720. [CrossRef] [PubMed]
109. Kimura, B.; Marshall, F.; Beja-Pereira, A.; Mulligan, C. Donkey domestication. *Afr. Archaeol. Rev.* **2013**, *30*, 83–95. [CrossRef]
110. Qiu, Q.; Wang, L.; Wang, K.; Yang, Y.; Ma, T.; Wang, Z.; Zhang, X.; Ni, Z.; Hou, F.; Long, R.; et al. Yak whole-genome resequencing reveals domestication signatures and prehistoric population expansions. *Nat. Commun.* **2015**, *6*, 10283. [CrossRef] [PubMed]
111. Wheeler, J.C. Evolution and present situation of the South American Camelidae. *Biol. J. Linn. Soc.* **1995**, *54*, 271–295. [CrossRef]
112. Kadwell, M.; Fernandez, M.; Stanley, H.F.; Baldi, R.; Wheeler, J.C.; Rosadio, R.; Bruford, M.W. Genetic analysis reveals the wild ancestors of the llama and the alpaca. *Proc. R. Soc. Lond. B* **2001**, *268*, 2575–2584. [CrossRef] [PubMed]
113. Revedin, A.; Aranguren, B.; Becattini, R.; Longo, L.; Marconi, E.; Lippi, M.M.; Skakun, N.; Sinitsyn, A.; Spiridonova, E.; Svoboda, J. Thirty thousand-year-old evidence of plant food processing. *Proc. Natl. Acad. Sci. USA* **2010**, *107*, 18815–18819. [CrossRef] [PubMed]
114. Lippi, M.M.; Foggi, B.; Aranguren, B.; Ronchitelli, A.; Revedin, A. Multistep food plant processing at Grotta Paglicci (Southern Italy) around 32,600 cal B.P. *Proc. Natl. Acad. Sci. USA* **2015**, *112*, 12075–12080. [CrossRef] [PubMed]
115. Bar-Yosef, O. The Natufian culture in the Levant, threshold to the origins of agriculture. *Evolut. Anthropol. Issues News Rev.* **1998**, *6*, 159–177. [CrossRef]
116. Nadel, D.; Piperno, D.R.; Holst, I.; Snir, A.; Weiss, E. New evidence for the processing of wild cereal grains at Ohalo II, a 23 000-year-old campsite on the shore of the Sea of Galilee, Israel. *Antiquity* **2012**, *86*, 990–1003. [CrossRef]
117. Balter, M. The tangled roots of agriculture. *Science* **2010**, *327*, 404–406. [CrossRef] [PubMed]
118. Nesbitt, M. Wheat evolution: Integrating archaeological and biological evidence. In *Wheat Taxonomy: The Legacy of John Percival*; Special Issue 3; Caligari, P.D.S., Brandham, P.E., Eds.; Linnean Society: London, UK, 2001; pp. 37–59.
119. Lazaridis, I.; Nadel, D.; Rollefson, G.; Merrett, D.C.; Rohland, N.; Mallick, S.; Fernandes, D.; Novak, M.; Gamarra, B.; Sirak, K.; et al. Genomic insights into the origin of farming in the ancient Near East. *Nature* **2016**, *536*, 419–424. [CrossRef] [PubMed]
120. Flad, R.; Li, S.; Wu, X.; Zhao, Z. Early wheat in China: Results from new studies at Donghuishan in the Hexi Corridor. *Holocene* **2010**, *20*, 955–965. [CrossRef]
121. Huang, X.; Kurata, N.; Wei, X.; Wang, Z.-X.; Wang, A.; Zhao, Q.; Zhao, Y.; Liu, K.; Lu, H.; Li, W.; et al. A map of rice genome variation reveals the origin of cultivated rice. *Nature* **2012**, *490*, 497–501. [CrossRef] [PubMed]
122. Rosen, A.M. The impact of environmental change and human land use on alluvial valleys in the Loess Plateau of China during the Middle Holocene. *Geomorphology* **2008**, *101*, 298–307. [CrossRef]
123. Harvey, E.L.; Fuller, D.Q. Investigating crop processing using phytolith analysis: The example of rice and millets. *J. Archaeol. Sci.* **2005**, *32*, 739–752. [CrossRef]
124. Fuller, D.Q. Agricultural origins and frontiers in South Asia: A working synthesis. *J. World Prehist.* **2006**, *20*, 1–86. [CrossRef]
125. Glaser, B.; Haumaier, L.; Guggenberger, G.; Zech, W. The "Terra Preta" phenomenon: A model for sustainable agriculture in the humid tropics. *Naturwissenschaften* **2001**, *88*, 37–41. [CrossRef] [PubMed]

126. Stinchcomb, G.E.; Messner, T.C.; Driese, S.G.; Nordt, L.C.; Stewart, R.M. Pre-colonial (A.D. 1100-1600) sedimentation related to prehistoric maize agriculture and climate change in eastern North America. *Geology* **2011**, *39*, 363–366. [CrossRef]

127. Downie, A.E.; Van Zwieten, L.; Smernik, R.J.; Morris, S.; Munroe, P.R. Terra Preta Australis: Reassessing the carbon storage capacity of temperate soils. *Agric. Ecosyst. Environ.* **2011**, *140*, 137–147. [CrossRef]

128. Tarolli, P.; Preti, F.; Romano, N. Terraced landscapes: From an old best practice to a potential hazard for soil degradation due to land abandonment. *Anthropocene* **2014**, *6*, 10–25. [CrossRef]

129. Hillel, D. *Rivers of Eden*; Oxford University Press: New York, NY, USA, 1994; 355p.

130. Wilkinson, T.J. Soil erosion and valley fills in the Yemen Highlands and southern Turkey: Integrating settlement, geoarchaeology, and climate change. *Geoarchaeology* **2005**, *20*, 169–192. [CrossRef]

131. Evershed, R.P.; Payne, S.; Sherratt, A.G.; Copley, M.S.; Coolidge, J.; Urem-Kotsu, D.; Kotsakis, K.; Özdoğan, M.; Özdoğan, A.E.; Nieuwenhuyse, O.; et al. Earliest date for milk use in the Near East and southeastern Europe linked to cattle herding. *Nature* **2008**, *455*, 528–531. [CrossRef] [PubMed]

132. Bogaard, A.; Fraser, R.; Heaton, T.H.E.; Wallace, M.; Vaiglova, P.; Charles, M.; Jones, G.; Evershed, R.P.; Styring, A.K.; Andersen, N.H.; et al. Crop maturing and intensive land management by Europe's first farmers. *Proc. Natl. Acad. Sci. USA* **2013**, *110*, 12589–12594. [CrossRef] [PubMed]

133. Mays, L.W. A very brief history of hydraulic technology during antiquity. *Environ. Fluid Mech.* **2008**, *8*, 471–484. [CrossRef]

134. Gillmore, G.K.; Coningham, R.A.E.; Fazeli, H.; Young, R.L.; Magshoudi, M.; Batt, C.M.; Rushworth, G. Irrigation on the Tehran Plain, Iran: Tepe Pardis—The site of a possible Neolithic irrigation feature? *Catena* **2009**, *78*, 285–300. [CrossRef]

135. Wilkinson, T.J. *Archaeological Landscapes of the Near East*; The University of Arizona Press: Tucson, AZ, USA, 2003; 260p.

136. Huckleberry, G. Assessing Hohokam canal stability through stratigraphy. *J. Field Archaeol.* **1999**, *26*, 1–18.

137. Nichols, D.L.; Spence, M.W.; Borland, M.D. Watering the fields of Teotihuacan: Early irrigation at the ancient city. *Anc. Mesoam.* **1991**, *2*, 119–129. [CrossRef]

138. Khayat, S.; Hotzl, H.; Geyer, S.; Ali, W. Hydrochemical investigation of water from the Pleistocene wells and springs, Jericho area, Palestine. *Hydrol. J.* **2006**, *14*, 192–202. [CrossRef]

139. Cullen, H.M.; deMenocal, P.B. North Atlantic influence on Tigris-Euphrates streamflow. *Int. J. Climatol.* **2000**, *20*, 853–863. [CrossRef]

140. Weiss, H.; Courty, M.-A.; Wetterstrom, W.; Guichard, F.; Senior, L.; Meadow, R.; Curnow, A. The genesis and collapse of third millennium North Mesopotamian Civilization. *Science* **1993**, *261*, 995–1004. [CrossRef] [PubMed]

141. Jacobsen, T.; Adams, R.M. Salt and silt in ancient Mesopotamian agriculture. *Science* **1958**, *128*, 1251–1258. [CrossRef] [PubMed]

142. Butzer, K.W. *Early Hydraulic Civilization in Egypt*; The University of Chicago Press: Chicago, IL, USA, 1976; 134p.

143. Eyre, C.J. The agriculture cycle, farming, and water management. In *Civilizations of the Ancient Near East*; Sasson, J.M., Ed.; Charles Scribner's Sons: New York, NY, USA, 1995; Volume I, pp. 175–189.

144. Viollet, P.-L. *Water Engineering in Ancient Civilizations: 5000 Years of History*; Taylor and Francis Group: Boca Rotan, FL, USA, 2007; 322p.

145. Will, P.-E. The Zheng-Bai irrigation system of Shaanxi Province in the Late-Imperial period. In *Sediments of Time, Environment and Society in Chinese History*; Elvin, M., Ts'ui-Jung, L., Eds.; Cambridge University Press: Cambridge, UK, 1998; pp. 282–343.

146. Pearsall, D.M. Plant domestication. In *Encyclopedia of Archaeology*; Pearsall, D.M., Ed.; Academic Press: Oxford, UK, 2008; pp. 1822–1842.

147. McNamee, G. *Gila: The Life and Death of an American River*; Orion Books: New York, NY, USA, 1994; 215p.

148. Masse, W.B. Prehistoric irrigation systems in the Salt River Valley, Arizona. *Science* **1981**, *214*, 408–415. [CrossRef] [PubMed]

149. McCully, P. *Silenced Rivers*; Zed Books: London, UK, 1996; 350p.

150. Tanchev, L. *Dams and Appurtenant Hydraulic Structures*, 2nd ed.; CRC Press, Taylor and Francis: Abingdon, UK, 2014.

151. Kamash, Z. Irrigation technology, society and environment in the Roman Near East. *J. Arid Environ.* **2012**, *86*, 65–74. [CrossRef]
152. Barnett, J.; Rogers, S.; Webber, M.; Finlayson, B.; Wang, M. Transfer project cannot meet China's water needs. *Nature* **2015**, *527*, 295–297. [CrossRef] [PubMed]
153. Weinberger, R.; Sneh, A.; Shalev, E. Hydrogeological insights in antiquity as indicated by Canaanite and Israelite water systems. *J. Archaeol. Sci.* **2008**, *35*, 3035–3042. [CrossRef]
154. Frumkin, A.; Shimron, A. Tunnel engineering in the Iron Age: Geoarcheology of the Siloam Tunnel, Jerusalem. *J. Archaeol. Sci.* **2006**, *33*, 227–237. [CrossRef]
155. Ward, C. Ships and shipbuilding. In *The Oxford Encyclopedia of Ancient Egypt*; Redford, D.B., Ed.; Oxford University Press: New York, NY, USA, 2001; Volume III, pp. 281–284.
156. Bass, G.F. Sea and river craft in the ancient Near East. In *Civilizations of the Ancient Near East*; Sasson, J.M., Ed.; Charles Scribner's Sons: New York, NY, USA, 1995; Volume III, pp. 1421–1431.
157. Ball, P. *The Water Kingdom: A Secret History of China*; The University of Chicago Press: Chicago, IL, USA, 2017; 341p.
158. Erickson-Gini, T. Nabatean agriculture: Myth and reality. *J. Arid Environ.* **2012**, *86*, 50–54. [CrossRef]
159. Nordt, L.C.; Hayashida, F.; Hallmark, T.; Crawford, C. Late prehistoric soil fertility, irrigation management, and agricultural production in northwest coastal Peru. *Geoarchaeology* **2004**, *19*, 21–46. [CrossRef]
160. Ertsen, M.W.; van der Spek, J. Modeling an irrigation ditch opens up the world. Hydrology and hydraulics of an ancient irrigation system in Peru. *Phys. Chem. Earth* **2009**, *34*, 176–191. [CrossRef]
161. Elvin, M.; Ninghu, S. The influence of the Yellow River on Hangzhou Bay since A.D. 1000. In *Sediments of Time, Environment and Society in Chinese History*; Elvin, M., Ts'ui-Jung, L., Eds.; Cambridge University Press: Cambridge, UK, 1998; pp. 344–407.
162. Morozova, G.S. A review of Holocene avulsions of the Tigris and Euphrates rivers and possible effects on the evolution of civilizations in Lower Mesopotamia. *Geoarchaeology* **2005**, *20*, 401–423. [CrossRef]
163. Dietrich, O.; Heun, M.; Notroff, J.; Schmidt, K.; Zarnkow, M. The role of cult and feasting in the emergence of Neolithic communities. New evidence from Göbekli Tepe, south-eastern Turkey. *Antiquity* **2012**, *86*, 674–695. [CrossRef]
164. Carter, T.; Shackley, M.S. Sourcing obsidian from Neolithic Çatalhöyük (Turkey) using energy dispersive X-ray fluorescence. *Archaeometry* **2007**, *3*, 437–454. [CrossRef]
165. Lawrence, D.; Philip, G.; Wilkinson, K.; Buylaert, J.P.; Murray, A.S.; Thompson, W.; Wilkinson, T.J. Regional power and local ecologies: Accumulated population trends and human impacts in the northern Fertile Crescent. *Quat. Int.* **2017**, *437*, 60–81. [CrossRef]
166. McMahon, A. Mesopotamia, Sumer, and Akkad. In *Encyclopedia of Archaeology*; Pearsall, D.M., Ed.; Academic Press: Oxford, UK, 2008; pp. 854–865.
167. Nichols, D.L.; Cover, R.A.; Abdi, K. Rise of civilization and urbanism. In *Encyclopedia of Archaeology*; Pearsall, D.M., Ed.; Academic Press: Oxford, UK, 2008; pp. 1003–1015.
168. Khan, A.M.; Lemmen, C. Bricks and urbanism in the Indus Civilization. *Hist. Philos. Phys.* **2014**, *24*, 1–11.
169. Kenoyer, J.M. Indus Civilization. In *Encyclopedia of Archaeology*; Pearsall, D.M., Ed.; Academic Press: Oxford, UK, 2008; pp. 715–733.
170. Zhang, R.; Pian, H.; Santosh, M.; Zhang, S. The history and economics of gold mining in China. *Ore Geol. Rev.* **2015**, *65*, 718–727. [CrossRef]
171. Lacovara, P. Bricks and brick architecture. In *The Oxford Encyclopedia of Ancient Egypt*; Redford, D.B., Ed.; Oxford University Press: New York, NY, USA, 2001; Volume I, pp. 198–200.
172. Wu, X.; Zhang, C.; Goldberg, P.; Cohen, D.; Pan, Y.; Arpin, T.; Bar-Yosef, O. Early pottery at 20,000 years ago in Xianrendong Cave, China. *Science* **2012**, *336*, 1696–1700. [CrossRef] [PubMed]
173. Craig, O.E.; Saul, H.; Lucquin, A.; Nishida, Y.; Tache, K.; Clarke, L.; Thompson, A.; Altoft, D.T.; Uchiyama, J.; Ajimoto, M.; et al. Earliest evidence for the use of pottery. *Nature* **2013**, *496*, 351–354. [CrossRef] [PubMed]
174. Nieuwenhuyse, O.P.; Akkermans, P.M.M.G.; van der Plicht, J. Not so coarse, nor always plain—The earliest pottery of Syria. *Antiquity* **2010**, *84*, 71–85. [CrossRef]
175. Roosevelt, A.C.; Housley, R.A.; Imazio da Silveira, M.; Maranca, S.; Johnson, R. Eighth millennium pottery from a prehistoric shell midden in the Brazilian Amazon. *Science* **1991**, *254*, 1621–1624. [CrossRef] [PubMed]
176. Kalicki, T.; Fraczek, M.; Przepiora, P. *Evolution of River Valleys in Central Europe: Field Guide*; Fluvial Archives Group Biennial Meeting, Kielce, Poland, 2016; FLAG: Amsterdam, The Netherlands, 2016; 132p.

177. Schild, R.; Królik, H. Rydno: A final Paleolithic ochre mining complex. *Prz. Archeol.* **1981**, *29*, 53–100.

178. Roaf, M. Palaces and temples in ancient Mesopotamia. In *Civilizations of the Ancient Near East*; Sasson, J.M., Ed.; Charles Scribner's Sons: New York, NY, USA, 1995; Volume I, pp. 423–441.

179. Hikade, T. Egypt, Pharaonic. In *Encyclopedia of Archaeology*; Pearsall, D.M., Ed.; Academic Press: Oxford, UK, 2008; pp. 31–45.

180. Leach, B.; Tait, J. Papyrus. In *The Oxford Encyclopedia of Ancient Egypt*; Redford, D.B., Ed.; Oxford University Press: New York, NY, USA, 2001; Volume III, pp. 22–24.

181. Tallet, P.; Marouard, G. The harbour of Khufu on the Red Sea coast at Wadi al-Jarf, Egypt. *Near East. Archaeol.* **2014**, *77*, 4–14. [CrossRef]

182. Roosevelt, A.C.; Lima da Costa, M.; Lopes Machado, C.; Michab, M.; Mercier, N.; Valladas, H.; Feathers, J.; Barnett, W.; Imazio da Silveira, M.; Henderson, A.; et al. Paleoindian cave dwellers in the Amazon: The peopling of the Americas. *Science* **1996**, *272*, 373–384. [CrossRef]

183. Butler, V.L.; O'Connor, J.E. 9000 years of salmon fishing on the Columbia River, North America. *Quat. Res.* **2004**, *62*, 1–8. [CrossRef]

184. Jing, Y.; Flad, R.; Yunbing, L. Meat-acquisition patterns in the Neolithic Yangzi river valley, China. *Antiquity* **2008**, *82*, 351–366. [CrossRef]

185. Maritan, L.; Iacumin, P.; Zerboni, A.; Venturelli, G.; Dal Sasso, G.; Linseele, V.; Talamo, S.; Salvatori, S.; Usai, D. Fish and salt: The successful recipe of White Nile Mesolithic hunter-gatherer-fishers. *J. Archaeol. Sci.* **2018**, *92*, 48–62. [CrossRef]

186. Vermeesch, P.M. Fishing along the Nile. In *Before Food Production in North Africa*; di Lernia, S., Manzi, G., Eds.; ABACO Edizioni: La Spezia, Italy, 1998; pp. 103–111.

187. Van Neer, W. Evolution of prehistoric fishing in the Nile Valley. *J. Afr. Archaeol.* **2004**, *2*, 251–269. [CrossRef]

188. deMenocal, P.; Ortiz, J.; Guilderson, T.; Adkins, J.; Sarnthein, M.; Baker, L.; Yarusinsky, M. Abrupt onset and termination of the African Humid Period: Rapid climate responses to gradual insolation forcing. *Quat. Sci. Rev.* **2000**, *19*, 347–361. [CrossRef]

189. Macklin, M.G.; Woodward, J.C.; Welsby, D.A.; Duller, G.A.T.; Williams, F.M.; Williams, M.A.J. Reach-scale river dynamics moderate the impact of rapid Holocene climate change on floodwater farming in the desert Nile. *Geology* **2013**, *41*, 695–698. [CrossRef]

190. De Bruhl, M. *The River Sea*; Counterpoint: Berkeley, CA, USA, 2010; 235p.

191. Bono, P.; Boni, C. Water supply of Rome in antiquity and today. *Environ. Geol.* **1996**, *27*, 126–134. [CrossRef]

192. Leo, A.D.; Tallini, M. Irrigation, groundwater exploitation and cult of water in the rural settlements of Sabina, Central Italy, in Roman times. *Water Sci. Technol. Water Supply* **2007**, *7*, 191–199. [CrossRef]

193. Caracuta, V.; Fiorentino, G. *Wood for Fuel in Roman Hypocaust Baths: New Data from the Late-Roman Villa of Faragola (SE Italy)*; Saguntum, Paleles del Laboratorio de Arqueología de Valencia, Extra-13; Universitat de València: València, Spain, 2011; pp. 167–168.

194. Astour, M.C. Overland trade routes in ancient western Asia. In *Civilizations of the Ancient Near East*; Sasson, J.M., Ed.; Charles Scribner's Sons: New York, NY, USA, 1995; Volume III, pp. 1401–1420.

195. Lech, J. Flint mining among the early farming communities of Poland. *Staringia* **1981**, *6*, 39–45.

196. Frankopan, P. *The Silk Roads*; Bloomsbury: London, UK, 2015; 636p.

197. Parker, A.G.; Lucas, A.S.; Walden, J.; Goudie, A.S.; Robinson, M.A.; Allen, T.G. Late Holocene geoarchaeological investigation of the Middle Thames floodplain at Dorney, Buckinghamshire, UK: An evaluation of the Bronze Age, Iron Age, Roman and Saxon landscapes. *Geomorphology* **2008**, *101*, 471–483. [CrossRef]

198. Lisé-Pronovost, A.; Salomon, F.; Goiran, J.-P.; St-Onge, G.; Herries, A.I.R.; Montero-Serrano, J.-C.; Heslop, D.; Roberts, A.P.; Levchenko, V.; Zawadzki, A.; et al. Dredging and canal gate technologies in Portus, the ancient harbour of Rome, reconstructed from event stratigraphy and multi-proxy sediment analysis. *Quat. Int.* **2018**. [CrossRef]

199. Rothenberg, B. Archaeo-metallurgical researches in the southern Arabah 1959-1990. Part 1: Late Pottery Neolithic to Early Bronze IV. *Palest. Explor. Q.* **1999**, *131*, 68–89. [CrossRef]

200. Ottaway, B.S. Innovation, production and specialization in early prehistoric copper metallurgy. *Eur. J. Archaeol.* **2001**, *4*, 87–112. [CrossRef]

201. Kaufman, B. Copper alloys from the 'Enot Shuni Cemetery and the origins of bronze metallurgy in the EB IV–MB II Levant. *Archaeometry* **2013**, *55*, 663–690. [CrossRef]

202. Zhang, S.; Yang, Y.; Storozum, M.J.; Li, H.; Cui, Y.; Dong, G. Copper smelting and sediment pollution in Bronze Age China: A case study in the Hexi corridor, Northwest China. *Catena* **2017**, *156*, 92–101. [CrossRef]

203. Balter, M. Early start for human art? Ochre may revise timeline. *Science* **2009**, *323*, 569. [CrossRef] [PubMed]

204. Walter, R.C.; Merritts, D.J. Natural streams and the legacy of water-powered mills. *Science* **2008**, *319*, 299–304. [CrossRef] [PubMed]

205. Constante, A.; Pena-Monne, J.L.; Munoz, A. Alluvial geoarchaeology of an ephemeral stream: Implications for Holocene landscape change in the central part of the Ebro Depression, northeast Spain. *Geoarchaeology* **2010**, *25*, 475–496. [CrossRef]

206. Beach, T.P.; Luzzadder-Beach, S. Geoarchaeology and aggradation around Kinet Hoyuk, an archaeological mound in the Eastern Mediterranean, Turkey. *Geomorphology* **2008**, *101*, 416–428. [CrossRef]

207. Casana, J. Mediterranean valleys revisited: Linking soil erosion, land use and climate variability in the Northern Levant. *Geomorphology* **2008**, *101*, 429–442. [CrossRef]

208. Cordova, C.E. Floodplain degradation and settlement history in Wadi al-Wala and Wadi ash-Shallalah, Jordan. *Geomorphology* **2008**, *101*, 443–457. [CrossRef]

209. Maghsoudi, M.; Simpson, I.A.; Kourampas, N.; Nashli, H.F. Archaeological sediments from settlement mounds of the Sagzabad Cluster, central Iran: Human-induced deposition on an arid alluvial plain. *Quat. Int.* **2014**, *324*, 67–83. [CrossRef]

210. Sharma, G.R.; Misra, V.D.; Mandal, D.; Misra, B.B.; Pal, J.N. *Beginnings of Agriculture*; Abinash Prakashan: Allahabad, India, 1980; 320p.

211. Milàn, J.; Clemmensen, L.B.; Buchardt, B.; Noe-Nygaard, N. Tracking the Bronze Age fauna: Preliminary investigations of a new Late Holocene tracksite, Lodbjerg dune system, northwest Jylland, Denmark. *Hantkeniana* **2006**, *5*, 42–45.

212. Goswami, D.C. Brahmaputra River, Assam, India: Physiography, basin denudation, and channel aggradation. *Water Resour. Res.* **1985**, *21*, 959–978. [CrossRef]

213. Li, X.; Shang, X.; Dodson, J.; Zhou, X. Holocene agriculture in the Guanzhong Basin in NW China indicated by pollen and charcoal evidence. *Holocene* **2009**, *19*, 1213–1220. [CrossRef]

214. Zhuang, Y.; Kidder, T.R. Archaeology of the Anthropocene in the Yellow River region, China, 8000–2000 cal. BP. *Holocene* **2014**, *24*, 1602–1623. [CrossRef]

215. Dearing, J.A. Landscape change and resilience theory: A palaeoenvironmental assessment from Yunnan, SW China. *Holocene* **2008**, *18*, 117–127. [CrossRef]

216. Wang, H.; Yang, Z.; Saito, Y.; Liu, J.P.; Sun, X.; Wang, Y. Stepwise decreases of the Huanghe (Yellow River) sediment load (1950-2005): Impacts of climate change and human activities. *Glob. Planet. Chang.* **2007**, *57*, 331–354. [CrossRef]

217. Marchetti, M. Environmental changes in the central Po Plain (northern Italy) due to fluvial modifications and anthropogenic activities. *Geomorphology* **2002**, *44*, 361–373. [CrossRef]

218. Bruno, L.; Amorosi, A.; Curina, R.; Severi, P.; Bitelli, R. Human-landscape interactions in the Bologna area (northern Italy) during the mid-late Holocene, with focus on the Roman period. *Holocene* **2013**, *23*, 1560–1571. [CrossRef]

219. Goman, M.; Joyce, A.; Mueller, R. Stratigraphic evidence for anthropogenically induced coastal environmental change from Oaxaca, Mexico. *Quat. Res.* **2005**, *63*, 250–260. [CrossRef]

220. Lenders, H.J.R.; Chamuleau, T.P.M.; Hendriks, A.J.; Lauwerier, R.C.G.M.; Leuven, R.S.E.W.; Verberk, W.C.E.P. Historical rise of waterpower initiated the collapse of salmon stocks. *Sci. Rep.* **2016**, *6*, 29269. [CrossRef] [PubMed]

221. Crook, D.; Elvin, M. Bureaucratic control of irrigation and labour in late-imperial China: The uses of administrative cartography in the Miju catchment, Yunnan. *Water Hist.* **2013**, *5*, 287–305. [CrossRef]

222. Lehner, B.; Liermann, C.R.; Revenga, C.; Vörösmarty, C.; Fekete, B.; Crouzet, P.; Döll, P.; Endejan, M.; Frenken, K.; Magone, J.; et al. High-resolution mapping of the world's reservoirs and dams for sustainable river-flow management. *Front. Ecol. Environ.* **2011**, *9*, 494–502. [CrossRef]

223. Ruyters, S.; Mertens, J.; Vassilieva, E.; Dehandschutter, B.; Poffijn, A.; Smolders, E. The red mud accident in Ajka (Hungary): Plant toxicity and trace metal bioavailability in red mud contaminated soil. *Environ. Sci. Technol.* **2011**, *45*, 1616–1622. [CrossRef] [PubMed]

224. Segura, F.R.; Nunes, E.A.; Paniz, F.P.; Paulelli, A.C.C.; Rodrigues, G.B.; Braga, G.U.L.; Dos Reis Pedreira Filho, W.; Barbosa, F., Jr.; Cerchiaro, G.; Silva, F.F.; et al. Potential risks of the residue from Samarco's mine dam burst (Bento Rodrigues, Brazil). *Environ. Pollut.* **2016**, *218*, 813–825. [CrossRef] [PubMed]

225. Kalis, A.J.; Merkt, J.; Wunderlich, J. Environmental changes during the Holocene climatic optimum in central Europe—Human impact and natural causes. *Quat. Sci. Rev.* **2003**, *22*, 33–79. [CrossRef]

226. Reale, O.; Shukla, J. Modeling the effects of vegetation on Mediterranean climate during the Roman Classical Period: Part II. Model simulation. *Glob. Planet. Chang.* **2000**, *25*, 185–214. [CrossRef]

227. Huang, C.C.; Pang, J.; Chen, S.; Su, H.; Han, J.; Cao, Y.; Zhao, W.; Tan, Z. Charcoal records of fire history in the Holocene loess-soil sequences over the southern Loess Plateau of China. *Palaeogeogr. Palaeoclimatol. Palaeoecol.* **2006**, *239*, 28–44. [CrossRef]

228. Brown, A.G. Learning from the past: Palaeohydrology and palaeoecology. *Freshw. Biol.* **2002**, *47*, 817–829. [CrossRef]

229. Słowik, M. Transformation of a lowland river from a meandering and multi-channel pattern into an artificial canal: Retracing a path of river channel changes (the Middle Obra River, W Poland). *Reg. Environ. Chang.* **2013**. [CrossRef]

230. Blott, S.J.; Pye, K.; van der Wal, D.; Neal, A. Long-term morphological change and its causes in the Mersey Estuary, NW England. *Geomorphology* **2006**, *81*, 185–206. [CrossRef]

231. Gastaldo, R.A.; Degges, C.W. Sedimentology and paleontology of a Carboniferous log jam. *Int. J. Coal Geol.* **2007**, *69*, 103–118. [CrossRef]

232. Nelson, A.D.; Church, M. Placer mining along the Fraser River, British Columbia: The geomorphic impact. *Geol. Soc. Am. Bull.* **2012**, *124*, 1212–1228. [CrossRef]

233. Knighton, A.D. River adjustment to changes in sediment load: The effects of tin mining on the Ringarooma River, Tasmania, 1875-1984. *Earth Surf. Process. Landf.* **1989**, *14*, 333–359. [CrossRef]

234. Sreebha, S.; Padmalal, D. Environmental impact assessment of sand mining from the small catchment rivers in the southwestern coast of India: A case study. *Environ. Manag.* **2011**, *47*, 130–140. [CrossRef] [PubMed]

235. Wohl, E. Wilderness is dead: Whither critical zone studies and geomorphology in the Anthropocene? *Anthropocene* **2013**, *2*, 4–15. [CrossRef]

236. Herodotus. *The History*; The University of Chicago Press: Chicago, IL, USA, 1988; 699p.

237. Dotterweich, M. The history of human-induced soil erosion: Geomorphic legacies, early descriptions and research, and the development of soil conservation—A global synopsis. *Geomorphology* **2013**, *201*, 1–34. [CrossRef]

238. Dubois, N.; Saulnier-Talbot, E.; Mills, K.; Gell, P.; Battarbee, R.; Bennion, H.; Chawchai, S.; Dong, X.; Francus, P.; Flower, R.; et al. First human impacts and responses of aquatic systems: A review of palaeolimnological records from around the world. *Anthr. Rev.* **2018**, *5*, 28–68. [CrossRef]

239. Geobel, T.; Waters, M.R.; O'Rourke, D.H. The Late Pleistocene dispersal of modern humans in the Americas. *Science* **2008**, *319*, 1497–1502. [CrossRef] [PubMed]

240. McWethy, D.B.; Whitlock, C.; Wilmshurst, J.M.; McGlone, M.S.; Fromont, M.; Li, X.; Dieffenbacher-Krall, A.; Hobbs, W.O.; Fritz, S.C.; Cook, E.R. Rapid landscape transformation in South Island, New Zealand, following initial Polynesian settlement. *Proc. Natl. Acad. Sci. USA* **2010**, *107*, 21343–21348. [CrossRef] [PubMed]

241. Kaplan, J.O.; Krumhardt, K.M.; Gaillard, M.-J.; Sugita, S.; Trondman, A.-K.; Fyfe, R.; Marquer, L.; Mazier, F.; Nielsen, A.B. Constraining the deforestation history of Europe: Evaluation of historical land use scenarios with pollen-based land cover reconstructions. *Land* **2017**, *6*, 91. [CrossRef]

242. Notebaert, B.; Verstraeten, G.; Ward, P.; Renssen, H.; Van Rompaey, A. Modeling the sensitivity of sediment and water runoff dynamics to Holocene climate and land use changes at the catchment scale. *Geomorphology* **2011**, *126*, 18–31. [CrossRef]

243. Syvitski, J.P.M.; Vörösmarty, C.J.; Kettner, A.J.; Green, P. Impact of humans on the flux of terrestrial sediment to the global coastal ocean. *Science* **2005**, *308*, 376–380. [CrossRef] [PubMed]

244. Trimble, S.W. The fallacy of stream equilibrium in contemporary denudation studies. *Am. J. Sci.* **1977**, *277*, 876–887. [CrossRef]

245. Reusser, L.; Bierman, P.; Rood, D. Quantifying human impacts on rates of erosion and sediment transport at a landscape scale. *Geology* **2015**, *43*, 171–174. [CrossRef]

246. Goodbred, S.L., Jr. Response of the Ganges dispersal system to climate change: A source-to-sink view since the last interstade. *Sediment. Geol.* **2003**, *162*, 83–104. [CrossRef]

247. Jain, V.; Tandon, S.K. Conceptual assessment of (dis)connectivity and its application to the Ganga River dispersal system. *Geomorphology* **2010**, *118*, 349–358. [CrossRef]

248. Blöthe, J.H.; Korup, O. Millennial lag times in the Himalayan sediment routing system. *Earth Planet. Sci. Lett.* **2013**, *382*, 38–46. [CrossRef]

249. Riehl, S. Archaeobotanical evidence for the interrelationship of agricultural decision-making and climate change in the ancient Near East. *Quat. Int.* **2009**, *197*, 93–114. [CrossRef]

250. Roberts, N.; Eastwood, W.J.; Kuzucuoğlu, C.; Fiorentino, G.; Caracuta, V. Climatic, vegetation and cultural change in the eastern Mediterranean during the mid-Holocene environmental transition. *Holocene* **2011**, *21*, 147–162. [CrossRef]

251. Wu, Q.; Zhao, Z.; Liu, L.; Granger, D.E.; Wang, H.; Cohen, D.J.; Wu, X.; Ye, M.; Bar-Yosef, O.; Lu, B.; et al. Outburst flood at 1920 BCE supports historicity of China's Great Flood and the Xia dynasty. *Science* **2016**, *353*, 579–582. [CrossRef] [PubMed]

252. Smith, D.E.; Harrison, S.; Firth, C.R.; Jordan, J.T. The early Holocene sea level rise. *Quat. Sci. Rev.* **2011**, *30*, 1846–1860. [CrossRef]

253. Stanley, D.J.; Chen, Z. Neolithic settlement distributions as a function of sea level-controlled topography in the Yangtze delta, China. *Geology* **1996**, *24*, 1083–1086. [CrossRef]

254. Stanley, D.J.; Warne, A.G. Sea level and initiation of Predynastic culture in the Nile delta. *Nature* **1993**, *363*, 435–438. [CrossRef]

255. Day, J.W., Jr.; Gunn, J.D.; Folan, W.J.; Yanez-Arancibia, A.; Horton, B.P. Emergence of complex societies after sea level stabilized. *Eos Trans. Am. Geophys. Union* **2007**, *88*, 169–176. [CrossRef]

256. Wang, Z.; Zhuang, C.; Saito, Y.; Chen, J.; Zhan, Q.; Wang, X. Early mid-Holocene sea-level change and coastal environmental response on the southern Yangtze delta plain, China: Implications for the rise of Neolithic culture. *Quat. Sci. Rev.* **2012**, *35*, 51–62. [CrossRef]

257. Ricklis, R.A.; Blum, M.D. The geoarchaeological record of Holocene sea level change and human occupation of the Texas Gulf Coast. *Geoarchaeology* **1997**, *12*, 287–314. [CrossRef]

MDPI

St. Alban-Anlage 66

4052 Basel

Switzerland

Tel. +41 61 683 77 34

Fax +41 61 302 89 18

www.mdpi.com

Quaternary Editorial Office

E-mail: quaternary@mdpi.com

www.mdpi.com/journal/quaternary